Fuzzy Analy

Hierarchy Process

Fuzzy Analytic Hierarchy Process

By
Ali Emrouznejad and William Ho

CRC Press
Taylor & Francis Group
Boca Raton London New York

CRC Press is an imprint of the
Taylor & Francis Group, an **informa** business

A CHAPMAN & HALL BOOK

CRC Press
Taylor & Francis Group
6000 Broken Sound Parkway NW, Suite 300
Boca Raton, FL 33487-2742

First issued in paperback 2020

ISBN-13: 978-0-367-57297-6 (pbk)
ISBN-13: 978-1-4987-3246-8 (hbk)

Library of Congress Cataloging-in-Publication Data

Names: Emrouznejad, Ali. | Ho, William.
Title: Fuzzy analytic hierarchy process / [edited by] Ali Emrouznejad,
William Ho.
Description: Boca Raton : CRC Press, 2017. | Includes bibliographical
references.
Identifiers: LCCN 2017011269 | ISBN 9781498732468 (hardback)
Subjects: LCSH: Fuzzy mathematics. | Fuzzy decision making. | Approximation
algorithms. | Decision making.
Classification: LCC QA248.5 .F8827 2017 | DDC 658.4/038011--dc23
LC record available at https://lccn.loc.gov/2017011269

Visit the Taylor & Francis Web site at
http://www.taylorandfrancis.com

and the CRC Press Web site at
http://www.crcpress.com

Contents

Preface

The analytic hierarchy process (AHP) introduced by Thomas Saaty (1980) is a modern tool for dealing with complex decision-making and may help the decision maker to set priorities to make the best decision. Due to its simplicity, ease of use, and great flexibility, the AHP has been studied extensively and used in nearly all applications related to multiple criteria decision-making (MCDM) since its development. Besides being applied to the finance sector, the AHP has been adopted in education; engineering; government; industry; management; manufacturing; personal, political, and social systems; and sports.

While the vast majority of applications of MCDM use crisp data, it is quite possible that the data in many real applications are not crisp. Hence, imprecision may arise for a variety of reasons: unquantifiable information, incomplete information, unobtainable information, and partial ignorance. Conventional multiple attribute decision-making methods cannot effectively handle problems with such imprecise information. Hence, fuzzy AHP (FAHP) has been developed to deal with uncertain data and the imprecision in assessing the relative importance of attributes and the performance ratings of alternatives with respect to attributes.

The main objective of this book is to provide the necessary background to work with FAHP by introducing some novel methods as well as to introduce some useful applications of FAHP for both interested academics and practitioners, and to benefit society, industry, academia, and government.

To facilitate this goal, Chapter 1 provides background information about fuzzy sets and FAHP. In this chapter, **Ali Emrouznejad** and **William Ho** investigate the research areas that are the most influenced by the AHP and discuss the use of fuzzy sets in uncertain situations along with the AHP. This chapter highlights a major effort involved in handling FAHP, which has led to several applications that have proven to be very useful when there is a notion of uncertainty in the data.

This is followed by a comprehensive discussion and review literature on risk assessment methodologies that incorporate FAHP. In Chapter 2, **Peh Sang Ng**, **Joshua Ignatius**, **Mark Goh**, **Amirah Rahman**, and **Feng Zhang** classify FAHP applications into two main categories: fuzzy singular analytic hierarchy processes and fuzzy hybrid analytic hierarchy processes. This chapter aims to uncover the recent trends and applications of FAHP over the past two decades. Pragmatically, this work serves as a guide to researchers who are seeking research gaps in the area of risk assessment and FAHP models.

In Chapter 3, **Nimet Yapici Pehlivan**, **Turan Paksoy**, and **Ahmet Çalik** compare methods in FAHP with application in supplier selection. As discussed in this chapter, there has recently been a growing pressure on the managers of global supply chains to be environmentally conscious with regard to their supply chain network. The main aim of this chapter is to provide a comparative analysis of FAHP methods for the supplier selection problems when dealing with green issues, such as carbon emissions, corporate social responsibility, green packaging, green handling, and waste management policy.

In the same area, in Chapter 4, **Mohammad Hasan Aghdaie** focuses on an application in supplier evaluation using data mining group decision-making with FAHP. This chapter explains that incorporation of imprecise information and vagueness is an unavoidable requirement of fine decision-making models. Specifically, this chapter provides a case study in the fast-moving consumer goods (FMCG) industry using a novel integrated fuzzy group analytic hierarchy process (FGAHP), simple additive weighting (SAW), and two-stage cluster analysis as a data mining tool for supplier evaluation and segmentation when the data are uncertain.

Further to this, the fundamental concepts of intuitionistic and hesitant fuzzy sets and proposed AHP approaches based on these extensions are presented by **Cengiz Kahraman** and **Fatih Tüysüz** in Chapter 5. This chapter also discusses, in detail, two different approaches of AHP methods based on interval-valued intuitionistic fuzzy sets (IVIFS) and hesitant fuzzy linguistic term sets (HFLTS).

The use of FAHP in combination with the technique for order of preference by similarity to ideal solution (TOPSIS) is the subject of Chapter 6 in which **Antonio Rodríguez** presents an overview of the implementation of MCDM methods based on the integration of FAHP and TOPSIS. In addition, this chapter presents a new method that incorporates variable weights analysis (VWA) to implement evaluation weights, a very useful feature for those evaluation models in which the evaluation formulas have to be fixed in advance.

Delphi-AHP-TOPSIS with an application to logistics service quality evaluation when using a multistakeholder multiperspective approach is the subject of Chapter 7. In this chapter, **Anjali Awasthi** and **Hassan Mukhtar** propose a three-step analysis where, in the first step, a fuzzy Delphi (FDELPHI) technique is used to identify the requirements (criteria) for logistics service quality evaluation. The second step involves attributing weights to these requirements using FAHP and, in the third step, fuzzy TOPSIS is used to rank the alternatives for logistics service quality improvement.

The inclusion of logical interaction between criteria in FAHP is discussed in Chapter 8 by **Ksenija Mandić**, **Vjekoslav Bobar**, and **Boris Delibašić**. This chapter proposes a hybrid model, by integrating interpolative Boolean algebra (IBA) into FAHP, for supporting the selection of suppliers in the telecommunications sector by using fuzzy MCDM and logical interactions between criteria. A comparative study of the traditional FAHP, where all criteria are standalone and assumed to be independent, against FAHP, which uses logical interactions set by decision makers as decision criteria, is also presented in this chapter.

Extending the ordinary FAHP to interval type-2 FAHP is the subject of Chapter 9 as discussed by **Cengiz Kahraman, Başar Öztayşi**, and **Sezi Çevik Onar**. Specifically, this chapter focuses on technology selection for wind energy using an interval type-2 FAHP method.

In Chapter 10, **Vinodh S** and **Vimal KEK** present a decision support system for prioritizing process sustainability tools using FAHP. This chapter first reviews literature on sustainable manufacturing criteria, process sustainability tools, and AHP applications in the sustainability domain. This is followed by a case study to improve the sustainable performance of the manufacturing processes using five tools: energy modeling, waste minimization, carbon footprint analysis, parametric optimization, and water footprint analysis.

The rest of this book provides guidance on using several applications of FAHP in different areas.

In Chapter 11, **Muhammet Gul** and **Ali Fuat Guneri** develop an application in the aluminum extrusion industry by using FAHP for occupational safety risk assessment. Specifically, FAHP is proposed to deal with shortcomings of a precise risk score measurement and decrease the inconsistency in decision-making, enabling occupational health and safety (OHS) experts to use linguistic terms for evaluating three parameters of the proportional risk assessment (PRA) method. First, these parameters are weighted by using FAHP. Then the orders of priority for hazard groups are determined by using fuzzy TOPSIS. In order to show the applicability of the proposed approach, a case application in an aluminum extrusion workplace is discussed.

An application of hybrid FDELPHI and FAHP as a tool for assessing the management of electronic scientific journals is given by **Nara Medianeira Stefano, Raul Otto Laux, Nelson Casarotto Filho, Lizandra Lupi Garcia Vergara**, and **Izabel Cristina Zattar** in Chapter 12. This is done in a two-stage analysis where in the first stage the FDELPHI is used to raise the critical factors (criteria/subcriteria) present in the management of electronic scientific journal, and in the second stage the FAHP is applied to calculate the relative weights of the selected criteria/subcriteria that affect management.

In Chapter 13, **Sonal K. Thengane, Andrew Hoadley, Sankar Bhattacharya, Sagar Mitra**, and **Santanu Bandyopadhyay** present the novel application of FAHP in analyzing energy systems. More specifically, this chapter discusses the application of different types of FAHPs in some major areas concerning energy systems with examples explaining the choice of criteria and methodology of analysis. The main focus in this study is on the applications of FAHP in four sections, namely energy sources, renewable energy technology, energy policy and site selection, and alternative fuels.

Applications of FAHP based on a decision-making framework for construction projects are discussed in Chapter 14. In this chapter, **Long D. Nguyen** and **Dai Q. Tran** explain the use of FAHP to quantify the complexity of transportation construction projects. The illustrative example shows that FAHP is an effective technique to deal with uncertain, subjective, and linguistic data in construction management.

In Chapter 15, **Theagarajan Padma**, **S. P. Shantharajah**, and **Shabir Ahmad Mir** present an application of plantation land segmentation for an orchard establishment using FAHP. This chapter first determines the major criterion and critical characteristics of each of the criterion in establishing an orchard and their contributions in planting a variety of fruits. Then FAHP is adopted to analyze the relative weight and the ranking of significance of individual factors.

FAHP has also been used in Medical studies. In Chapter 16, **Lazim Abdullah** discusses the use of FAHP in identifying the most likely type of prostate cancer using weighted-based approach of decision-making technique. The decision model is utilized to overcome medical experts' predicaments for identifying the type of prostate cancer while FAHP is used in a seven-step computation with the ultimate aim to establish the relative weights of alternatives.

Finally, this book is concluded by providing another application of FAHP in assessing environmental actions in the Natura 2000 network area. In Chapter 17, **J. M. Sánchez-Lozano** and **J. A. Bernal-Conesa** discuss the use of FAHP to obtain the weights of the criteria that have influence in an emerging decision problem such as preserving the natural habitats and wild fauna and flora of protected areas. Normally, the information provided by the criteria is of a different types, with quantitative criteria (distance to settlements, main roads, crops, etc.) coexisting with qualitative criteria (desertification risk and level of protection). Therefore, this chapter explains how linguistic labels and numerical values are employed to obtain the importance coefficients of the criteria. FAHP is used in order to compare different models for extracting the knowledge based on heterogeneous and homogeneous aggregations.

We hope this book will add value to business, government, and professional readers in various sectors of society, including business leaders, policymakers, and professionals/practitioners in agriculture, banking, economic, education, environment, energy, finance, government, health care, regulation, transport, sustainability, social, and medicine departments. We have similar hope that this book will be useful for researchers and students who are specializing in the areas of multicriteria decision-making, AHP, and FAHP.

Ali Emrouznejad
Aston Business School, Aston University, Birmingham, UK
William Ho
Department of Management and Marketing, University of Melbourne, Australia

Editors

Ali Emrouznejad is a professor and chair in business analytics at Aston Business School, United Kingdom. His areas of research interest include performance measurement and management, and efficiency, and productivity analysis, as well as data mining. Dr. Emrouznejad is editor of *Annals of Operations Research*, associate editor of *RAIOR-Operations Research,* associate editor of *Socio-Economic Planning Sciences*, associate editor of *IMA Journal of Management Mathematics*, senior editor of *Data Envelopment Analysis Journal*, and a member of editorial boards or guest editor for several other scientific journals. He has published more than 100 articles in top-ranked journals; he is author of the book *Applied Operational Research with SAS* (Chapman & Hall/CRC Press), editor of the books *Big Data Optimization, Performance Measurement with Fuzzy Data Envelopment Analysis* (Springer), *Managing Service Productivity* (Springer), and *Handbook of Research on Strategic Performance Management and Measurement* (IGI Global). He is also cofounder of Performance Improvement Management Software (PIM-DEA).

William Ho is a senior lecturer in the Department of Management and Marketing at the University of Melbourne, Australia. William's research interests include strategic sourcing and supplier performance management, supply chain risk management, and sustainable supply chain management. He has published more than 50 articles in various international journals, such as *Computers and Operations Research, Decision Sciences, European Journal of Operational Research, International Journal of Production Economics, International Journal of Production Research, Journal of the Operational Research Society, Supply Chain Management: An International Journal*, and so on. He published two authored books: *Optimal Production Planning for PCB Assembly* (Springer) and *Applied Operational Research with SAS* (Chapman & Hall/CRC Press) in 2006 and 2011, respectively. He serves as an editorial board member for *IEEE Transactions on Engineering Management*.

Contributors

Lazim Abdullah is a professor of computational mathematics at the School of Informatics and Applied Mathematics, Universiti Malaysia Terengganu. He received a BSc (Hons) in mathematics from the University of Malaya, Kuala Lumpur in June 1984 and an MEd in mathematics from Universiti Sains Malaysia, Penang, in 1999. He received his PhD from the Universiti Malaysia Terengganu (Information Technology Development) in 2004. His research focuses on the mathematical theory of fuzzy sets and its diverse applications. His research findings have been published in more than 250 publications, including refereed journals, conference proceedings, chapters in books, and research books. He is a member of the editorial boards of several international journals related to computing and applied mathematics. Dr. Abdullah is an associate member of the IEEE Computational Intelligence Society, a member of the Malaysian Mathematical Society, and a member of the International Society on MCDM.

Mohammad Hasan Aghdaie received both his bachelor's and master's degrees in industrial engineering from Shomal University. His current research interests include operations research, decision analysis, multiple criteria decision analysis, system dynamics, business analytics, data mining, data science and big data, application of fuzzy sets and systems, and water resource management. Currently, he is an MSc student in the area of industrial engineering at Concordia University.

Anjali Awasthi is an associate professor in the Concordia Institute of Information Systems Engineering Department at Concordia University, Montreal, Canada. Her areas of research are city logistics, sustainable supply chains, information technology and decision-making, and quality assurance in supply chain management. She is the author of several journal and conference papers in these areas.

Santanu Bandyopadhyay is an institute chair professor in the Department of Energy Science and Engineering, IIT Bombay, India. He worked as an assistant professor in energy systems engineering in the Department of Mechanical Engineering, IIT Bombay from 2001 to 2005, and as an associate professor from 2005 to 2009. His research interests are process integration, pinch analysis, industrial energy conservation, modeling and simulation of energy systems, energy integration of distillation processes, and renewable energy systems.

J. A. Bernal-Conesa was born in 1974 in Cartagena, Spain. He received his bachelor's degree in industrial engineering from the Technical University of Cartagena (TUCT), Spain, in 2003. In 2010, he finished his master's degree in renewable energies. He received his PhD from the TUCT in 2016. He was a senior consultant and lead auditor of internal audits of management systems from 2003 until 2009. He is currently a full-time lecturer and researcher at the University Centre of Defence at the Spanish Air Force Academy. He has published papers in scientific journals and has a large number of publications in national and international conferences. His main research interests are production and operations management, quality and environmental systems, corporate social responsibility, and human resources.

Sankar Bhattacharya started his career in India as a design and commissioning engineer for coal-fired power stations. He then worked in Thailand on utilization of agro-forestry residues for gaseous fuel production, as a principal research engineer with Lignite CRC in Australia, as a principal process engineer with Anglo Coal Australia, and as a senior energy analyst with the International Energy Agency in Paris managing their Cleaner Fossil Fuels program. His research area involves advanced coal and biomass utilization for power and fuel production through gasification and combustion, and biofuels including algae. He commissioned the first CFBC pilot plant in Australia and managed the first pressurized oxygen-blown gasification trials of brown coal in Australia and the United States. With his strong background in industry and now academia, he believes in tangible outcomes from research.

Vjekoslav Bobar is an Information and Communication Department director at Electric Power Industry of Serbia, Belgrade, Serbia. He received his PhD in 2015 at Faculty of Organizational Sciences, University of Belgrade, Serbia. He has worked as an Information and Communication Department director for the Government of Serbia for the last 12 years. Also, he has worked as assistant professor at the Faculty of Organizational Sciences, University of Belgrade, Serbia. His areas of expertise are in the domains of multiattribute decision-making, decision support systems, fuzzy set theory, information and communications technologies, data mining, and public electronic procurement.

Ahmet Çalik is an assistant professor at KTO Karatay University, Karatay, Turkey, in the Department of Logistics Management. He was a research assistant at Selçuk University from 2011 to 2016. He graduated from Gazi University in the Department of Statistics in 2010. He received a master's degree in statistics in 2012 and a PhD in statistics in 2016 from the Selçuk University Institute of Science and Technology. His research interests are supply chain management, fuzzy sets theory, mathematical programming, and multicriteria decision-making.

Boris Delibašić is full professor at the University of Belgrade, Faculty of Organizational Sciences (School of Business), Serbia. His research interests lie in business intelligence, data mining, machine learning, multicriteria decision analysis, and decision support systems. He serves on the editorial board of several international journals. He is a coordinator of the EURO working group on Decision Support Systems. His research profile is available at https://www.researchgate.net/profile/Boris_Delibasic.

Nelson Casarotto Filho has a bachelor degree in chemical engineering from the Federal University of Rio Grande do Sul (1974), master's degree in production engineering from the Federal University of Santa Catarina (1977), and PhD in production engineering from the Federal University of Santa Catarina (1995) with sandwich at the University of Minho, Portugal. He has experience in the area of production engineering, with emphasis on project evaluation, working mainly on the following topics: regional development, industrial competitiveness, business networks, competitiveness, and investment analysis.

Mark Goh is a former Colombo Plan Scholar and holds a PhD from the University of Adelaide, South Australia. In the National University of Singapore, he holds the appointments of director (Industry Research) at the Logistics Institute-Asia Pacific, a joint venture with Georgia Tech, and principal researcher at the Centre for Transportation Research, and he was a program director of the Penn-State NUS Logistics Management Program. He also was director of Supply Chain Solutions for Asia/Middle East with APL Logistics, crafting logistics engineering solutions for major multinational corporations (Dell, Nike, Lenovo, VW, BMW, Roche, BBraun) in this part of the world. He was on the board of Rigelsoft, a reverse logistics company. He is a professor at NUS Business School, National University of Singapore.

Muhammet Gul works as an assistant professor in the Department of Industrial Engineering at Munzur University, Tunceli, Turkey. He received his MSc and PhD in industrial engineering from Yildiz Technical University, Turkey. His research interests are simulation modeling, health-care system management, occupational safety and risk assessment, multicriteria decision-making, and fuzzy sets. His papers have appeared in international high-cited journals such as *Computers & Industrial Engineering, Knowledge-Based Systems, Applied Soft Computing, Journal of Loss Prevention in the Process Industries*, and *European Journal of Industrial Engineering*.

Ali Fuat Guneri is now a full professor and has been working at the Department of Industrial Engineering, Yildiz Technical University, Turkey, since 1990. He received his MSc and PhD in industrial engineering from Yildiz Technical University. His research interests are production management, supply chain management, and occupational safety. His papers have appeared in international high-cited journals such as *European Journal of Operations Research, Computers & Industrial Engineering, Knowledge-Based Systems, Applied Soft Computing, Journal of Loss Prevention in the Process Industries,* and *European Journal of Industrial Engineering.*

Andrew Hoadley has been a faculty member of Monash University, Clayton, Australia, since 1999. He obtained his PhD (chemical engineering) from the University of Cambridge, UK, in 1988 and worked as a research fellow at the Laboratory of Physical Metallurgy, EPFL, Switzerland from 1988 to 1991, and as senior engineer at Comalco Research and Technology from 1991 to 1997. His research interests are process design and process integration, dewatering and drying processes, multiobjective optimization, and biofuels.

Joshua Ignatius is an associate professor of Operations Research at the School of Mathematical Sciences, Universiti Sains Malaysia. His research is transdisciplinary and focuses on resource allocation and performance evaluation models across a variety of industries. He has published in numerous high-impact SCI/SSCI journals such as *European Journal of Operational Research, TECHNOVATION, Knowledge-Based Systems, Journal of Intelligent & Fuzzy Systems, Experts Systems with Applications,* and *Group Decision & Negotiation.* Dr. Ignatius is the recipient of the Endeavour Executive Award for high-achieving professionals in 2010–2011 from the Australian Government under then Prime Minister Julia Gillard's administration.

Cengiz Kahraman earned his BSc, MSc, and PhD degrees in industrial engineering from Istanbul Technical University (ITU), Turkey. His main research areas are engineering economics, quality management and control, statistical and multicriteria decision-making, and fuzzy sets applications. He has published about 180 papers in international journals, 140 papers in the proceedings of international conferences, and 10 edited books from Springer. He has guest-edited many special issues of international journals and organized international conferences. He is on the editorial boards of 20 international journals. He was the vice dean of the ITU Management Faculty between 2004 and 2007 and the head of the ITU Industrial Engineering Department between 2010 and 2013.

Vimal KEK is an assistant professor, Department of Mechanical Engineering, Sri Venkateswara College of Engineering, Sriperumbudur, Tamil Nadu, India. He completed his PhD and MTech in the Production Engineering Department, National Institute of Technology, Tiruchirappalli, Tamil Nadu. He received his bachelor's degree in production engineering (Sandwich Program) from the PSG College of Technology, Coimbatore, India. He has published 15 papers in international journals and 15 papers in international conferences. His areas of research interest include lean manufacturing, sustainable manufacturing, neural network, and fuzzy logic.

Raul Otto Laux has a solid academic background in Business Administration and University Administration, ranging from undergraduate to PhD, with post-doctorate in University Administration of the Programa de Pós-Graduação em Administração, Universidade Federal de Santa Catarina, Brasil. He has professional experience at the national and international level in the areas of innovation and entrepreneurship, marketing, strategic marketing, executive education, sales, negotiation processes, competitive intelligence, promotion and merchandising, trade marketing, business strategies, market development, products and services, competitiveness and foreign trade, new developments, incubators, business plans, and analysis of entrepreneur and startup profiles.

Ksenija Mandić received her BSc, MSc, and PhD degrees from the Faculty of Organizational Science at the University of Belgrade, Serbia, in 2006, 2008, and 2015, respectively. She has worked at the telecommunication company Crony since 2007. Her research interests are decision-making theory, supply chain management, multicriteria decision-making methods, fuzzy logic, and interpolative Boolean algebra.

Shabir Ahmad Mir is a program assistant (Computer) at the Sher-e-Kashmir, University of Agricultural Sciences and Technology of Kashmir, India, and is pursuing a PhD (Information Technology) at AMIT University, Chennai, India. He received his Master of Computer Applications degree at Indira Gandhi National Open University, New Delhi, India, in 2002 and MPhil (computer science) at Madurai Kamaraj University, Tamil Nadu, India in 2005. He has published more than 20 papers and book chapters in national and international journals in the fields of decision support systems, fuzzy logic, computational intelligence, and blended learning. He has developed different software solutions for academics, accountancy, and human resource development on client–server architecture. He has developed ONVAREF DSS for onion.

Sagar Mitra joined the Department of Energy Science and Engineering at IIT Bombay, India, and has set up an Electrochemical Energy Laboratory doing extensive research in the field of energy storage. He has received a prestigious award from the Swedish Govt. Tax Authority (Skatteverket) as an Expert in Electrochemistry for the electronics industry to work for the Replisaurus Technology R&D group. His research interests are nanostructured materials-synthesis, lithium ion batteries, hybrid vehicles, Na-ion and magnesium ion batteries, and hybrid capacitors. His teaching interests are chemistry for energy science, electrochemistry, introduction to energy systems engineering, and thermodynamics and energy conversion.

Hassan Mukhtar is a PhD student at the Concordia Institute for Information Systems Engineering at Concordia University, Montreal, Canada. His thesis addresses the theme of supplier quality management in global supply chains. He has published one journal and one conference paper in this area.

Peh Sang Ng is a lecturer at the Department of Physical and Mathematical Sciences, Faculty of Science, Universiti Tunku Abdul Rahman (UTAR), Petaling Jaya, Malaysia. She received her master's degree in multicriteria decision-making from Universiti Sains Malaysia (USM). Her research interest is multicriteria decision-making.

Long D. Nguyen is an associate professor in the Department of Environmental and Civil Engineering, U.A. Whitaker College of Engineering at Florida Gulf Coast University (FGCU), Fort Myers, Florida. Prior to his tenure at FGCU, he had more than 6 years of professional experience with full-time positions in the United States and Vietnamese construction industries. He is a professional engineer registered in California. He earned an MS and PhD in civil engineering from the University of California, Berkeley, an MEng in construction engineering and management from the Asian Institute of Technology, and a BEng in civil engineering from Ho Chi Minh City University of Technology. His teaching and research interests include project planning and controls, quantitative methods in construction, and resilient and sustainable built environment. He has served as an assistant specialty editor of the *Journal of Construction Engineering and Management* and an editorial board member of the *International Journal of Project Management*.

Sezi Çevik Onar is a full-time associate professor at the Industrial Engineering Department of Istanbul Technical University (ITU) Management Faculty, Turkey. She earned her BSc in industrial engineering and MSc in engineering management, both from ITU. She completed her PhD studies at ITU and visited Copenhagen Business School and Eindhoven Technical University during these studies. Her PhD was on strategic options. Her research interests include strategic management, multicriteria decision-making, and fuzzy decision-making. She has published many papers and an international book from Springer. She has taken part as a researcher in many privately funded projects such as organization design, intelligent debt collection system design, and human resource management system design. Her refereed articles have appeared in a variety of journals.

Başar Öztayşi is a full-time associate professor at the Industrial Engineering Department of Istanbul Technical University (ITU), Turkey. After earning his BSc in industrial engineering and receiving his MSc in management engineering, he finished his PhD in industrial engineering program at the ITU Institute of Science and Technology in 2009. He has published many papers and an international book from Springer. His research interests include fuzzy sets, multicriteria decision-making, intelligent systems, and data mining.

Theagarajan Padma received her master's degree in computer applications from the University of Madras, India, in 1992, an MTech degree in information technology from AAU, India, in 2004 and MPhil and PhD degrees in computer science from Mother Teresa Women's University, India, in 2003 and 2010, respectively. Currently she is working as a professor in the Department of Computer Applications at the Sona College of Technology, Tamil Nadu, India. Her research interests include artificial intelligence, data analytics, and knowledge-based systems. Dr. Padma is a fellow of the Computer Society of India and life member of the Indian Society for Technical Education. She was the recipient of the Shayesta Akhtar Memorial National Award of ISTE in 2015. Her biography was included in the 2010 Edition of the Marquis *Who's Who in the World*, USA. She serves as an editor and reviewer for well-known national and international journals. She has more than 30 publications in journals and chapters in books.

Turan Paksoy is currently full professor in industrial engineering at Selçuk University, Konya, Turkey. He graduated from Gazi University, Department of Industrial Engineering in 1998. He has continued to Gazi University (Industrial Engineering Department), Ankara, Turkey, and received a his master's degree from the Selçuk University Institute of Science and Technology. He received his PhD degree from Selçuk University, Department of Business Administration (Production Management) in 2004. His current academic interests include lean manufacturing, logistics, supply chain management, and fuzzy set theory. Dr. Paksoy has several published international conference and journal papers.

Amirah Rahman is a PhD graduate from the University of New South Wales, Australia, and is currently serving as a senior lecturer at the School of Mathematical Sciences, Universiti Sains Malaysia. Her research interest is in discrete optimization. She is actively pursuing advancement of mathematical applications across various industries, in particular, transportation and construction.

Antonio Rodríguez holds a PhD in project management from the University of Oviedo, Spain. Certified as MS in telecommunication engineering by the Spanish Ministry of Education, he obtained a degree in electronic engineering from the Universidad Simón Bolívar (USB), Caracas, Venezuela. Certified as PMP and as PRINCE2 Practitioner, he works in IT projects at SELAE, Spain, and works as an independent consultant in the IT sector. Simultaneously, he is a professor at the Universidad Alfonso X El Sabio (UAX) in Madrid, Spain. His research involves quantitative methods for managerial decision-making in information technology projects. The work of Dr. Rodríguez has been recently published in *Expert Systems with Applications* and *Information Sciences*.

Vinodh S is an assistant professor in the Production Engineering Department of the National Institute of Technology, Tiruchirappalli, Tamil Nadu. Dr. Vinodh completed his bachelor's degree in mechanical engineering at Bharathiar University, Coimbatore, India, and earned a master's degree in production engineering and PhD in mechanical engineering at Anna University, Chennai, India. He was awarded a National Doctoral Fellowship for pursuing doctoral research by the All India Council for Technical Education, New Delhi, India, during 2006–2008. He was awarded a Highly Commended Paper Award and Outstanding Paper Award by Emerald Publishers, United Kingdom, in 2009 and 2011, respectively. He has published/presented more than 100 papers at various international journals/conferences. He is serving on the editorial advisory board of the *International Journal of Lean Six Sigma* and *Journal of Engineering, Design and Technology*. His research interests include agile manufacturing, lean manufacturing, sustainable manufacturing, Lean Six Sigma, additive manufacturing, product development, and multicriteria decision-making.

J. M. Sánchez-Lozano received a BS in industrial engineering, an MSc in renewable energy, and a PhD in renewable energy from the Technical University of Cartagena, Murcia, Spain, in 2004, 2010, and 2013, respectively. Currently, he works in the field of engineering projects and is also a full-time senior lecturer and researcher at the University Centre of Defence at the Spanish Air Force Academy. He has published papers in top-quality scientific journals and has contributed a large number of studies in international conferences. His research is focused on decision-making in renewable energy, environmental problems, geographic information systems and cartography, multicriteria decision methods in engineering projects, and fuzzy logic applied to multicriteria decision-making.

S. P. Shantharajah received his a master's degree in computer applications from Bharadhidasan University, Tiruchirappalli, India, in 1998 and a PhD degree in computer science from Periyar University, Salem, India, in 2007. He has more than 18 years of experience in teaching and is currently works as a professor at the School of Information Technology and Engineering, VIT University, Vellore, India. His specializations are network security, database systems, and decision support systems. His main areas of research include interdisciplinary research on various domains with information technology. He is a life member of the Computer Society of India and Indian Society for Technical Education. He serves as editor and reviewer for reputed national and international journals and has published more than 45 journals and book chapters. Five research candidates have been awarded a their PhD under his supervision. He was the recipient of the Professional Society Activities Award for the year 2015–2016.

Nara Medianeira Stefano is a professor and researcher, working acting mainly on the following research themes: improving business performance using artificial intelligence (fuzzy logic) and performance evaluation; evaluation and diagnosis of technological learning in incubators, disclosure of knowledge management and intellectual capital in medium and large companies; implementation of strategic planning and quality systems; cost management; management in services; multicriterial analysis; and management of innovation; and entrepreneurship. She has business experience in economics and administration with emphasis on industrial economics, quantitative methods, and costs.

Sonal K. Thengane earned his PhD from IIT Bombay Monash Research Academy, India, and has been working as a postdoctoral fellow in the Department of Chemical Engineering, Tata Centre for Technology and Design, IIT Bombay since May 2016. His research interests are the analytic hierarchy process, hydrogen production, chemical looping, biomass gasification, and process modeling and simulation.

Dai Q. Tran is an assistant professor in the Department of Civil, Environmental and Architectural Engineering at the University of Kansas, Lawrence, Kansas. His teaching focuses on design/construction interface, construction engineering including value engineering and constructability, risk and decision analysis, cost engineering and project controls, construction quality, productivity, and safety. His research focuses on transportation project planning and development, performance-based decision-making, alternative project delivery, asset management, work zone design and safety, and the application of statistics, modeling, and technology to solve problems in civil infrastructure projects. He received his PhD in construction engineering and management and master's degree in statistics and probability from the University of Colorado at Boulder. He also received his master's degree in structural engineering at Georgia Institute of Technology. He has 5 years of professional experience in the design and construction industry.

Fatih Tüysüz received his BSc in industrial engineering in 2002, MSc in engineering management in 2004, and PhD in industrial engineering in 2010 from İstanbul Technical University. He has been a full-time faculty member at the Department of Industrial Engineering in İstanbul University since , Turkey, 2012. He has been giving lectures relating decision theory, system simulation, engineering economics, manufacturing systems modeling, and lean production at both undergraduate and graduate levels. His research interests include fuzzy logic and applications, gray system theory, system modeling, and modern manufacturing systems.

Lizandra Lupi Garcia Vergara is an architect and urban planner with a master's degree and PhD in production engineering. She is a professor in postgraduate courses in occupational safety engineering and architecture. She is the leader of the GMETTA research group (UFSC-CNPq): Multidisciplinary Group Ergonomics of Labor and Applied Technologies and effective member of the Brazilian Association of Ergonomics (ABERGO), participating in the following technical groups: Ergonomics and Built Environment and Integral Accessibility. She develops research in the following areas: ergonomics, occupational health and safety, products with technological innovation, usability, architecture, accessibility, and assistive technologies.

Nimet Yapici Pehlivan is currently an associate professor at the Department of Statistics of Selçuk University, Turkey. She graduated from Hacettepe University, Ankara, Turkey, Department of Statistics in 1997. She received her master's degree in statistics in 2000 and PhD degree in mathematics in 2005 from the Selçuk University, Institute of Science and Technology. She has published several international conference papers and journal papers. Her current research interests include operations research, optimization, applied statistics, fuzzy set theory, and multicriteria decision-making.

Izabel Cristina Zattar earned a degree in technology in mechanics from the Educational Society of Santa Catarina in 2002, master's degree in mechanical engineering and PhD in mechanical engineering from the Federal University of Santa Catarina, Florianópolis, Brazil, in 2004 and 2008, respectively. She has experience in manufacturing processes, planning and process improvements, and productivity indicators. Currently she works on projects with several companies, including publishers, incubators, and manufacturing industries, as well as development agencies.

Feng Zhang is a professor at the College of Mathematics and Computer Science, Hebei University, Baoding, China. She specializes in soft computing and decision-making methods. Her work has been published in *Information Sciences, Knowledge-Based Systems*, and *European Journal of Operational Research*.

1

Analytic Hierarchy Process and Fuzzy Set Theory

Ali Emrouznejad and William Ho

CONTENTS

1.1 Introduction

Because of its simplicity, ease of use, and great flexibility, the analytic hierarchy process (AHP) has been studied extensively and used in nearly all applications related to multiple criteria decision-making (MCDM) since its development (Saaty, 1980). Besides being applied in the finance sector (Steuer and Na, 2003), the AHP has been adopted in the education, engineering, government, industry, management, manufacturing, personal, political, social, and sports sectors (Vaidya and Kumar, 2006).

The three major elements of AHP are hierarchy construction, priority analysis, and consistency verification. First, the decision makers need to break down complex multiple criteria decision problems into their component parts of which every possible attribute is arranged into multiple hierarchical levels. Second, the decision makers have to compare each cluster in the same level in a pairwise manner based on their own judgments. Specifically, two criteria in the second level are compared at a time with respect to the goal, whereas two attributes of the same criteria in the third level are compared at a time with respect to the corresponding criterion. Since the comparisons are carried out through personal or subjective judgments, some degree of inconsistency may occur. To guarantee that the judgments are consistent, the third element, called consistency verification, which is regarded as one of the most significant advantages of the AHP, is incorporated to measure the degree of consistency among the pairwise comparisons by computing the consistency ratio. If the consistency ratio exceeds the limit, the decision makers should review and revise the pairwise comparisons. Once all pairwise comparisons are carried out at every level and are proved to be consistent, the judgments can then be synthesized to find out the priority ranking of each criterion and its attributes. The overall procedure of the AHP is shown in Figure 1.1 (Ho, 2008).

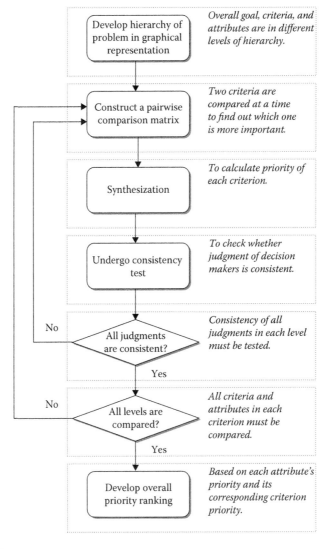

FIGURE 1.1
The flowchart of the analytic hierarchy process.

1.2 Fuzzy Set Theory

The fuzzy set theory has been developed to deal with the concept of partial truth values ranging from absolutely true to absolutely false. Fuzzy set theory has become the prominent tool for handling imprecision or vagueness aiming at tractability, robustness, and low-cost solutions for real-world problems. According to Zadeh (1975), it is very difficult for conventional quantification to reasonably express complex situations and it is necessary to use linguistic variables whose values are words or sentences in a natural or artificial language. The potential of working with linguistic variables, low computational cost, and ease of understanding are characteristics that have contributed to the popularity of this approach. Fuzzy set algebra, developed by Zadeh (1965), is the formal body of theory that allows the treatment of imprecise and vague estimates in uncertain environments.

Zadeh (1965, p. 339) states "The notion of a fuzzy set provides a convenient point of departure for the construction of a conceptual framework which parallels in many respects the framework used in the case of ordinary sets, but is more general than the latter and, potentially, may prove to have a much wider scope of applicability." The application of fuzzy set theory in multiattribute decision-making (MADM) became possible when Bellman and Zadeh (1970) and Zimmermann (1978) introduced fuzzy sets into the field of MADM. They cleared the way for a new family of methods to deal with problems that had been unapproachable and unsolvable with standard techniques (see Chen and Hwang, 1992 for a numerical comparison of fuzzy and classical MADM models). Bellman and Zadeh's (1970) framework was based on the maximin principle and the simple additive weighing model of Yager and Basson (1975) and Bass and Kwakernaak (1977). Bass and Kwakernaak's (1977) method is widely known as the classic work of fuzzy MADM methods.

In 1992, Chen and Hwang (1992) proposed an easy-to-use and easy-to-understand approach to reduce some of the cumbersome computations in the previous MADM methods. Their approach includes two steps: (1) converting fuzzy data into crisp scores and (2) introducing some comprehensible and easy methods. In addition, Chen and Hwang (1992) made distinctions between fuzzy ranking methods and fuzzy MADM methods. Their first group contained a number of methods for finding a ranking: degree of optimality, Hamming distance, comparison function, fuzzy mean and spread, proportion to the ideal, left and right scores, area measurement, and linguistic ranking methods. Their second group was built around methods for assessing the relative importance of multiple attributes: fuzzy simple additive weighting methods, FAHP, fuzzy conjunctive/disjunctive methods, fuzzy outranking methods, and fuzzy maximin methods. The group with the most frequent contributions is fuzzy mathematical programming. Inuiguchi et al. (1990) have provided a useful survey of fuzzy mathematical programming applications including flexible programming, possibilistic programming, possibilistic programming with fuzzy preference relations, possibilistic linear programming (LP) using fuzzy max, possibilistic LP with fuzzy goals, and robust programming.

Recently, fuzzy set theory has been applied in a wide range of fields such as management science, decision theory, artificial intelligence, computer science, expert systems, logic, control theory, and statistics, among others (Chen, 2001; Chen and Tzeng, 2004; Chiou et al., 2005; Ding and Liang, 2005; Figueira et al., 2004; Geldermann et al., 2000; Ho et al., 2010; Triantaphyllou, 2000; Ölçer and Odabaşi, 2005; Wang and Lin, 2003; Wang et al., 2009; Xu and Chen, 2007; Emrouznejad and Tavana, 2014).

1.3 Fuzzy Set Theory and FAHP

The observed values in real-world problems are often imprecise or vague. Imprecise or vague data may be the result of unquantifiable, incomplete, and nonobtainable information. They are often expressed with bounded intervals, ordinal (rank order) data, or fuzzy numbers.

To effectively handle subjective perceptions and impreciseness, fuzzy numbers are integrated with AHP, allowing the appropriate expression of linguistic evaluation (Calabrese et al., 2016). Fuzzy numbers are also used to deal with uncertainties affecting subjective preferences in assessing real-world decision-making problems.

Despite the convenience of AHP in handling both quantitative and qualitative criteria of MCDM problems based on decision makers' judgments, FAHP can reduce or even eliminate the fuzziness; vagueness existing in many decision-making problems may contribute to the imprecise judgments of decision makers in conventional AHP approaches. The field of FAHP has rapidly grown. As seen in this book, in recent years, many researchers have formulated FAHP models in many applications to deal with situations where some of the data are imprecise or vague. The FAHP method is thus suited to solving decision-making problems concerning subjective evaluations and is currently among the most widely used MCDM methods in the fields of business, management, manufacturing, industry, and government.

The first FAHP method was proposed by Van Laarhoven and Pedrycz (1983) using triangular fuzzy numbers (TFNs) in the pairwise comparison matrix. Later, many other methods were proposed, using various types of fuzzy numbers such as the trapezoidal membership function (e.g., see Onar et al., 2016) or the bell-shaped/Gaussian membership function (e.g., see Paul, 2015). In more recent years, FAHP has mostly been applied in the areas of selection and evaluation with a significant amount of literature on combining/integrating

FAHP with other tools, particularly with the technique for order of preference by similarity to ideal solution (TOPSIS), quality function deployment (QFD), Delphi, and analytic network process (ANP).

1.4 Integrated AHP and Fuzzy Set Theory—Recent Literature (2007 to 2016)

Fuzzy set theory has been widely used in conjunction with AHP because fuzzy set theory enables decision makers to render interval judgments and consider uncertainty or fuzziness.

We reviewed journal articles published between 2007 and 2016. The research methodology is as follows. First, the search terms were defined. A title/abstract/keyword search with the terms "analytic hierarchy process" and "fuzzy set theory" was used in the search process. Second, Scopus was utilized to identify the journal articles. To achieve the highest level of relevance, only peer-reviewed articles written in English and published in international journals were selected; conference papers, dissertations, textbooks, book chapters, and notes were excluded. Third, several criteria were determined and used to filter the articles. Regarding the criteria, abstracts of articles were examined to determine whether the article addressed integrated AHP and fuzzy set theory approaches. In addition, because thousands of articles have been published in the last decade, we focused on those articles published in international journals with high ranking, including A/A[*] journals on the Australian Business Deans Council (ABDC) journal list and/or 3/4/4[*] journals on the Association of Business Schools (ABS) journal list in the United Kingdom. Articles that did not meet one of these filtration criteria were excluded. Fourth, the reference lists of the shortlisted articles were carefully evaluated to ensure that no other articles of relevance were omitted in the search. Finally, the content of each article was thoroughly reviewed to ensure that the article fit into the context of integrated AHP and fuzzy set theory approaches. This analysis resulted in 20 journal articles.

Twenty articles applied the integrated FAHP approach, as shown in Table 1.1. The articles can be classified into five categories: supplier evaluation and selection, product/process evaluation and selection, business information system assessment, project assessment, and others.

A number of researchers applied the integrated FAHP approach to supplier evaluation and selection. Chan and Kumar (2007) and Chan et al. (2008) applied the integrated FAHP approach to evaluate and select global suppliers with risk considerations for a hypothetical case. Other groups of researchers demonstrated the applications of the integrated FAHP approach with real cases, including the textile industry (Wang et al.,

TABLE 1.1

The Integrated FAHP Set Theory Approach and Its Applications

Approach	Authors	Application Areas	Specific Problems
FAHP	Akarte and Ravi (2007)	Steel casting manufacturing	Casting process and producer evaluation and selection
	Chan and Kumar (2007)	General manufacturing	Supplier evaluation and selection
	Chu et al. (2007)	Multiple industries	Organizational transformation performance assessment
	Kreng and Wu (2007)	Stone manufacturing	Knowledge portal system evaluation and selection
	Ma et al. (2007)	Furniture manufacturing	Sofa design evaluation and selection
	Nagahanumaiah et al. (2007)	Mould manufacturing	Rapid tooling process evaluation and selection
	Wang and Chang (2007)	LCD manufacturing	Knowledge management project evaluation and selection
	Wang et al. (2007)	Thermal power	Maintenance strategy evaluation and selection
	Chan et al. (2008)	General manufacturing	Supplier evaluation and selection
	Huang et al. (2008)	Government	Industrial technology development program evaluation and selection
	Wang et al. (2008)	Textiles and fashion	Catering firm evaluation and selection
	Arslan (2009)	Government	Transportation project evaluation and selection
	Huang et al. (2009)	LCD manufacturing	SPC implementation strategy evaluation and selection
	Lee (2009)	LCD manufacturing	Supplier evaluation and selection
	Kahraman et al. (2010)	Automotive	ERP supplier evaluation and selection
	Sarfaraz et al. (2012)	General manufacturing	ERP customization technique evaluation and selection
	Wang et al. (2012)	Textiles and fashion	Green initiative risk assessment
	Rezaei and Ortt (2013)	Food	Supplier evaluation and selection
	van de Kaa et al. (2014)	Telecommunications	Technology standard assessment
	Liang (2015)	Telecommunications	Information system assessment

AHP, analytic hierarchy process; ERP, enterprise resource planning; LCD, liquid crystal display; SPC, statistical process control.

2008), liquid crystal display (LCD) manufacturing (Lee, 2009), the automotive industry (Kahraman et al., 2010), and the food industry (Rezaei and Ortt, 2013).

Several scholars applied the integrated FAHP approach for product/process evaluation and selection in the manufacturing industry, including in the casting process and in producer evaluation and selection (Akarte and Ravi, 2007), sofa design evaluation and selection (Ma et al., 2007), rapid tooling process evaluation and selection (Nagahanumaiah et al., 2007), equipment maintenance strategy evaluation and selection (Wang et al., 2007), and statistical process control (SPC) implementation strategy evaluation and selection (Huang et al., 2009).

Regarding business information system assessment, Kreng and Wu (2007) applied the integrated FAHP approach to evaluate the performance of a knowledge portal system to improve the competitiveness of the Taiwanese stone industry. Sarfaraz et al. (2012) adopted the integrated FAHP approach to assess the customization options of an enterprise resource planning (ERP) system. Liang (2015) applied the integrated FAHP approach to measure the performance of interorganizational information systems for the Taiwanese telecommunications industry.

With respect to project assessment, three groups of researchers applied the integrated FAHP approach to evaluate and select projects, including knowledge management projects for a Taiwanese LCD manufacturing company (Wang and Chang, 2007), a government-sponsored industrial technology development program assessment in Taiwan (Huang et al., 2008), and a transportation project assessment in the United States (Arslan, 2009).

Three additional applications of integrated FAHP approach cannot be classified in the earlier categories. Chu et al. (2007) assessed the performance of organizational transformation by communities of practice using the integrated FAHP approach. Wang et al. (2012) developed a risk assessment model based on the integrated FAHP approach that can analyze the aggregative risk of implementing green initiatives in the fashion supply chain. van de Kaa et al. (2014) proposed an integrated FAHP approach to model the process of technology standard selection.

For a full list of publications on FAHP, see Kubler et al. (2016).

1.5 Conclusion

This chapter provides the basic details of FAHP. The FAHP method is suited to solving decision-making problems concerning subjective evaluations and is currently among the most widely used MCDM methods in the fields of business, management, manufacturing, industry, and government. FAHP is mostly applied in the areas of selection and evaluation with a significant amount of literature on combining/integrating FAHP with other tools, particularly with TOPSIS, QFD, Delphi, and ANP.

References

Akarte, M.M., Ravi, B., 2007. Casting product–process–producer compatibility evaluation and improvement. *International Journal of Production Research* 45 (21), 4917–4936.

Arslan, T., 2009. A hybrid model of fuzzy and AHP for handling public assessments on transportation projects. *Transportation* 36 (1), 97–112.

Bass SJ, Kwakernaak H., 1977. Rating and ranking of multi-aspects alternatives using fuzzy sets. *Automatica* 13:47–58.

Bellman, R.E., Zadeh, L.A. 1970. Decision-making in a fuzzy environment. *Management Science* 17(4), 141–164.

Calabrese A., Costa, R., Levialdi, N., Menichini, T., 2016. A fuzzy analytic hierarchy process method to support materiality assessment in sustainability reporting. *Journal of Cleaner Production* 121, 248–264.

Chan, F.T.S., Kumar, N., 2007. Global supplier development considering risk factors using fuzzy extended AHP-based approach. *Omega* 35 (4), 417–431.

Chan, F.T.S., Kumar, N., Tiwari, M.K., Lau, H.C.W., Choy, K.L., 2008. Global supplier selection: A fuzzy-AHP approach. *International Journal of Production Research* 46 (14), 3825–3857.

Chen, C.-T. 2001. A fuzzy approach to select the location of the distribution center. *Fuzzy Sets and Systems* 118(1), 65–73.

Chen, S.J., Hwang, C.L. 1992. *Fuzzy Multi-Attribute Decision-Making: Methods and Applications.* Berlin: Springer.

Chen, M.F., Tzeng, G.H. 2004. Combining grey relation and TOPSIS concepts for selecting an expatriate host country, *Mathematical and Computer Modelling* 40(13), 1473–1490.

Chiou, H.K., Tzeng, G.H., Cheng, D.C. 2005. Evaluating sustainable fishing development strategies using fuzzy MCDM approach. *Omega* 33(3), 223–234.

Chu, M.T., Shyu, J.Z., Tzeng, G.H., Khosla, R., 2007. Using nonadditive fuzzy integral to assess performances of organizational transformation via communities of practice. *IEEE Transactions on Engineering Management* 54 (2), 327–339.

Ding, J.F., Liang, G.S. 2005. Using fuzzy MCDM to select partners of strategic alliances for liner shipping. *Information Sciences* 173(1–3), 197–225.

Emrouznejad, A., Tavana, M., 2014. *Performance Measurement with Fuzzy Data Envelopment Analysis.* In the series "Studies in Fuzziness and Soft Computing," New York/Dordrecht/London: Springer-Verlag, ISBN 978-3-642-41371-1.

Figueira, J., Greco, S., Ehrgott, M. (Eds.) 2004. *Multiple Criteria Decision Analysis: State of the Art Surveys*, New York: Springer, pp. 229–241.

Geldermann, J., Spengler, T., Rentz, O. 2000. Fuzzy outranking for environmental assessment. Case study: Iron and steel making industry. *Fuzzy Sets and Systems* 115(1), 45–65.

Ho, W., 2008. Integrated analytic hierarchy process and its applications—A literature review. *European Journal of Operational Research* 186 (1), 211–228.

Ho, W., Xu, X., Dey, P.K. 2010. Multi-criteria decision making approaches for supplier evaluation and selection: A literature review. *European Journal of Operational Research* 202(1), 16–24.

Huang, C.C., Chu, P.Y., Chiang, Y.H., 2008. A fuzzy-AHP application in government-sponsored R&D project selection. *Omega* 36 (6), 1038–1052.

Huang, C.T., Yeh, T.M., Lin, W.T., Lee, B.T., 2009. A fuzzy AHP-based performance evaluation model for implementing SPC in the Taiwanese LCD industry. *International Journal of Production Research* 47 (18), 5163–5183.

Inuiguchi, M., Ichihashi, H., Tanaka, H. 1990. Fuzzy programming: A survey of recent developments, in R Slowinski and J Teghem (Eds.), *Stochastic versus Fuzzy Approaches to Multiobjective Mathematical Programming under Uncertainty*, Dordrecht: Kluwer Academic Publishers, pp. 45–68.

Kahraman, C., Beskese, A., Kaya, I., 2010. Selection among ERP outsourcing alternatives using a fuzzy multi-criteria decision making methodology. *International Journal of Production Research* 48 (2), 547–566.

Kreng, V.B., Wu, C.Y., 2007. Evaluation of knowledge portal development tools using a fuzzy-AHP approach: The case of Taiwanese stone industry. *European Journal of Operational Research* 176 (3), 1795–1810.

Kubler S., Robert, J., Derigent, J., Voisin, A., Traon, Y.L., 2016. A state-of the-art survey & testbed of fuzzy-AHP (FAHP) applications. *Expert Systems with Applications* 65, 398–422.

Kwakernaak H. 1978. Fuzzy random variables. I: Definitions and theorems. *Information Sciences* 15, 1–29.

Lee, A.H.I., 2009. A fuzzy-AHP evaluation model for buyer-supplier relationships with the consideration of benefits, opportunities, costs and risks. *International Journal of Production Research* 47 (15), 4255–4280.

Liang, Y.H., 2015. Performance measurement of interorganizational information systems in the supply chain. *International Journal of Production Research* 53 (18), 5484–5499.

Ma, M.Y., Chen, C.Y., Wu, F.G., 2007. A design decision-making support model for customized product color combination. *Computers in Industry* 58 (6), 504–518.

Nagahanumaiah, Ravi, B., Mukherjee, N.P., 2007. Rapid tooling manufacturability evaluation using fuzzy-AHP methodology. *International Journal of Production Research* 45 (5), 1161–1181.

Ölçer, A.İ., Odabaşi, A.Y. 2005. A new fuzzy multiple attributive group decision making methodology and its application to propulsion/maneuvering system selection problem. *European Journal of Operational Research* 166(1), 93–114.

Onar, S.Ç., Büyüközkan, G., Öztay, B., Kahraman, C., 2016. A new hesitant fuzzy QFD approach: An application to computer workstation selection. *Applied Soft Computing* 46, 1–16.

Paul, S.K., 2015. Supplier selection for managing supply risks in supply chain: A fuzzy approach. *International Journal of Advanced Manufacturing Technology* 79 (1–4), 657–664.

Rezaei, J., Ortt, R., 2013. Multi-criteria supplier segmentation using a fuzzy preference relations based AHP. *European Journal of Operational Research* 225 (1), 75–84.

Saaty, T.L., 1980. *The Analytic Hierarchy Process*. New York, NY: McGraw-Hill.

Sarfaraz, A., Jenab, K., D'Souza, A.C., 2012. Evaluating ERP implementation choices on the basis of customisation using fuzzy AHP. *International Journal of Production Research* 50 (23), 7057–7067.

Steuer, R.E., Na, P., 2003. Multiple criteria decision making combined with finance: A categorized bibliographic study. *European Journal of Operational Research* 150 (3), 496–515.

Triantaphyllou, E. 2000. *Multi-Criteria Decision Making Methods: A Comparative Study*, London: Kluwer Academic Publishers.

Vaidya, O.S., Kumar, S., 2006. Analytic hierarchy process: An overview of applications. *European Journal of Operational Research* 169 (1), 1–29.

van de Kaa, G., van Heck, E., de Vries, H.J., van den Ende, J., Rezaei, J., 2014. Supporting decision making in technology standards battles based on a fuzzy analytic hierarchy process. *IEEE Transactions on Engineering Management* 61 (2), 336–348.

Van Laarhoven, P.J.M., Pedrycz, W. 1983. A fuzzy extension of Saaty's priority theory. *Fuzzy Sets and Systems* 11(1–3), 229–241.

Wang, J., Lin, Y.T. 2003. Fuzzy multicriteria group decision making approach to select configuration items for software development, *Fuzzy Sets and Systems* 134(3), 343–363.

Wang, L., Chu, J., Wu, J., 2007. Selection of optimum maintenance strategies based on a fuzzy analytic hierarchy process. *International Journal of Production Economics* 107 (1), 151–163.

Wang, T.C., Chang, T.C., 2007. Application of consistent fuzzy preference relations in predicting the success of knowledge management implementation. *European Journal of Operational Research* 182 (3), 1313–1329.

Wang, X., Chan, H.K., Yee, R.W.Y., Diaz-Rainey, I., 2012. A two-stage fuzzy-AHP model for risk assessment of implementing green initiatives in the fashion supply chain. *International Journal of Production Economics* 135 (2), 595–606.

Wang, Y.M., Luo, Y., Hua, Z., 2008. On the extent analysis method for fuzzy-AHP and its applications. *European Journal of Operational Research* 186 (2), 735–747.

Wang, J.J., Jing, Y.Y., Zhang, C.F., Zhao, J.H. 2009. Review on multi-criteria decision analysis aid in sustainable energy decision-making, *Renewable and Sustainable Energy Reviews* 13(9), 2263–2278.

Xu, Z.-S., Chen, J. 2007. An interactive method for fuzzy multiple attribute group decision making, *Information Sciences* 177(1), 248–263.

Yager, R.R., Basson, D. 1975. Decision making with fuzzy sets, *Decision Sciences* 6(3), 590–600.

Zadeh, L. 1965. Fuzzy sets, *Information and Control* 8(3), 338–353.

Zadeh, L.A. 1975. The concept of a linguistic variable and its application to approximate reasoning, *Information Sciences* 8(3), 199–249.

Zimmermann, H.J. 1978. Fuzzy programming and linear programming with several objective functions, *Fuzzy Sets and Systems* 1(1), 45–55.

2

The State of the Art in FAHP in Risk Assessment

Peh Sang Ng, Joshua Ignatius, Mark Goh,
Amirah Rahman, and Feng Zhang

CONTENTS

2.1 Introduction

According to the Commission on Risk Assessment and Risk Management (Christopher Frey and Patil, 2002), risk assessment addresses the social question, "What is unsafe?" Unfortunate events can occur at any time and place, with no specificity. Given sufficient preparation and evaluation, such events can be mitigated. What constitutes an effective risk assessment depends on the situation. Different risks call for different assessments. A crucial element in risk assessment is the risk framework. This framework, which precedes the risk assessment, not only provides the basic measurements of risk estimation, but also fundamental information for risk management.

Klinke and Renn (2002) suggest that there is no simple recipe for evaluating and managing risks, as it involves many stakeholders who possess diverse objectives and criteria that may conflict with each other. In addition, the subjectivity in the decision-making process often leads to ambiguity due to the lack of information or the information being compromised. These factors compound the list of challenges that decision makers face when evaluating risks. Managers need a formalized framework for better and faster risk evaluation that ensures the integration of stakeholder opinions through a wide spectrum of measures. One option is to introduce a method to determine risk performance from information. Multicriteria decision-making (MCDM) methods readily provide this means of capturing the judgment of the decision makers to weigh, rank, and select the parameters that reflect the true value of the decision maker when accurate and complete risk data are unavailable.

This chapter surveys risk assessment methodologies that were integrated with the fuzzy analytic hierarchy process (FAHP) model and aims to uncover the recent trends of their applications over the past two decades. This includes the attributes of (1) year of publication, (2) frequency distribution by two main categories of FAHP, (3) main objective, (4) frequency distribution by application area, (5) types of decision-making, (6) sensitivity analysis, and (7) validation methods.

To conduct this review, we select relevant articles (up to November 10, 2016) from the ISI Web of Knowledge database. The articles were found using the keywords of "fuzzy analytic hierarchy process," "risk assessment," and "risk evaluation," where publications such as book chapters and proceeding papers were not considered. We begin the screening process by considering each abstract and excluding irrelevant articles. Then, articles with the following issues were further excluded: (1) articles with significantly overlapping content that appeared in different journals (e.g., Levary, 2008), (2) articles that had a relatively poor relationship to risk assessment such as confusing and indirect issues of risk assessment, (3) articles that focused on other risk issues such as risk attitudes without any relevance to the application context of evaluation, and (4) articles that assessed the risk criterion as part of the evaluation criteria in the proposed model but the main objective was not risk assessment. Only 82 articles from 1996 to 2016 were found to be relevant to the topic of interest.

The remainder of this chapter is organized as follows: Section 2.2 presents the classification scheme for the reviewed articles, which utilizes two subcategories, fuzzy singular analytic hierarchy process (FSAHP) and fuzzy hybrid analytic hierarchy process (FHAHP). Within each subcategory, reviewed articles are tabulated based on the five requirements in Table 2.1. Section 2.3 discusses the observations drawn from the classification. The overall summary of the findings and future research directions for the FAHP

TABLE 2.1

Five Fields of Information Gathered from Each of the Review Articles

No.	Field
1	Did the study perform group decision-making?
2	Did the study differentiate each expert's competence according to their background and experience in evaluating the model?
3	Did the study model the decision maker's risk attitude in the evaluation model?
4	Did the study consider sensitivity analysis?
5	Did the study present a validation method for the proposed model in risk evaluation?

model in relation to risk assessment are presented in Section 2.4. Finally, conclusions are drawn in Section 2.5.

2.2 Classification

For the purpose of classification and to avoid confusion, we also group the methods that evaluate uncertain data, such as interval AHP (Zhang et al., 2013a) and rough AHP (Song et al., 2013), under the category of FAHP. The selected 82 articles were then classified into two subcategories: FSAHP and FHAHP. The singular approach is a stand-alone approach to evaluate the risk level or make decisions on the risk problem without combining with other tools. This means that it does not incorporate other methods. When compared with the hybrid approach, the singular approach is easier to conduct, does not need much computation, and saves time due to its relative simplicity. Aside from the two subcategories, each of the reviewed articles is tabulated across the five fields as shown in Table 2.1. Then, each article in each subcategory is summarized according to the format of Table 2.2.

2.2.1 Fuzzy Singular Analytic Hierarchy Process

A total of 49 articles used FSAHP in risk assessment. We treat the following fuzzy methods as fuzzy singular models: fuzzy set theory, fuzzy logic, fuzzy linguistic, fuzzy reasoning, fuzzy inference system, fuzzy comprehensive, and fuzzy synthesis evaluation. Table 2.3 shows the articles that use FSAHP in risk assessment.

2.2.2 Fuzzy Hybrid Analytic Hierarchy Process

There are 33 articles that applied FHAHP in risk assessment. Table 2.4 shows the articles that use fuzzy hybrid models in risk assessment.

TABLE 2.2

Classification Scheme for the Reviewed Articles

	Author (s)	Proposed Method	Reason for Using Hybrid Method	Application Area	Research Objective	1	2	3	4	5	Future Work
1											
2											
...											
82											

TABLE 2.3

Articles Based on Fuzzy Singular Analytic Hierarchy Process (49 Articles or 59.8% out of 82 Articles)

Author(s)	Method	Area	Objective of Study (S, R, W)	1	2	3	4	5	Future Work
Abdul-Rahman et al. (2013)	FAHP	CCE	R-Rank 3 building project objectives with respect to project risks.	x	x	–	–	x	Develop decision support tool to enhance efficiency of problem evaluation in terms of time constraint; develop methodology to simplify fuzzy computations.
An et al. (2016)	FAHP	CCE	W-Evaluate railway risk at Waterloo depot.	x	–	–	–	–	–
Arikan et al. (2013)	FAHP	IE	R-Rank 7 strategic risks in a public institution.	x	–	–	–	–	Consider interdependence relationships among the evaluated criteria.
Bayrakdaroglu and Yalcin (2013)	FAHP	FBA	R-Rank 10 operational risk factors for two commercial banks in Turkey.	x	–	–	–	–	Extend model to banking sector; consider use of other fuzzy MCDM methods such as fuzzy PROMETHEE, fuzzy ELECTRE, and fuzzy_ TOPSIS.
Bell and Crumpton (1997)	Fuzzy linguistic and AHP	HSM	W-Evaluate carpal tunnel syndrome risk in occupational setting.	x	–	–	x	x	–
Bi et al. (2015)	FCE and AHP	MA	W-Evaluate the low-carbon technological innovation risks in a manufacturing company.	x	–	–	–	–	Consider the innovation risks throughout the life cycle of low-carbon products; extend the model to incorporate the impact of stakeholders on low-carbon technological innovation risk and strategic planning with respect to the risk response.
Cebi (2011)	FAHP	CCE	W-Evaluate the risk level of a housing project	x	x	–	–	–	–

(Continued)

TABLE 2.3 (*Continued*)

Articles Based on Fuzzy Singular Analytic Hierarchy Process (49 Articles or 59.8% out of 82 Articles)

Author (s)	Method	Area	Objective of Study (S, R, W)	1	2	3	4	5	Future Work
Chen and Zhang (2009)	FCE and AHP	SE	W-Evaluate emergency planning and system of petrochemical storage depot.	x	–	–	–	–	
Dagdeviren and Yuksel (2008)	FAHP	HSM	W-Evaluate two work systems-based faulty behavior risk factors in work systems.	x	–	–	–	x	
Fan et al. (2016)	FCE and AHP	SE	R-Rank the risk level for 17 items of maglev train bogie.	x	–	–	–	–	Extend the research to include risk sources, effects, and FAHP model to address the drawback of using exact numbers for linguistic judgments.
Hamidi et al. (2008)	FAHP	CCE	S-Select rock tunnel boring machine with respect to geotechnical risk assessment (three alternatives).	–	–	–	–	–	
Hsueh et al. (2013)	Fuzzy logic and AHP	CCE	W-Evaluate the investment risks for redeveloping derelict public construction.	x	–	–	–	–	
Kang et al. (2014)	FCE and AHP	PEM	W-Evaluate the risk level for the oil storage tank zone.	–	–	–	–	–	Program the functions of the proposed model.
Keramati et al. (2013)	FAHP	BM	R-Rank 22 customer relationship management risk factors.	x	–	x	–	x	Consider the proposed methodology for other fields/countries; extend the model to incorporate life-cycle customer relationship management and other models such as fuzzy TOPSIS and fuzzy DEA.
Khan and Sadiq (2005)	Fuzzy synthetic and AHP	EM	R-Rank 35 air pollution monitoring locations.	x	–	x	–	–	

(*Continued*)

(Continued)

TABLE 2.3 (Continued)

Articles Based on Fuzzy Singular Analytic Hierarchy Process (49 Articles or 59.8% out of 82 Articles)

Author (s)	Method	Area	Objective of Study (S, R, W)	1	2	3	4	5	Future Work
Khan et al. (2004)	Fuzzy logic and AHP	SE	W-Evaluate risk performance for four different industries.	x	–	–	x	–	–
Khazaeni et al. (2012)	FAHP and Delphi method	CCE	W-Evaluate and determine risk allocation between owner and contractor in contract project.	x	–	–	–	x	–
Lee et al. (2009)	FSDM and AHP	CCE	S-Select optimal risk-sharing decision for construction contract (three alternatives).	x	–	–	–	–	–
Li et al. (2015)	FCE and AHP	EM	W-Evaluate the water inrush risk for two karst tunnels.	x	–	–	–	x	–
Li et al. (2016)	FCE and AHP	SCT	W-Evaluate the risk of using polysorbate 80 in supply chain.	x	–	–	–	–	–
Li and Zou (2011)	FAHP	BM	R-Rank 25 risk factors for Public–Private Partnership project.	x	–	x	–	x	–
Li et al. (2012a)	FCE and AHP	HSM	R-Rank four risk indicators and 29 subrisk indicators for mass crowd stampede–trampling accidents in stadium.	x	–	–	–	–	–
Luo et al. (2013)	Fuzzy set theory and AHP	EM	R-Rank coastal erosion risks for 23 regions.	x	–	–	–	–	–
Ma et al. (2010)	FCE and AHP	CCE	W-Evaluate the risk level for horizontal directional drilling crossing project.	x	–	–	–	–	–

TABLE 2.3 (*Continued*)

Articles Based on Fuzzy Singular Analytic Hierarchy Process (49 Articles or 59.8% out of 82 Articles)

Author(s)	Method	Area	Objective of Study (S, R, W)	1	2	3	4	5	Future Work
Mccauley-Bell and Badiru (1996)	Fuzzy rule-based system and AHP	HSM	W-Evaluate cumulative trauma disorder risk performance at local assembly plant.	x	–	–	–	x	Conduct further analysis to validate preliminary system.
Nieto-Morote and Ruz-Vila (2011)	FAHP	CCE	R-Rank 13 risk factors for a building rehabilitation project.	x	–	–	–	–	
Padma and Balasubramanie (2011)	FAHP	HSM	R-Rank 3 occupational settings-based shoulder and neck pain risks.	x	–	–	–	–	
Radivojević and Vladimir (2014)	FAHP	SCT	R-Rank five risk categories in a supply chain.	x	–	–	–	–	Develop decision support system for the proposed application case, extend model to include more evaluation criteria.
Sadiq and Husain (2005)	Fuzzy synthetic and AHP	EM	W-Estimate environmental risk level for three discharge alternatives of drilling waste disposal.	–	–	–	x	–	Consider sociopolitical, financial, and technical risks with the existing criteria in model for comprehensive evaluation.
Sadiq et al. (2004)	Fuzzy synthetic and AHP	EM	S-Select best discharge options-based risk concept for drilling waste disposal (three alternatives).	–	–	x	x	–	
Sadiq and Rodriguez (2004)	Fuzzy synthetic and AHP	HSM	W-Evaluate cancerous and noncancerous risks resulting from chlorinated DBPs.	–	–	–	x	x	Extend model to incorporate risks related to nonchlorinated DBPs and microbiological water quality to improve risk index; include expert panels and selection processes for aggregation operators, other chlorinated DBPs and nonchlorinated in evaluation.

(Continued)

TABLE 2.3 (Continued)

Articles Based on Fuzzy Singular Analytic Hierarchy Process (49 Articles or 59.8% out of 82 Articles)

Author(s)	Method	Area	Objective of Study (S, R, W)	1	2	3	4	5	Future Work
Sari et al. (2012)	FAHP	CCE	R-Rank risk governance for four urban rail systems in Istanbul.	x	–	–	x	–	Incorporate interdependency of risk criteria using the Choquet integral or ANP methods in model.
Song et al. (2013)	Rough AHP	MA	R-Rank 17 risk factors of customer integration in the development of new product.	x	–	x	–	x	Develop rough group ANP to consider interactive of risk criteria.
Sun et al. (2013)	Fuzzy logic and AHP	EM	W-Evaluate the risk of pollution from phosphogypsum tailing dams	x	x	–	–	–	–
Tesfamariam and Sadiq (2006)	FAHP	DE	S-Select the best drilling fluids of offshore oil and gas operations under each activities' risks (three alternatives).	–	–	x	x	–	–
Tian and Yan (2013)	FAHP	IT	W-Evaluate the risk level of general-assembling process	–	–	–	–	–	–
Topuz et al. (2011)	Fuzzy logic and AHP	EM	R-Rank six environmental and health risk sources for industrial hazardous materials.	x	–	–	–	–	–
Topuz and Van Gestel (2016)	Fuzzy inference rules and AHP	EM	W-Evaluate the environmental risk level due to the usage of engineered nanoparticles.	x	–	–	–	–	Extend the research by considering the correlation of environmental parameters with toxicity of engineered nanoparticles.

(Continued)

TABLE 2.3 (Continued)

Articles Based on Fuzzy Singular Analytic Hierarchy Process (49 Articles or 59.8% out of 82 Articles)

Author (s)	Method	Area	Objective of Study (S, R, W)	1	2	3	4	5	Future Work
Tsaur et al. (1997)	Fuzzy synthetic and AHP	TA	R-Rank six risk criteria for tour itineraries.	x	–	–	–	–	Consider tourists that experience full set of tour itinerary alternatives as respondents of study evaluation; incorporate other risk criteria such as financial risk, satisfaction risk, and time risk if alternative of foreign tour itinerary is considered in model evaluation.
Tuysuz and Kahraman (2006)	FAHP	IT	W-Evaluate risk level of information technology project.	x	–	–	–	–	Consider other FAHP models (except Chang's FAHP) and different fuzzy models such as ORESTE, PROMETHEE, ELECTRE, TOPSIS, and DEA to compare with results obtained.
Mangla et al. (2015)	FAHP	SCT	R-Rank 25 supply chain risk factors.	x	–	–	x	–	Extend the research to include risk management. Extend proposed methodology to developing/developed countries.
Ustundag et al. (2011)	Fuzzy rule based system and AHP	BM	W-Evaluate the risk for the real estate investment.	x	–	–	–	x	Develop decision support system for the proposed methodology.
Wang et al. (2012b)	FAHP	SCT	S-Select best green initiative alternative for three different case scenarios of fashion supply chain (three alternatives).	–	–	–	–	–	Evaluate model using real application case for more practical results.
Wang et al. (2012d)	FCE and AHP	ME	W-Evaluate risk level of floor water inrush in Huaibei Permian mining area, China.	x	–	–	–	x	–
Wang et al. (2016)	FAHP	SE	W-Evaluate the risk level of coal mine.	x	–	–	–	x	Consider partial correlation between the risk factors in the assessment.

(Continued)

TABLE 2.3 (*Continued*)

Articles Based on Fuzzy Singular Analytic Hierarchy Process (49 Articles or 59.8% out of 82 Articles)

Author (s)	Method	Area	Objective of Study (S, R, W)	1	2	3	4	5	Future Work
Zhang and Zou (2007)	FAHP	CCE	W-Evaluate risk in joint-venture construction project.	x	–	–	–	–	
Zhang et al. (2012)	FCE and AHP	DE	W-Evaluate risk of relief well project in South China offshore exploration.	x	–	–	–	–	Consider database in evaluation so as to avoid subjectivity and impersonal judgment; develop other more advanced methods such as BP neutral networks in risk assessment.
Zhang et al. (2013b)	Fuzzy set theory and AHP	EM	W-Evaluate the risk of groundwater contamination.	–	–	–	–	–	
Zheng et al. (2012)	FAHP	HSM	W-Evaluate the safety grade and the early warning grade in a coal mine environment.	x	–	–	–	x	–

AHP, analytic hierarchy process; ANP, analytic network process; BM, Business Management; BP, back propagation; CCE, Construction and Civil Engineering; DBPs, disinfection of by-products; DE, Design Engineering; DEA, data envelopment analysis; EM, Environmental Management; ELECTRE, elimination et choix traduisant la realité; FAHP, fuzzy analytic hierarchy process; FBA, Finance, Banking, and Accounting; FCE, fuzzy comprehensive evaluation; FSAHP, fuzzy singular analytic hierarchy process; FSDM, fuzzy synthetic decision method; HSM, Health and Safety Management; IE, Industrial Engineering; IT, Information Technology; MA, Manufacturing; ME, Mining Engineering; ORESTE, organization; rangement et synthese de donnes relationnelles; PEM, Power and Energy Management; PROMETHEE, preference ranking organization method for enrichment evaluations; SCT, Supply Chain and Transportation; SE, Safety Engineering; TA, Tourism; TOPSIS, technique for order of preference by similarity to ideal solution.

TABLE 2.4

Articles Based on Fuzzy Hybrid Analytic Hierarchy Process (33 Articles or 40.2% out of 82 Articles)

Author (s)	Method	Purpose of Using Hybrid Method	Area	Objective of study (S, R, W)	1	2	3	4	5	Future Work
Fuzzy hybrid AHP (26 articles)										
Abdelgawad and Fayek (2010)	FAHP and FMEA expert system	Use fuzzy FMEA expert system to provide valuable insight in analyzing effect of failure modes based on risk score rating and FAHP's importance weights of parameters of probability of occurrence, detection, and severity.	CCE	W-Evaluate risk for the construction company.	–	–	–	–	x	Develop relevant membership functions for adjusted number of linguistic terms; calibrate membership functions by avoiding full reelicitation; construct database for corrective actions toward associated risks.
Aliahmadi et al. (2011)	FAHP and game theory	Use game theory to analyze return of players by evaluating and study possible alternative decisions made by them.	CCE	S-Select best strategy in minimizing risk of Resalat tunnel project (nine alternatives).	x	–	–	–	–	–
An et al. (2007)	FAHP and fuzzy reasoning	Use fuzzy reasoning approach to evaluate incomplete or redundant of data parameters.	CCE	W-Evaluate railway risk at Waterloo depot.	x	–	–	–	–	–
An et al. (2011)	FAHP and fuzzy reasoning	Refer to An et al. (2007).	CCE	W-Evaluate railway risk at Hammersmith depot.	x	x	–	–	–	–

(Continued)

TABLE 2.4 (*Continued*)

Articles Based on Fuzzy Hybrid Analytic Hierarchy Process (33 Articles or 40.2% out of 82 Articles)

Author (s)	Method	Purpose of Using Hybrid Method	Area	Objective of study (S, R, W)	1	2	3	4	5	Future Work
Aydi et al. (2013)	Fuzzy logic, AHP, WLC approach, ELECTRE III, and GIS	Use WLC to aggregate criteria weight and standardized score information to obtain overall score of suitability index in selecting most appropriate scenario for landfill evaluation. Apply GIS to derive suitability map for alternative landfill site selection using ranking information form ELECTRE III.	EM	R-Rank five landfill sites to minimize environmental risk for siting municipal solid waste landfill in Ariana Region, Tunisia.	x	–	–	–	–	–
Chiang and Che (2010)	FAHP, fuzzy data envelopment analysis, Bayesian belief network model, cost and return analysis	Use Bayesian belief network model to consider causal relationships in evaluating input risk dimension of projects. Use cost and return model to evaluate output of cost and return dimension. Apply information in an integrated FAHP and FDEA ranking method.	MA	R-Rank 12 projects based on risks of new product development projects.	x	–	–	–	–	Develop resource allocation model of research and development.

(*Continued*)

TABLE 2.4 (Continued)

Articles Based on Fuzzy Hybrid Analytic Hierarchy Process (33 Articles or 40.2% out of 82 Articles)

Author (s)	Method	Purpose of Using Hybrid Method	Area	Objective of study (S, R, W)	1	2	3	4	5	Future Work
Debnath et al. (2016)	FIS, subtractive clustering technique, and AHP	Use subtractive clustering to reduce the number of if-then rules in FIS, whereas FIS-based AHP is used to refine the fuzzy if-then rules from the experts to describe the system's behavior.	HSM	W-Evaluate risk level for 17 body parts and the overall occupational risk level for construction industry.	x–	–	–	–	–	–
Gul and Guneri (2016)	FAHP and fuzzy TOPSIS	Incorporate the results from FAHP in fuzzy TOPSIS to rank the hazard types of an aluminum manufacturing facility.	MA	R-Rank the risk level associated with the hazards of the 12 departments in an aluminum factory.	x	–	–	x	–	Develop other MCDM models such as ANP, VIKOR, and PROMETHEE for evaluating hazards in an aluminum factory.
Hayaty et al. (2014)	FAHP, TOPSIS	Apply the results from FAHP into fuzzy TOPSIS to rank the risk of pollution from metals.	EM	W-Evaluate the pollution level of 10 metals in mine's tailings dam sediments.	–	–	–	–	–	–

(Continued)

TABLE 2.4 (*Continued*)

Articles Based on Fuzzy Hybrid Analytic Hierarchy Process (33 Articles or 40.2% out of 82 Articles)

Author (s)	Method	Purpose of Using Hybrid Method	Area	Objective of study (S, R, W)	1	2	3	4	5	Future Work
Hu et al. (2009)	FAHP and FMEA with RPN method	Multiply the criteria weighting from AHP with RPN values to determine the risk ranking for each alternative.	EM	R-Rank 17 risk factors for green components.	x	–	–	–	x	Consider FAHP and make adjustments to criteria setting of occurrence and severity indices to improve the model; develop compliances other than restriction of hazardous substance compliance in the criteria testing report; extend risk scope to include polybrominated biphenyls, polybrominated diphenyl ethers and hexavalent chromium products.
John et al. (2014)	FAHP, fuzzy set theory, belief degree concept, evidential reasoning, expected utility	Use evidential reasoning to derive the risk synthesis results by combining the results from FAHP and belief relief concept and fuzzy set theory. Expected utility is used to obtain the crisp value of risk synthesis.	SE	W-Evaluate the risk of disruption for seaport operations.	x	–	–	x	–	–

(*Continued*)

TABLE 2.4 (Continued)

Articles Based on Fuzzy Hybrid Analytic Hierarchy Process (33 Articles or 40.2% out of 82 Articles)

Author (s)	Method	Purpose of Using Hybrid Method	Area	Objective of study (S, R, W)	1	2	3	4	5	Future Work
Khan et al. (2002)	Fuzzy composite programming, AHP, and RBLCA method	Use fuzzy composite programming and AHP in RBLCA method to overcome large-scale computation and data requirements.	EM	S-Select best management alternative by considering risk-based life cycle for each urea production process.	x	–	–	x	–	–
Kutlu and Ekmekcioglu (2012)	FAHP and fuzzy TOPSIS	Model fuzziness of human feeling into mathematical rating and consider importance of risk factors in potential failure mode prioritization.	MA	R-Rank eight failure modes for four cases in manufacturing facility.	x	–	–	x	–	Compare proposed MCDM model with other MCDM models such as fuzzy VIKOR, fuzzy PROMETHEE, and fuzzy ELECTRE.
Lai et al. (2015)	FCE, AHP, entropy theory, game theory, and GIS	To integrate the subjective weights and objective weights of the evaluation criteria.	EM	W-Evaluate the flood risk areas in Dongjiang River Basin.	x	–	–	–	x	–
Lari et al. (2009)	Fuzzy logic, AHP, and budgetary allocation	Use AHP when number of variables is limited. Otherwise, apply budgetary allocation to determine importance weights of criteria.	EM	W-Evaluate technological and natural risk for the Lombardy Region, Italy.	x	–	x	x	–	–

(Continued)

TABLE 2.4 (Continued)

Articles Based on Fuzzy Hybrid Analytic Hierarchy Process (33 Articles or 40.2% out of 82 Articles)

Author (s)	Method	Purpose of Using Hybrid Method	Area	Objective of study (S, R, W)	1	2	3	4	5	Future Work
Lavasani et al. (2011)	FAHP and evidential reasoning	Use evidential reasoning approach to update evaluation result in any level of the hierarchy by merging newly obtained data with old data.	EM	W-Evaluate hydrocarbon leakage from offshore well.	x	–	–	x	–	Distinguish quantitative and qualitative data by assigning different weights in process of aggregation.
Li et al. (2012b)	VFS theory, AHP, and information diffusion technique	Use information diffusion to optimize evaluation sample into fuzzy sets to offset information deficiency.	EM	W-Evaluate flood disaster assessment in China.	–	–	–	–	x	–
Li et al. (2012c)	FAHP and neural networks	Select high–weight criteria as input in ANN method to reduce problem complexity and increase accuracy and speed of simulated assessment.	PEM	W-Evaluate risk in large power transformer.	x	–	x	–	x	–
Liu et al. (2014)	FAHP, entropy method, and fuzzy VIKOR	Both FAHP and entropy methods were used to derive the importance weights of risk factors subjectively and objectively in fuzzy VIKOR.	HSM	R-Rank seven failure modes in a university hospital.	x	x	–	x	x	Extend model to include other risk factors; develop an optimization model that can generate influential parameters objectively.

(Continued)

TABLE 2.4 (Continued)

Articles Based on Fuzzy Hybrid Analytic Hierarchy Process (33 Articles or 40.2% out of 82 Articles)

Author (s)	Method	Purpose of Using Hybrid Method	Area	Objective of study (S, R, W)	1	2	3	4	5	Future Work
Rodriguez et al. (2016)	FAHP and FIS	Use modified FAHP to simplify the implementation of FAHP in FIS for the purpose of integrating different groups of factors.	IT	R-Rank the risk level for three different IT projects.	–	–	–	–	–	Extend model to other applications; consider the concept of implementing classic pairwise comparison, adjustable weighting procedure, and other types of membership functions in the assessment; introduce historical data from postmortem studies of former projects in the evaluation; compare the proposed model with other fuzzy MCDM models.
Sadiq et al. (2007)	Fuzzy logic, AHP, evidential reasoning and exponential OWA	Merge newly obtained data with old data (evidential reasoning) and integrate decision maker's attitude in defuzzification (OWA operator).	HSM	W-Evaluate risk of water quality failure.	–	–	×	–	–	–

(Continued)

TABLE 2.4 (Continued)

Articles Based on Fuzzy Hybrid Analytic Hierarchy Process (33 Articles or 40.2% out of 82 Articles)

Author (s)	Method	Purpose of Using Hybrid Method	Area	Objective of study (S, R, W)	1	2	3	4	5	Future Work
Samuel et al. (2016)	FAHP and ANNs	Use FAHP to derive the weight for the attributes in an ANN approach.	HSM	W-Evaluate and predict the heart failure risk in patients.	–	–	–	–	x	Extend model to include prediction time and optimization techniques such as ant colony or genetic algorithm in the ANN approach.
Samvedi et al. (2013)	FAHP and fuzzy TOPSIS	Apply the results from FAHP into fuzzy TOPSIS to rank the risks of a supply chain.	SCT	R-Rank 17 supply chain risk factors in Indian textile and steel industry.	x	–	–	–	–	Consider customized risk index; extend the risk index by incorporating artificial intelligence modeling techniques, graph theory, multiagent, Petri net, etc.; develop interactive risk criteria by using fuzzy ANP.
Sharma et al. (2012)	FAHP and GIS	Use GIS to depict risk level of different areas in the map by gathering, processing, and integrating different types of data, including criteria weight from MADM method.	EM	W-Evaluate risk of fire for Taradevi forest range of Shimla Division in India.	–	–	–	–	x	–

(Continued)

TABLE 2.4 (Continued)

Articles Based on Fuzzy Hybrid Analytic Hierarchy Process (33 Articles or 40.2% out of 82 Articles)

Author (s)	Method	Purpose of Using Hybrid Method	Area	Objective of study (S, R, W)	1	2	3	4	5	Future Work
Sohn et al. (2001)	Fuzzy set theory, AHP, and MAUA	Integrate fuzzy set theory with MAUA to enable aggregation data from two separate pools (experts' and public's opinions) in determining total utility of each evaluation alternative.	EM	R-Rank six options for fuel research.	x	–	–	–	x	–
Taylan et al. (2014)	FAHP and fuzzy TOPSIS	Apply the results from FAHP into fuzzy TOPSIS to evaluate risks in construction projects.	CCE	R-Rank 30 construction projects based on risk factors.	x	–	–	–	–	–
Vadrevu et al. (2010)	Fuzzy set theory, AHP, and GIS	Use GIS to depict risk level of different areas in the map by gathering, processing and integrating different types of data, including criteria weight from MADM method.	EM	W-Evaluate forest fire risk in Andhra Pradesh, India.	x	–	–	–	–	Consider web-based GIS methodology in study.
Wang, et al. (2012c)	Fuzzy set theory, AHP, and risk parameter analysis	Use risk parameter analysis to determine risk score for each hazard. Obtain overall risk score by multiplying risk score and importance weight of each hazard.	SCT	W-Evaluate food safety risks for raw material batches in food supply chain.	x	–	–	–	–	Consider effects on risk transmission of supply chain processes such as removal, partitioning, growth, cross-contamination, and inactivation.

(Continued)

TABLE 2.4 (Continued)

Articles Based on Fuzzy Hybrid Analytic Hierarchy Process (33 Articles or 40.2% out of 82 Articles)

Author (s)	Method	Purpose of Using Hybrid Method	Area	Objective of study (S, R, W)	1	2	3	4	5	Future Work
Wang et al. (2011)	FAHP and GIS	Use GIS to depict risk level of different areas in the map by gathering, processing, and integrating different types of data, including criteria weight from MADM method.	EM	W-Evaluate flood risk level in Dongting Lake region.	x	–	–	–	–	–
Yang et al. (2013)	Triangular FAHP and GIS	Use GIS to depict risk level of different areas in the map by gathering, processing, and integrating different types of data, including criteria weight from MADM method.	EM	W-Evaluate flood risk level in Yangtze River.	x	–	–	–	x	–
Zhang et al. (2013a)	Interval AHP and interval TOPSIS	Use interval AHP to deal with uncertain data and determine criteria weights that fulfill condition of acceptable consistency. Apply results to interval TOPSIS analysis.	PEM	R-Rank 15 potential failure modes for hydropower project.	x	x	–	–	–	Develop new models that can justify quality of evaluation.

(Continued)

TABLE 2.4 (*Continued*)

Articles Based on Fuzzy Hybrid Analytic Hierarchy Process (33 Articles or 40.2% out of 82 Articles)

Author (s)	Method	Purpose of Using Hybrid Method	Area	Objective of study (S, R, W)	1	2	3	4	5	Future Work
Zheng et al. (2016)	FAHP and interval-valued intuitionistic fuzzy sets cross entropy	Use the concept of interval intuitionistic fuzzy sets to address the degrees of membership, nonmembership, and unknown membership.	CCE	R-Rank 18 failure modes for an earth-rock dam.	x	–	–	–	–	Extend the research to include the probability failure of dam risks.
Zou et al. (2013)	FAHP, SPA, VFS theory, and GIS	Use SPA to determine VFS membership degree function, which can contribute to less tedious and complicated computational process.	EM	W-Evaluate flood hazard and vulnerability assessment in Jingjiang flood diversion district, China.	x	–	–	–	x	–

AHP, analytic hierarchy process; ANN, artificial neural networks; BM, Business Management; CCE, Construction and Civil Engineering; DE, Design Engineering; EM, Environmental Management; FAHP, fuzzy analytic hierarchy process; FBA, Finance, Banking, and Accounting; FDEA, fuzzy data envelopment analysis; FIS, fuzzy inference system; FMEA, fuzzy mode effects analysis; GIS, Geographical Information System; HSM, Health and Safety Management; IE, Industrial Engineering; IT, Information Technology; MA, Manufacturing; MADM, multiattribute decision-making; MAUA, multiat-tribute utility analysis; MCDM, multicriteria decision-making; ME, Mining Engineering; OWA, ordered weighted averaging; PEM, Power and Energy Management; RBLCA, risk-based life cycle assessment; RoHS, restriction of hazardous substance; RPN, risk priority number; SCT, Supply Chain and Transportation; SE, Safety Engineering; SPA, set pair analysis; TA, Tourism; TOPSIS, technique for order of preference by similarity to ideal solution; VFS, variable fuzzy set; VIKOR, vlse kriterijumska optimizacija kompromisno resenje; WLC, weighted linear combination.

2.3 The Classification Scheme and Recommendations

An analysis of the overall growth of FAHP and each subcategory of FAHP as applied to risk assessment from 1996 to 2016 was conducted. The distribution by year of publication, frequency distribution by FAHP, the research objective of FAHP, and the application areas were analyzed. Following that a distribution of sensitivity analysis and summary of validation methods were organized.

Table 2.5 suggests that FSAHP models (49 articles) are preferred over FHAHP models (33 articles). This may be due to the simpler computational steps involved as well as the ease of applicability of FSAHP. However, the use of hybrid methods in decision analysis is expected to increase in the future across diverse application areas, given their "mix-and-match" ability to cancel out the weaknesses of their respective techniques when used singularly (see Vaidya and Kumar, 2006; Ho, 2008; Sipahi and Timor, 2010; Behzadian et al., 2010).

Furthermore, Table 2.5 shows that there is a growth in risk assessment using FAHP models, in which the two subcategories of FAHP exhibit a gradual upward trend since 2006. Specifically, both the FSAHP and FHAHP models surge in output after 2011 and 2008, respectively. From Figure 2.1, we can term the years 2009–2011 as the fast follower period where researchers found publishing opportunities and cues from the pioneering articles in deterministic hybrid models.

Each article was then categorized based on the main purpose of the FAHP application in addressing the risk context problem. Most of the risk problems applied the FAHP models for weighting (W) and ranking of alternatives (R), with coverage of 56.1% and 35.4%, respectively. The applications of FAHP models in dealing with decision problems (S) were not significant, with only 8.5% coverage. The extensive use of FAHP models to rank and evaluate the risk problems shows that the FAHP models have practical

TABLE 2.5

FAHP Categories

FAHP Category	<1997	1997–1999	2000–2002	2003–2005	2006–2008	2009–2011	2012–2014	2015–2016	Total	Total Percentage (%)
Fuzzy singular	1	2	0	5	5	9	19	8	49	59.8
Fuzzy hybrid	0	0	2	0	2	9	15	5	33	40.2
Total	1	2	2	5	7	18	34	13	82	100.0
Total percentage (%)	1.2	2.4	2.4	6.1	8.5	22.0	41.5	15.9		

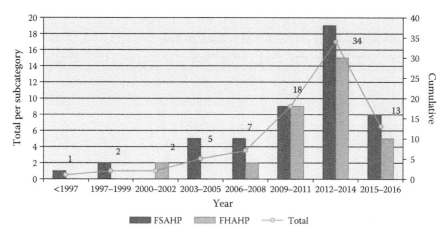

FIGURE 2.1
Total number of FAHP articles for each interval.

applicability in aiding the decision maker to predict and estimate risk and thus undertake proper precaution strategies to avoid the occurrence of adverse events.

The reviewed articles are further segregated into 11 main application areas, namely Environmental Management (EM), Construction and Civil Engineering (CCE), Health and Safety Management (HSM), Business Management (BM), Manufacturing (MA), Power and Energy Management (PEM), Supply Chain and Transportation (SCT), Safety Engineering (SE), Information Technology (IT), Design Engineering (DE), and other areas. The other areas cover the fields of Finance, Banking and Accounting (FBA), Mining Engineering (ME), Industrial Engineering (IE), and Tourism (TA). The distribution of the number of FAHP articles by application area is presented in Table 2.6.

Among the 11 application areas, the most commonly studied risk application in the review is in the EM area (23 articles). This tendency can be supported by the number of review articles that discussed the growth and risk assessment issues for the environmental field (Butt et al., 2014; Koornneef et al., 2012; Wang et al., 2012a). Fourteen out of these 23 articles employed hybrid methods in addressing the risk problems of EM. This may be due to the fact that environmental problems involve a number of subjective and unforeseen circumstances that are difficult to precisely measure. Singular methodology approaches pose some limitations when one considers the complexity and fuzziness in the evaluation of various disasters, such as flood (Li et al., 2012b; Wang et al., 2011; Yang et al., 2013; Zou et al., 2013) and grassland/forest fire (Sharma et al., 2012; Vadrevu et al., 2010). Thus, usually the FAHP model is combined with a Geographical Information System (GIS) methodology to gather and process different data in determining the distribution of risk performance through risk maps.

TABLE 2.6

The Distribution of FAHP Articles by Application Area

	EM	CCE	HSM	BM	MA	PEM	SCT	SE	IT	DE	Others	Total
1996			1									1
1997			1								1	2
2001	1											1
2002	1											1
2003												0
2004	1		1					1				3
2005	2											2
2006									1	1		2
2007		2	1									3
2008		1	1									2
2009	2	1						1				4
2010	1	2			1							4
2011	3	4	1	2								10
2012	2	2	2		1	1	2			1	1	12
2013	6	2		1	1	1	1		1		2	15
2014	2	1	1			1	1	1				7
2015	1				1		1					3
2016	1	2	2		1		1	2	1			10
Total	23	17	11	3	5	3	6	5	3	2	4	82

BM, Business Management; CCE, Construction and Civil Engineering; DE, Design Engineering; EM, Environmental Management; HSM, Health and Safety Management; IT, Information Technology; MA, Manufacturing; PEM, Power and Energy Management; SCT, Supply Chain and Transportation; SE, Safety Engineering.

Following that, we observed that the most common type of decision-making applied in the literature is group decision-making. Group decision-making accounts for 80.5% of the articles in risk assessment. This reflects the emphasis on avoiding biases of a single person's opinion in the risk assessment literature. To strive for an effective group evaluation, the concern should also be placed in the profile of the experts as the accuracy of the evaluation results is partially dependent on the judgment information provided by expert knowledge, with the other half relying on the structure as well as variables and measures of the model. Therefore, recent research has focused on developing FAHP models that take into account the weight of each expert by differentiating each expert's competence according to their background and experience (see Abdul-Rahman et al., 2013; An et al., 2011; Cebi, 2011; Sun et al., 2013; Liu et al., 2014; Zhang et al., 2013a).

In addition, FAHP models are limited to the consideration of the decision maker's risk attitude in the evaluation. The ability to identify the decision maker's personality and preference in risk evaluation would allow for a higher degree of confidence toward the selected alternatives. The issue of overestimating (underestimating) the utility of positive outcomes relative to the utility of negative outcomes received may occur if the risk-seeking risk-averse continuum of a decision maker is not accounted for properly, rendering the ranking outcome questionable (Tsaur, 2011). Some research has incorporated the modeling of the risk attitude (tolerance) in the risk framework. For instance, Sadiq et al. (2007) account for the decision maker's attitude and preferences by incorporating the aggregation method of exponential ordered weighted averaging (OWA) into the proposed risk analysis. Most of the articles reviewed considered the risk attitude of the decision maker for AHP models by introducing different modeling frameworks such as the weighted average approach (Khan and Sadiq, 2005; Sadiq et al., 2004), α-method (Li and Zou, 2011), Deng's risk attitude index (Tesfamariam and Sadiq, 2006), possibility degree matrix (Li et al., 2012c), budgetary allocation approach (Lari et al., 2009), and optimistic indicators (Song et al., 2013).

Of the articles reviewed, 15 out of 82 included sensitivity analysis to revise estimation of the current scenario in risk assessment. Sensitivity analysis provides the flexibility for the analyst to revise the key variable weights to identify the most significant exposure or risk factors, which directly aids in developing priorities for risk reduction. This process helps in outcomes validation, especially in risk-based decisions where an expert's judgment involves a certain degree of uncertainty (Christopher Frey and Patil, 2002).

To address the issue of "Are the evaluation results trustworthy?" 25 articles conducted validation methods such as comparison methods, analyzing a previous study's results, reviewing the actual situation, further test analysis, and feedback from the experts to verify their evaluation results. Table 2.7 shows that various types of validation methods applied for model justification. In terms of percentage, the majority of the reviewed articles, 64%, validate their evaluation results by comparing their proposed method with other methods. The validation method of comparing the assessment results with actual case conditions is the second most common validation method (20%).

2.4 Findings and Future Research on Risk Assessment

Overall, the growth rate of FAHP applications involving risk assessment process has been on the rise for the period 1996–2016. The growth rate was relatively slow from 1996 to 2008 but rapidly expanded from 2009 to 2016, contributing to 65 out of the 82 reviewed articles (79.3%). The deployment of FAHP methods in risk applications is expected to grow further as topics in human health and environmental issues remain a growing concern (Butt et al., 2014; Wang et al., 2012a).

TABLE 2.7

Different Types of Validation Method

Author(s)					
Validation Method	Comparison Methods	Results of Previous Studies	Actual Case Condition/ Decision	Further Test Analysis/ Calibration	Feedback from Experts
Abdelgawad and Fayek (2010)					x
Abdul-Rahman et al. (2013)					x
Bell and Crumpton (1997)	x				
Dagdeviren and Yuksel (2008)	x	x			x
Hu et al. (2009)	x				
Keramati et al. (2013)		x			
Khazaeni et al. (2012)	x				
Lai et al. (2015)		x			
Li and Zou (2011)	x				
Li et al. (2012b)	x			x	
Li et al. (2012c)					x
Li et al. (2015)			x		
Liu et al. (2014)	x				
Mccauley-Bell and Badiru (1996)			x	x	
Sadiq and Rodriguez (2004)	x			x	
Samuel et al. (2016)	x				
Sharma et al. (2012)	x				
Sohn et al. (2001)	x				
Song et al. (2013)	x				
Ustundag et al. (2011)			x		
Wang et al. (2012d)			x		
Wang et al. (2016)	x		x		
Yang et al. (2013)	x				
Zheng et al. (2012)	x				
Zou et al. (2013)	x				
Total percent (%)	64	12	20	12	16.0

In this review, fuzzy hybrid models appeared after 1999, which is after the appearance of fuzzy singular models. One would expect this situation since fuzzy singular approaches are less complex than fuzzy hybrid methods. Nevertheless, we anticipate more risk assessment articles to bear the FHAHP model methodology in the future as risk is naturally complicated. This was highlighted by Kangas and Kangas (2005) stating that the efficacy of the planning process can be improved by proposing a hybrid method. We believe that hybrid methods will see consistent growth in the future as they allow the interaction of different methods to compensate for the weaknesses of singular methods when applied in isolation.

Of the 11 application areas grouped in risk assessment, the area that received the most significant attention of FAHP models is Environmental Management (23 articles). This confirms the observation made by Huang et al. (2011) that multicriteria decision analysis in the environmental field has been on the rise for the past decade. It may be an indication that the use of FAHP in environmental risk assessment will gain more attention in the future. However, the areas that did not receive significant attention included DE, FBA, ME, IE, and TA, with fewer than three articles per area. Future work should explore FAHP models on risk problems in these areas, as well as other areas of interest such as aircraft building and university labs.

Our summary statistics revealed that 61% of articles in the area of environment management applied FHAHP. It is noteworthy that 42.7% of the FHAHP models assessed risks associated with natural disasters (see Vadrevu et al., 2010; Wang et al., 2011; Sharma et al., 2012; Yang et al., 2013; Zou et al., 2013; Lai et al., 2015). The application of FHAHP to natural disaster risk assessment is scarce within the first 14 years of our review period, but this changed after 2010. Indeed, the FHAHP model appeared as a solution to some of the weaknesses presented by the FSAHP model. Overall, the trend of using FAHP models in environment management is popular among green and eco-friendly environmental studies.

More research could be done in terms of risk response strategy and risk allocation (e.g., loss and benefits). Most reviewed articles on risk assessment are limited to reporting the results for the contextual problems, without further investigation of potential improved treatments for the consequences associated with an unacceptable or risky risk level. We found relatively fewer number of applications that use the risk sharing concept. More empirical works on risk assessment with response actions are needed to capture realistic strategies in managing risks. This would allow researchers to gain context-specific knowledge for risk management in any particular decision situation.

In addition, a number of articles in group decision-making provide the final aggregated score without elaborating on the group decision-making process and assume all decision makers have the same personality. It would also be helpful if the articles could put more thought into selecting certain preference functions rather than assuming that all decision makers would behave similarly. For example, PROMETHEE remains a far superior method in risk management due to its well-conceived preference functions and flexibility in applying them. Alternatively, the integration of a decision maker's personality and risk attitude in FAHP applications, which has already been applied in some studies, should be considered during risk evaluation.

Finally, most articles do not design specific software for their risk frameworks. Thus, in order to enhance risk evaluation research, practitioners and decision makers are encouraged to translate their objectives through decision support software, which can facilitate sequential decision-making, visualization support, and documentation integration (Tan and Platts, 2003).

2.5 Conclusion

This study is a review on risk assessment using a range of FAHP models from 1996 to 2016. Our intention is to classify the FAHP models in various application areas of risk assessment. The objective was to cover recent trends of FAHP models in supporting risk assessment. Our screening process produced 82 papers, covering 11 application areas, which can be further categorized into two types of models: singular and fuzzy hybrid. From these, we determined the distribution by the objective of FAHP in the risk application, the frequency distribution of FAHP models used, the distribution by application area, a summary of the types of decision-making models, the extent of the popularity of sensitivity analysis, and whether validation methods were used. The methodology that was used in this review poses several limitations. The first limitation is that while the ISI Web of Knowledge database covers a wide range of journals, we faced the problem of obtaining the full set of articles. Some articles from this database cannot be accessed while some of the articles still in press in the ISI journals are yet to be included in the database. Nonetheless, this review provides a broad overview and some useful insights into the anatomy of the various FAHP models used in risk assessment.

References

Abdelgawad, M., Fayek, A. R., 2010. Risk management in the construction industry using combined fuzzy FMEA and fuzzy AHP. *Journal of Construction Engineering and Management* 136, 1028–1036.

Abdul-Rahman, H., Wang, C., Lee, Y. L., 2013. Design and pilot run of fuzzy synthetic model (FSM) for risk evaluation in civil engineering. *Journal of Civil Engineering and Management* 19, 217–238.

Aliahmadi, A., Sadjadi, S. J., Jafari-Eskandari, M., 2011. Design a new intelligence expert decision making using game theory and fuzzy AHP to risk management in design, construction, and operation of tunnel projects (case studies: Resalat tunnel). *International Journal of Advanced Manufacturing Technology* 53, 789–798.

An, M., Chen, Y., Baker, C. J., 2011. A fuzzy reasoning and fuzzy analytical hierarchy process based approach to the process of railway risk information: A railway risk management system. *Information Sciences* 181, 3946–3966.

An, M., Huang, S., Baker, C. J., 2007. Railway risk assessment—The fuzzy reasoning approach and fuzzy analytic hierarchy process approaches: A case study of shunting at Waterloo depot. *Proceedings of the Institution of Mechanical Engineers Part F—Journal of Rail and Rapid Transit* 221, 365–383.

An, M., Qin, Y., Jia, L. M., Chen, Y., 2016. Aggregation of group fuzzy risk information in the railway risk decision making process. *Safety Science* 82, 18–28.

Arikan, R., Dagdeviren, M., Kurt, M., 2013. A fuzzy multi-attribute decision making model for strategic risk assessment. *International Journal of Computational Intelligence Systems* 6(3), 487–502.

Aydi, A., Zairi. M., Ben Dhia, H., 2013. Minimization of environmental risk of landfill site using fuzzy logic, analytical hierarchy process, and weighted linear combination methodology in a geographic information system environment. *Environmental Earth Sciences* 68, 1375–1389.

Bayrakdaroglu, A., Yalcin, N., 2013. A fuzzy multi-criteria evaluation of the operational risk factors for the state-owned and privately-owned commercial banks in Turkey. *Human and Ecological Risk Assessment* 19, 443–461.

Behzadian, M., Kazemzadeh, R., Albadvi, A., Aghdasi, M., 2010. PROMETHEE: A comprehensive literature review on methodologies and applications. *European Journal of Operational Research* 200, 198–215.

Bell, P. M., Crumpton, L., 1997. A fuzzy linguistic model for the prediction of carpal tunnel syndrome risks in an occupational environment. *Ergonomics* 40, 790–799.

Bi, K., Huang, P., Hui, Y., 2015. Risk identification, evaluation and response of low-carbon technological innovation under the global value chain: A case of the Chinese manufacturing industry. *Technological Forecasting and Social Change* 100, 238–248.

Butt, T., Gouda, H., Baloch, M., Paul, P., Javadi, A., Alam, A., 2014. Literature review of baseline study for risk analysis—The landfill leachate case. *Environment International* 63, 149–162.

Cebi, S., 2011. Developing a fuzzy based decision making model for risk analysis in construction project. *Journal of Multiple-Valued Logic and Soft Computing* 17(4), 387–405.

Chen, G., Zhang, X., 2009. Fuzzy-based methodology for performance assessment of emergency planning and its application. *Journal of Loss Prevention in the Process Industries* 22, 125–132.

Chiang, T. A., Che, Z. H., 2010. A fuzzy robust evaluation model for selecting and ranking NPD projects using Bayesian belief network and weight-restricted DEA. *Expert Systems with Applications* 37, 7408–7418.

Christopher Frey, H., Patil, S. R., 2002. Identification and review of sensitivity analysis methods. *Risk Analysis* 22, 553–578.

Dagdeviren, M., Yuksel, I., 2008. Developing a fuzzy analytic hierarchy process (AHP) model for behavior-based safety management. *Information Sciences* 178, 1717–1733.

Debnath, J., Biswas, A., Sivan, P., Sen, K. N., Sahu, S., 2016. Fuzzy inference model for assessing occupational risks in construction sites. *International Journal of Industrial Ergonomics* 55, 114–128.

Fan, C., Dou, F., Tong, B., Long, Z., 2016. Risk analysis based on AHP and fuzzy comprehensive evaluation for maglev train bogie. *Mathematical Problems in Engineering* 2016, 10, Article ID 1718257.

Gul, M., Guneri, A. F., 2016. A fuzzy multi criteria risk assessment based on decision matrix technique: A case study for aluminum industry. *Journal of Loss Prevention in the Process Industries* 40, 89–100.

Hamidi, J. K., Shahriar, K., Rezai, B., Rostami, J., Bejari, H., 2008. Risk assessment based selection of rock TBM for adverse geological conditions using fuzzy-AHP. *Bulletin of Engineering Geology and the Environment* 69, 523–532.

Hayaty, M., Mohammadi, M. R. T., Rezaei, A., Shayestehfar, M. R., 2014. Risk assessment and ranking of metals using FDAHP and TOPSIS. *Mine Water and the Environment* 33(2), 157–164.

Ho, W., 2008. Integrated analytic hierarchy process and its applications—A literature review. European *Journal of Operational Research* 186, 211–228.

Hsueh, S. L., Lee, J. R., Chen, Y. L. 2013. DFAHP multicriteria risk assessment model for redeveloping derelict public buildings. *International Journal of Strategic Property Management* 17(4), 333–346.

Hu, A. H., Hsu, C. W., Kuo, T. C., Wu, W. C., 2009. Risk evaluation of green components to hazardous substance using FMEA and FAHP. *Expert Systems with Applications* 36, 7142–7147.

Huang, I. B., Keisler, J., Linkov, I., 2011. Multi-criteria decision analysis in environmental sciences: Ten years of applications and trends. *Science of the Total Environment* 409, 3578–3594.

John, A., Paraskevadakis, D., Bury, A., Yang, Z., Riahi, R., Wang, J., 2014. An integrated fuzzy risk assessment for seaport operations. *Safety Science* 68, 180–194.

Kang, J., Liang, W., Zhang, L., Lu, Z., Liu, D., Yin, W., Zhang, G., 2014. A new risk evaluation method for oil storage tank zones based on the theory of two types of hazards. *Journal of Loss Prevention in the Process Industries* 29, 267–276.

Kangas, J., Kangas, A., 2005. Multiple criteria decision support in forest management—The approach, methods applied, and experiences gained. *Forest Ecology and Management* 207, 133–143.

Keramati, A., Nazari-Shirkouhi, S., Moshki, H., Afshari-Mofrad, M., Maleki-Berneti, E., 2013. A novel methodology for evaluating the risk of CRM projects in fuzzy environment. *Neural Computing & Applications* 23(1), 29–53.

Khan, F. I., Sadiq, R., 2005. Risk-based prioritization of air pollution monitoring using fuzzy synthetic evaluation technique. *Environmental Monitoring and Assessment* 105, 261–283.

Khan, F. I., Sadiq, R., Haddara, M. M., 2004. Risk-based inspection and maintenance (RBIM): Multi-attribute decision-making with aggregative risk analysis. *Process Safety and Environmental Protection* 82, 398–411.

Khan, F. I., Sadiq, R., Husain, T., 2002. GreenPro I: A risk-based life cycle assessment and decision-making methodology for process plant design. *Environmental Modelling & Software* 17, 669–692.

Khazaeni, G., Khanzadi, M., Afshar, A., 2012. Fuzzy adaptive decision making model for selection balanced risk allocation. *International Journal of Project Management* 30, 511–522.

Klinke, A., Renn, A., 2002. A new approach to risk evaluation and management: Risk-based, precaution-based, and discourse-based strategies. *Risk Analysis* 22, 1071–1094.

Koornneef, J., Ramirez, A., Turkenburg, W., Faaij, A., 2012. The environmental impact and risk assessment of CO_2 capture, transport and storage—An evaluation of the knowledge base. *Progress in Energy and Combustion Science*, 38(1), 62–86.

Kutlu, A. C., Ekmekcioglu, M., 2012. Fuzzy failure modes and effects analysis by using fuzzy TOPSIS-based fuzzy AHP. *Expert Systems with Applications* 39, 61–67.

Lai, C., Chen, X., Chen, X., Wang, Z., Wu, X., Zhao, S., 2015. A fuzzy comprehensive evaluation model for flood risk based on the combination weight of game theory. *Natural Hazards* 77(2), 1243–1259.

Lari, S., Frattini, P., Crosta, G. B., 2009. Integration of natural and technological risks in Lombardy, Italy. *Natural Hazards and Earth System Sciences* 9, 2085–2106.

Lavasani, S. M. M., Yang, Z., Finlay, J., Wang, J., 2011. Fuzzy risk assessment of oil and gas offshore wells. *Process Safety and Environmental Protection* 89, 277–294.

Lee, T. C., Lee, T. H., Wang, C. H., 2009. Decision analysis for construction contract risk-sharing. *Journal of Marine Science and Technology* 17, 75–87.

Levary, R. R., 2008. Using the analytic hierarchy process to rank foreign suppliers based on supply risks. *Computers & Industrial Engineering* 55, 535–542.

Li, J., Zou, P. X. W., 2011. Fuzzy AHP-based risk assessment methodology for PPP projects. *Journal of Construction Engineering and Management* 137, 1205–1209.

Li, L., Lei, T., Li, S., Zhang, Q., Xu, Z., Shi, S., Zhou, Z., 2015. Risk assessment of water inrush in karst tunnels and software development. *Arabian Journal of Geosciences* 8(4), 1843–1854.

Li, M., Du, Y., Wang, Q., Sun, C., Ling, X., Yu, B., Tu, J., Xiong, Y., 2016. Risk assessment of supply chain for pharmaceutical excipients with AHP-fuzzy comprehensive evaluation. *Drug Development and Industrial Pharmacy* 42(4), 676–684.

Li, M., Peng, H., Zhang, X., Deng, L., 2012a. Research on risk assessment system of mass crowded stampede-trampling accidents in stadium. *Applied Mathematics & Information Sciences* 6, 9–14.

Li, Q., Zhou, J., Liu., D., Jiang, X., 2012b. Research on flood risk analysis and evaluation method based on variable fuzzy sets and information diffusion. *Safety Science* 50, 1275–1283.

Li, W., Yu., Q., Luo, R., 2012c. Application of fuzzy analytic hierarchy process and neural network in power transformer risk assessment. *Journal of Central South University of Technology* 19, 982–987.

Liu, H. C., You, J. X., You, X. Y., Shan, M. M., 2014. A novel approach for failure mode and effects analysis using combination weighting and fuzzy VIKOR method. *Applied Soft Computing* 28, 579–588.

Luo, S., Wang, H., Cai, F., 2013. An integrated risk assessment of coastal erosion based on fuzzy set theory along Fujian coast, southeast China. *Ocean & Coastal Management* 84, 68–76.

Ma, B., Najafi, M., Shen, H., Wu, L., 2010. Risk evaluation for maxi horizontal directional drilling crossing projects. *Journal of Pipeline Systems Engineering and Practice* 1(2), 91–97.

Mangla, S. K., P. Kumar, Barua, M. K., 2015. Risk analysis in green supply chain using fuzzy AHP approach: A case study. *Resources, Conservation and Recycling* 104, 375–390.

Mccauley-Bell, P., Badiru, A. B., 1996. Fuzzy modeling and analytic hierarchy processing—Means to quantify risk levels associated with occupational injuries—Part II: The development of a fuzzy rule-based model for the prediction of injury. *IEEE Transactions on Fuzzy Systems* 4, 132–138.

Nieto-Morote, A., Ruz-Vila, F., 2011. A fuzzy approach to construction project risk assessment. *International Journal of Project Management* 29, 220–231.

Padma, T., Balasubramanie, P., 2011. A fuzzy analytic hierarchy processing, decision support system to analyze occupational menace forecasting the spawning of shoulder and neck pain. *Expert Systems with Applications* 38, 15303–15309.

Radivojević, G., Gajovic, V., 2014. Supply chain risk modeling by AHP and fuzzy AHP methods. *Journal of Risk Research* 17(3), 337–352.

Rodriguez, A., Ortega, F., Concepcion, R., 2016. A method for the evaluation of risk in IT projects. *Expert Systems with Applications* 45, 273–285.

Sadiq, R., Husain, T., 2005. A fuzzy-based methodology for an aggregative environmental risk Assessment: A case study of drilling waste. *Environmental Modelling & Software* 20, 33–46.

Sadiq, R., Husain, T., Veitch, B., Bose, N., 2004. Risk-based decision-making for drilling waste discharges using a fuzzy synthetic evaluation technique. *Ocean Engineering* 31, 1929–1953.

Sadiq, R., Kleiner, Y., Rajani, B., 2007. Water quality failures in distribution networks—Risk analysis using fuzzy logic and evidential reasoning. *Risk Analysis* 27, 1381–1394.

Sadiq, R., Rodriguez, M. J., 2004. Fuzzy synthetic evaluation of disinfection by-product a risk based indexing system. *Journal of Environmental Management* 73, 1–13.

Samuel, O. W., Asogbon, G. M., Sangaiah, A. K., Fang, P., Li, G., 2016. An integrated decision support system based on ANN and Fuzzy_AHP for heart failure risk prediction. *Expert Systems with Applications* 68, 163–172.

Samvedi, A., Jain, V., Chan, F. T. S., 2013. Quantifying risks in a supply chain through integration of fuzzy AHP and fuzzy TOPSIS. *International Journal of Production Research* 51(8), 2433–2442.

Sari, I. U., Behret, H., Kahraman, C., 2012. Risk governance of urban rail systems using fuzzy AHP: The case of Istanbul. *International Journal of Uncertainty Fuzziness and Knowledge-Based Systems* 20, 67–79.

Sharma, L. K., Kanga, S., Nathawat, M. S., Sinha, S., Pandey, P. C., 2012. Fuzzy AHP for forest fire risk modeling. *Disaster Prevention and Management* 21, 160–171.

Sipahi, S., Timor, M., 2010. The analytic hierarchy process and analytic network process: An overview of applications. *Management Decision* 48, 775–808.

Sohn, K. Y., Yang, J. W., Kang, C. S., 2001. Assimilation of public opinions in nuclear decision-making using risk perception. *Annals of Nuclear Energy* 28, 553–563.

Song, W., Ming, X., Xu, Z., 2013. Risk evaluation of customer integration in new product development under uncertainty. *Computers & Industrial Engineering* 65, 402–412.

Sun, X., Ning, P., Tang, X., Yi, H., Li, K., Zhou, L., Xu, X., 2013. Environmental risk assessment system for phosphogypsum tailing dams. *Scientific World Journal* 2013, 13, Article ID 680798.

Tan, K. H., Platts, K., 2003. Linking objectives to actions: A decision support approach based on cause–effect linkages. *Decision Sciences* 34, 569–593.

Taylan, O., Bafail, A. O., Abdulaal, R. M. S., Kabli, M. R., 2014. Construction projects selection and risk assessment by fuzzy AHP and fuzzy TOPSIS methodologies. *Applied Soft Computing* 17, 105–116.

Tesfamariam, S., Sadiq, R., 2006. Risk-based environmental decision-making using fuzzy analytic hierarchy process (F-AHP). *Stochastic Environmental Research and Risk Assessment* 21, 35–50.

Tian, J., Yan, Z. F., 2013. Fuzzy analytic hierarchy process for risk assessment to general-assembling of satellite. *Journal of Applied Research and Technology* 11, 568–577.

Topuz, E., Talinli, I., Aydin, E., 2011. Integration of environmental and human health risk assessment for industries using hazardous materials: A quantitative multi criteria approach for environmental decision makers. *Environment International* 37, 393–403.

Topuz, E., Van Gestel, C. A., 2016. An approach for environmental risk assessment of engineered nanomaterials using analytical hierarchy process (AHP) and fuzzy inference rules. *Environment International* 92, 334–347.

Tsaur, R. C., 2011. Decision risk analysis for an interval TOPSIS method. *Applied Mathematics and Computation* 218, 4295–4304.

Tsaur, S. H., Tzeng, G. H., Wang, K. C., 1997. Evaluating tourist risks from fuzzy perspectives. *Annals of Tourism Research* 24, 796–812.

Tuysuz, F., Kahraman, C., 2006. Project risk evaluation using a fuzzy analytic hierarchy process: An application to information technology projects. *International Journal of Intelligent Systems* 21, 559–584.

Ustundag, A., Cevikcan, E., Kilinc, M. S., 2011. A hybrid fuzzy risk evaluation model for real estate investments. *Journal of Multiple-Valued Logic and Soft Computing* 17(4), 339–362.

Vadrevu, K. P., Eaturu, A., Badarinath, K. V. S., 2010. Fire risk evaluation using multicriteria analysis—A case study. *Environmental Monitoring and Assessment* 166, 223–239.

Vaidya, O. S., Kumar, S., 2006. Analytic hierarchy process: An overview of applications. *European Journal of Operational Research* 169, 1–29.

Wang, H., Yan, Z. G., Li, H., Yang, N. Y., Leung, K. M.,. Wang, Y. Z., Yu, R. Z., Zhang, L., Wang, W. H., Jiao, C. Y., Liu, Z. T., 2012a. Progress of environmental management and risk assessment of industrial chemicals in China. *Environmental Pollution* 165, 174–181.

Wang, Q., Wang, H., Zuoqiu, Q., 2016. An application of nonlinear fuzzy analytic hierarchy process in safety evaluation of coal mine. *Safety Science* 86, 78–87.

Wang, X., Chan, H. K., Yee, R. W. Y., Diaz-Rainey, I., 2012b. A two-stage fuzzy-AHP model for risk assessment of implementing green initiatives in the fashion supply chain. *International Journal of Production Economics* 135, 595–606.

Wang, X., Li, D., Shi, X., 2012c. A fuzzy model for aggregative food safety risk assessment in food supply chains. *Production Planning & Control* 23, 377–395.

Wang, Y., Li, Z., Tang, Z., Zeng, G., 2011. A GIS-based spatial multi-criteria approach for flood risk assessment in the Dongting Lake Region, Hunan, Central China. *Water Resources Management* 25, 3465–3484.

Wang, Y., Yang, W., Li, M., Liu, X., 2012d. Risk assessment of floor water inrush in coal mines based on secondary fuzzy comprehensive evaluation. *International Journal of Rock Mechanics and Mining Sciences* 52, 50–55.

Yang, X. I., Ding, J. H., Hou, H., 2013. Application of a triangular fuzzy AHP approach for flood risk evaluation and response measures analysis. *Natural Hazards* 68, 657–674.

Zhang, G., Zou, P. X. W., 2007. Fuzzy analytical hierarchy process risk assessment approach for joint venture construction projects in China. *Journal of Construction Engineering and Management* 133, 771–779.

Zhang, H., Gao, D., Liu, W., 2012. Risk assessment for Liwan relief well in South China Sea. *Engineering Failure Analysis* 23, 63–68.

Zhang, Q., Yang, X., Zhang, Y., Zhong, M., 2013b. Risk assessment of groundwater contamination: A multilevel fuzzy comprehensive evaluation approach based on DRASTIC model. *Scientific World Journal* 2013, 1–9.

Zhang, S., Sun, B., Yan, L., Wang, C., 2013a. Risk identification on hydropower project using the IAHP and extension of TOPSIS methods under interval-valued fuzzy environment. *Natural Hazards* 65, 359–373.

Zheng, G., Zhu, N., Tian, Z., Chen, Y., Sun, B., 2012. Application of a trapezoidal fuzzy AHP method for work safety evaluation and early warning rating of hot and humid environments. *Safety Science* 50(2), 228–239.

Zheng, X., Gu,. C., Qin, D., 2016. Dam's risk identification under interval-valued intuitionistic fuzzy environment. *Civil Engineering and Environmental Systems* 32(4), 351–363.

Zou, Q., Zhou, J., Zhou, C., Song, L., Guo, J., 2013. Comprehensive flood risk assessment based on set pair analysis-variable fuzzy sets model and fuzzy AHP. *Stochastic Environmental Research and Risk Assessment* 27, 525–546.

3

Comparison of Methods in FAHP with Application in Supplier Selection

Nimet Yapici Pehlivan, Turan Paksoy, and Ahmet Çalik

CONTENTS

3.1 Introduction

Green supply chain management (GSCM) is described as the combination of environmental thinking and supply chain management (SCM) encompassing product design, material sourcing and selection, manufacturing processes, delivery of the final product to the consumer, and end-of-life management of the product (Srivastava, 2007; Tseng and Chiu, 2013). Environmental sustainability of a supply chain depends on the purchasing strategy of the supply chain members. Most of the supply chain models focus on cost, quality, lead time, carbon emission, etc., issues for supplier selection. Shaw et al. (2012) presented an integrated approach for selecting the appropriate supplier in the supply chain, considering the factors of cost, quality rejection percentage, late delivery percentage, greenhouse gas emission, and demand, using fuzzy analytic hierarchy process (FAHP) and multiobjective linear programming (MOLP).

This chapter presents an integrated approach using FAHP and MOLP to develop a supplier selection model, including purchasing cost, late delivery, corporate social responsibility (CSR), carbon emission during production and handling, energy consumption, waste generation, and green packaging. FAHP methodologies are used to determine the weights of supplier selection criteria. By using a trade-off table, we calculate the membership function values of each objective. The weights of the criteria obtained from FAHP are multiplied with each membership function to formulate the single-objective linear programming.

The remainder of this chapter is organized as follows: Section 3.2 presents some of the FAHP methods, including Van Laarhoven and Pedrycz's fuzzy priority method, Chang's extent analysis method, Buckley's geometric mean method, Mikhailov's fuzzy preference programming method, and Mikhailov's fuzzy prioritization method. Section 3.3 gives a literature review on FAHP applications in supplier selection. In Section 3.4, a proposed supplier selection model based on FAHP and MOLP is presented and a numerical example is given. Finally, conclusions and suggestions are drawn in Section 3.5.

3.2 A Brief History of FAHP Methods

In this section, we present the five fundamental methods of FAHP proposed by Van Laarhoven and Pedrycz (1983), Buckley (1985), Chang (1996), and Mikhailov (2002, 2003).

In the FAHP methods, fuzzy pairwise comparison matrices have been constructed by using linguistic evaluations with respect to the decision makers' judgments. Linguistic variables for pairwise comparison of each criterion are shown in Table 3.1.

3.2.1 Van Laarhoven and Pedrycz (1983) Fuzzy Priority Method

Van Laarhoven and Pedrycz (1983) proposed a fuzzy method for choosing among the alternatives under conflicting criteria; this is a fuzzy version of the Saaty (1980) AHP method extended by de Graan (1980) and Lootsma (1981). In this method, the logarithmic regression method of Lootsma (1981) is used to derive fuzzy weights or fuzzy performance scores through AHP operations by using triangular fuzzy numbers (Van Laarhoven and Pedrycz, 1983).

TABLE 3.1

Linguistic Variables for Pairwise Comparison of Each Criterion

Linguistic Variables	Triangular Fuzzy Scale	Triangular Fuzzy Reciprocal Scale
Equally strong	(1, 1, 1)	(1, 1, 1)
Moderately strong	(2, 3, 4)	(1/4, 1/3, 1/2)
Strong	(4, 5, 6)	(1/6, 1/5, 1/4)
Very strong	(6, 7, 8)	(1/8, 1/7, 1/6)
Extremely strong	(9, 9, 9)	(1/9, 1/9, 1/9)
	(1, 2, 3)	(1/3, 1/2, 1)
Intermediate values	(3, 4, 5)	(1/5, 1/4, 1/3)
	(5, 6, 7)	(1/7, 1/6, 1/5)
	(7, 8, 9)	(1/9, 1/8, 1/7)

Source: Kannan, D. et al., *Journal of Cleaner Production*, 47, 355–367, 2013.

The algorithm of Van Laarhoven and Pedrycz (1983)'s method is summarized as follows:

Step 1: The fuzzy pairwise comparison matrix $\tilde{D} = [\tilde{a}_{ij}]$ including a group of decision makers is constituted as follows:

$$\tilde{D} = \begin{bmatrix} (1,1,1) & \tilde{a}_{12}{}^{\delta_{ij}} & \cdots & \tilde{a}_{1n}{}^{\delta_{ij}} \\ \tilde{a}_{21}{}^{\delta_{ij}} & (1,1,1) & \cdots & \tilde{a}_{2n}{}^{\delta_{ij}} \\ \vdots & \vdots & (1,1,1) & \vdots \\ \tilde{a}_{n1}{}^{\delta_{ij}} & \tilde{a}_{n2}{}^{\delta_{ij}} & \cdots & (1,1,1) \end{bmatrix}$$

where $\tilde{a}_{ij}{}^{\delta_{ij}}$ denotes fuzzy judgments evaluated by multiple decision makers. Note that δ_{ij} may be 0 when no decision maker expresses his or her comparisons or greater than 1 when more than one decision maker expresses his or her comparisons.

Step 2: The linear equations for triangular fuzzy numbers $\tilde{x}_i = (l_i, m_i, u_i)$ and $\tilde{y}_{ijk} = \ln \tilde{a}_{ijk} = (l_{ijk}, m_{ijk}, u_{ijk})$ are solved by

$$l_i \sum_{\substack{j=1 \\ j \neq i}}^{n} \delta_{ij} - \sum_{\substack{j=1 \\ j \neq i}}^{n} \delta_{ij} u_j = \sum_{\substack{j=1 \\ j \neq i}}^{n} \sum_{k=1}^{\delta_{ij}} l_{ijk}, \quad i, j = 1, 2, \ldots, n; \ k = 1, 2, \ldots, \delta_{ij} \qquad (3.1)$$

$$m_i \sum_{\substack{j=1 \\ j \neq i}}^{n} \delta_{ij} - \sum_{\substack{j=1 \\ j \neq i}}^{n} \delta_{ij} m_j = \sum_{\substack{j=1 \\ j \neq i}}^{n} \sum_{k}^{\delta_{ij}} m_{ijk}, \quad i, j = 1, 2, \ldots, n; \ k = 1, 2, \ldots, \delta_{ij} \qquad (3.2)$$

$$u_i \sum_{\substack{j=1 \\ j\neq i}}^{n} \delta_{ij} - \sum_{\substack{j=1 \\ j\neq i}}^{n} \delta_{ij}l_j = \sum_{\substack{j=1 \\ j\neq i}}^{n} \sum_{k}^{\delta_{ij}} u_{ijk}, \quad i,j=1,2,...,n; \ k=1,2,...,\delta_{ij} \qquad (3.3)$$

In Equations 3.1 through 3.3, l_{ijk} and u_{ijk} denote lower and upper values of $\tilde{a}_{jik} = -\ln \tilde{a}_{ijk}$.

Generally, a solution of linear equations given in Equations 3.1 through 3.3 is obtained as

$$x_i = (l_i + p_1, m_i + p_2, u_i + p_1); \quad i=1,2,...,n \qquad (3.4)$$

where p_1 and p_2 are arbitrary values.

Step 3: The exponential of x_i in Equation 3.4 is taken as follows:

$$\beta_i = \exp(x_i) = (\exp(l_i + p_1), \exp(m_i + p_2), \exp(u_i + p_1)); \quad i=1,2,...,n \qquad (3.5)$$

Normalizing the β_i, the $\tilde{\alpha}_i$'s are calculated as

$$\tilde{\alpha}_i = \beta_i \left(\sum_{i=1}^{n} \beta_i \right)^{-1}, i=1,2,...,n \qquad (3.6)$$

where $\tilde{\alpha}_i$ denotes fuzzy estimations for the weights w_i, $i=1,2,...,n$ and can be computed as

$$\tilde{\alpha}_i = (\gamma_1 \exp(l_i), \gamma_2 \exp(m_i), \gamma_3 \exp(u_i)), \ i=1,2,...,n \qquad (3.7)$$

In Equation 3.7, γ_1, γ_2, and γ_3 are calculated by the following:

$$\gamma_1 = \left(\sum_{i=1}^{n} \exp(u_i) \right)^{-1}$$

$$\gamma_2 = \left(\sum_{i=1}^{n} \exp(m_i) \right)^{-1} \qquad (3.8)$$

$$\gamma_3 = \left(\sum_{i=1}^{n} \exp(l_i) \right)^{-1}$$

Step 4: The weights w_i, $i = 1, 2, ..., n$ are obtained by defuzzifying the $\tilde{\alpha}_i's$ using any defuzzification method; here we use the center of area (CoA) method, as follows:

$$w_i = \frac{\gamma_1 \exp(l_i) + \gamma_2 \exp(m_i) + \gamma_3 \exp(u_i)}{3}, i = 1, 2, ..., n \qquad (3.9)$$

(Van Laarhoven and Pedrycz, 1983).

3.2.2 Buckley (1985) Geometric Mean Method

The geometric mean method was first developed by Buckley (1985) to extend the AHP to the situation of using linguistic variables.

The steps for the geometric mean method of Buckley (1985) are summarized as follows:

Step 1: The fuzzy pairwise comparison matrix $\tilde{D} = [\tilde{a}_{ij}]$ is constructed as

$$\tilde{D} = \begin{bmatrix} (1,1,1) & \tilde{a}_{12} & \cdots & \tilde{a}_{1n} \\ \tilde{a}_{21} & (1,1,1) & \cdots & \tilde{a}_{2n} \\ \vdots & \vdots & \ddots & \vdots \\ \tilde{a}_{n1} & \tilde{a}_{n2} & \cdots & (1,1,1) \end{bmatrix}$$

where $\tilde{a}_{ij} \times \tilde{a}_{ji} \approx 1$ and $\tilde{a}_{ij} \cong w_i/w_j$, $i, j = 1, 2, ..., n$.

Step 2: The fuzzy geometric mean value $\tilde{r}_i$, for each criterion i is computed as

$$\tilde{r}_i = (\tilde{a}_{i1} \times \tilde{a}_{i2} \times ... \times \tilde{a}_{in})^{1/n} \qquad (3.10)$$

Step 3: The fuzzy weight $\tilde{w}_i$ for each criterion i is calculated as

$$\tilde{w}_i = \tilde{r}_i \times (\tilde{r}_1 + \tilde{r}_2 + ... + \tilde{r}_n)^{-1} \qquad (3.11)$$

where $\tilde{r}_k = (l_k, m_k, u_k)$ and $(\tilde{r}_k)^{-1} = (1/u_k, 1/m_k, 1/l_k)$.

Step 4: The fuzzy weights $\tilde{w}_i = (l_i, m_i, u_i)$ are defuzzified by any defuzzification method; here we use the CoA method as follows:

$$\tilde{w}_i = \frac{l_i + m_i + u_i}{3}$$

(Buckley, 1985; Tzeng and Huang, 2011).

3.2.3 Chang (1996) Extent Analysis Method

The extent analysis method proposed by Chang (1996) has been widely used to obtain crisp weights from a fuzzy comparison matrix. In the method, every criteria or alternative is evaluated by linguistic variables and then the extent analysis is performed.

The algorithm of the extent analysis method can be summarized as follows:

Step 1: The fuzzy pairwise comparison matrix $\tilde{D} = [\tilde{a}_{ij}]$ is set as follows:

$$
\tilde{D} = \begin{bmatrix}
(1,1,1) & \tilde{a}_{12} & \cdots & \tilde{a}_{1n} \\
\tilde{a}_{21} & (1,1,1) & \cdots & \tilde{a}_{2n} \\
\vdots & \vdots & \ddots & \vdots \\
\tilde{a}_{n1} & \tilde{a}_{n2} & \cdots & (1,1,1)
\end{bmatrix}
$$

where $\tilde{a}_{ij} \times \tilde{a}_{ji} \approx 1$ *ve* $\tilde{a}_{ij} \cong w_i/w_j$, $i,j = 1,2,...,n$ and all $\tilde{a}_{ij} = (l_{ij}, m_{ij}, u_{ij})$, $i,j = 1, 2,...,n$ are triangular fuzzy numbers.

Step 2: The value of fuzzy synthetic extent with respect to the criteria i is defined as

$$
S_i = \sum_{j=1}^{n} \tilde{a}_{ij} \left[\sum_{i=1}^{n} \sum_{j=1}^{n} \tilde{a}_{ij} \right]^{-1}
\tag{3.12}
$$

In Equation 3.12, $\displaystyle\sum_{j=1}^{n} \tilde{a}_{ij}$ and $\left[\displaystyle\sum_{i=1}^{n} \sum_{j=1}^{n} \tilde{a}_{ij} \right]^{-1}$ are calculated by using the fuzzy addition operation of n extent analysis for a fuzzy pairwise comparison matrix as follows:

$$
\sum_{j=1}^{n} \tilde{a}_{ij} = \left(\sum_{j=1}^{n} l_j, \sum_{j=1}^{n} m_j, \sum_{j=1}^{n} u_j \right)
\tag{3.13}
$$

$$
\sum_{i=1}^{n} \sum_{j=1}^{n} \tilde{a}_{ij} = \left(\sum_{i=1}^{n} l_i, \sum_{i=1}^{n} m_i, \sum_{i=1}^{n} u_i \right)
\tag{3.14}
$$

$$\left[\sum_{i=1}^{n}\sum_{j=1}^{n}\tilde{a}_{ij}\right]^{-1} = \left(\frac{1}{\sum_{i=1}^{n}u_i}, \frac{1}{\sum_{i=1}^{n}m_i}, \frac{1}{\sum_{i=1}^{n}l_i}\right) \tag{3.15}$$

The principles for the comparison of fuzzy numbers were introduced to derive the weight vectors of all elements for each level of hierarchy with the use of fuzzy synthetic values.

Step 3: To compare the fuzzy numbers, the degree of possibility of $M_2 \geq M_1$ is calculated as

$$V(M_2 \geq M_1) = \sup_{y \geq x}[\min(\mu_{M_1(x)}, \mu_{M_2(y)})] = hgt(M_1 \cap M_2) = \mu_{M_2(d)}$$

$$= \begin{cases} 1, \text{ if } m_2 \geq m_1 \\ 0, \ l_1 \geq u_2 \\ \dfrac{(l_1 - u_2)}{(m_2 - u_2) - (m_1 - l_1)}, \text{ otherwise} \end{cases} \tag{3.16}$$

where $M_1 = (l_1, m_1, u_1)$ and $M_2 = (l_2, m_2, u_2)$ and d is the ordinate of the highest intersection point D between μ_{M_1} and μ_{M_2} (see Figure 3.1). To compare M_1 and M_2, both $V(M_2 \geq M_1)$ and $V(M_1 \geq M_2)$ are needed.

Step 4: The degree of possibility for a fuzzy number to be greater than k fuzzy numbers $S_i \ (i = 1, 2, ..., k)$ can be defined by

$$V(S \geq S_1, S_2, ..., S_k) = \min V(S \geq S_i), \ i = 1, 2, ..., k \tag{3.17}$$

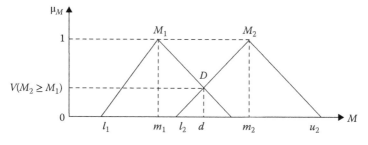

FIGURE 3.1
The intersection between M_1 and M_2.

Assume that

$$d'(A_i) = \min V(S_i \geq S_k), \quad i,k = 1,2,...,n; \ k \neq i \qquad (3.18)$$

Then, the weight vector is given by

$$W' = (d'(A_1),d'(A_2),...,d'(A_n))^T \qquad (3.19)$$

Step 5: Via normalization, the normalized nonfuzzy weight vector is computed as

$$W = (d(A_1),d(A_2),...,d(A_n))^T \qquad (3.20)$$

(Chang, 1996; Kahraman et al., 2003; Paksoy et al., 2012).

3.2.4 Mikhailov (2000) Fuzzy Preference Programming Method

The Fuzzy Preference Programming (FPP) method was proposed by Mikhailov (2000) for deriving weights from fuzzy comparison judgments. In the FPP method, initial fuzzy judgments are transformed into interval ones by using α-cuts and it is applied to derive crisp weights from the interval judgments. Then, the assessment of the weights is obtained by solving any optimization problem, maximizing the decision maker's satisfaction with a specific crisp weight vector (Mikhailov, 2002, 2003).

The algorithm of the FPP method is given as follows:

Step 1: The fuzzy pairwise comparison matrix $\tilde{D} = [\tilde{a}_{ij}]$ is constituted as follows:

$$\tilde{D} = \begin{bmatrix} (1,1,1) & \tilde{a}_{12} & \cdots & \tilde{a}_{1n} \\ \tilde{a}_{21} & (1,1,1) & \cdots & \tilde{a}_{2n} \\ \vdots & \vdots & \ddots & \vdots \\ \tilde{a}_{n1} & \tilde{a}_{n2} & \cdots & (1,1,1) \end{bmatrix}$$

where $\tilde{a}_{ij} = (l_{ij},m_{ij},u_{ij})$, $i,j = 1,2,...,n$ are fuzzy judgments provided by the decision maker considered as triangular fuzzy numbers.

Step 2: The maximin prioritization problem is considered as

$$\text{Max} \quad \lambda$$

$$\lambda \leq 1 - \frac{R_k w}{d_k} = d_k \lambda + R_k w \leq d_k \qquad (3.21)$$

$$\sum_{i=1}^{n} w_i = 1, \quad w_i > 0, \quad i = 1, 2, ..., n; \ k = 1, 2, ..., 2m$$

In Equation 3.21,

$$R_k w = \begin{cases} w_i - w_j u_{ij}(\alpha) \lessgtr 0 \\ -w_i + w_j l_{ij}(\alpha) \lessgtr 0 \end{cases} \tag{3.22}$$

and its membership function is defined as

$$\mu_k(R_k w) = \begin{cases} 1 - \dfrac{R_k w}{d_k}, & R_k w \leq d_k \\ 0, & R_k w \geq d_k \end{cases} \tag{3.23}$$

In Equation 3.22, the priority ratios at each α-cut level should satisfy $l_{ij}(\alpha) \leq w_i/w_j \leq u_{ij}(\alpha)$ and the bounds of the α-cut intervals are defined as

$$l_{ij}(\alpha) = \alpha(m_{ij} - l_{ij}) + l_{ij}$$

$$u_{ij}(\alpha) = \alpha(m_{ij} - u_{ij}) + u_{ij}$$

Step 3: By introducing a variable λ, taking into consideration the membership functions for the triangular fuzzy numbers, Equation 3.21 is transformed into a linear programming problem:

Max λ

$$\lambda d_k + w_i - u_{ij}(\alpha)\, w_j \leq d_k$$

$$\lambda d_k - w_i + l_{ij}(\alpha)\, w_j \leq d_k \tag{3.24}$$

$$\sum_{i=1}^{n} w_i = 1,\ w_i > 0;\ i = 1, 2, \ldots, n;\ k = 1, 2, \ldots, 2m$$

Step 4: The optimal solution (λ^*, w^*) to the linear programming problem given in Equation 3.24 is obtained. The consistency index λ^* measures the degree of satisfaction. If $\lambda^* \geq 1$, then interval judgments are consistent, whereas if $0 \leq \lambda^* \leq 1$ then interval judgments are inconsistent and it depends on the tolerance parameters d_k (Mikhailov, 2000, 2002, 2003; Mikhailov and Tsetinov, 2004).

3.2.5 Mikhailov (2003) Fuzzy Prioritization Method

The solution procedure of the fuzzy prioritization method proposed by Mikhailov (2003) is based on the maximin decision rule. The maximin rule was applied by Bellman and Zadeh (1970) for solving decision-making problems in uncertain environments and was applied by Zimmermann (1978) for fuzzy linear problems with constraints.

The steps of the fuzzy prioritization method are given as follows:

Step 1: The fuzzy pairwise comparison matrix $\tilde{D} = [\tilde{a}_{ij}]$ is set as follows:

$$\tilde{D} = \begin{bmatrix} (1,1,1) & \tilde{a}_{12} & \cdots & \tilde{a}_{1n} \\ \tilde{a}_{21} & (1,1,1) & \cdots & \tilde{a}_{2n} \\ \vdots & \vdots & \ddots & \vdots \\ \tilde{a}_{n1} & \tilde{a}_{n2} & \cdots & (1,1,1) \end{bmatrix}$$

where $\tilde{a}_{ij} = (l_{ij}, m_{ij}, u_{ij}), i, j = 1, 2, ..., n$ are triangular fuzzy numbers.

Step 2: The maximin prioritization problem is considered as

$$\text{Max } \lambda$$

$$\lambda \le \mu_{ij}(w); \ i = 1, 2, ..., n-1; \ j = 1, 2, ..., n; \ j > i \tag{3.25}$$

$$\sum_{l=1}^{n} w_l = 1, \ w_l > 0; \ l = 2, 3, ..., n-1$$

Step 3: Taking into consideration the membership functions for the triangular fuzzy numbers, Equation 3.25 is transformed into a nonlinear programming problem:

$$\text{Max } \lambda$$

$$(m_{ij} - l_{ij})\lambda w_j - w_i + l_{ij}w_j \le 0$$

$$(u_{ij} - m_{ij})\lambda w_j + w_i - u_{ij}w_j \le 0 \tag{3.26}$$

$$\sum_{k=1}^{n} w_k = 1, \ w_k > 0; \ k = 1, 2, ..., n$$

$$i = 1, 2, ..., n-1; \ j = 2, 3, ..., n-1; \ j > i$$

Step 4: The optimal solution (λ^*, w^*) to Equation 3.26 is obtained by some appropriate numerical method for nonlinear optimization.

The optimal value λ^* can be used for measuring the consistency of the initial set of fuzzy judgments. If the optimal value λ^* is positive, the initial set of fuzzy judgments is rather consistent. If the optimal value λ^* is negative, the fuzzy judgments are strongly inconsistent (Mikhailov, 2003; Mikhailov and Tsetinov, 2004).

3.2.6 Advantages and Disadvantages of FAHP Methods

In the literature, there are several FAHP methods proposed by various authors. These methods are systematic approaches to an alternative selection and justification problem using fuzzy set theory and hierarchical structure analysis. The earliest work on FAHP was by Van Laarhoven and Pedrycz (1983), which compared fuzzy ratios described by triangular fuzzy numbers. In the method, fuzzy weights are derived from fuzzy comparison matrices via a fuzzy logarithmic least squares method. Buckley (1985) determined fuzzy weights of comparison ratios with trapezoidal fuzzy numbers using the geometric mean method. Chang (1996) proposed an extent analysis method for FAHP, with the use of triangular fuzzy numbers for pairwise comparison matrices to derive crisp weights. The FPP method, proposed by Mikhailov (2000), aims to compute crisp weights from fuzzy comparison matrices through a linear programming problem. Mikhailov (2003) proposed a fuzzy prioritization method to obtain crisp weights from fuzzy comparison matrices via nonlinear optimization (Büyüközkan et al., 2004; Bozbura and Beskese, 2007; Javanbarg et al., 2012).

In this chapter, we considered the FAHP methods proposed by Van Laarhoven and Pedrycz (1983), Buckley (1985), Chang (1996), and Mikhailov (2002, 2003). FAHP methods have significant differences in their theoretical structures and computational simplicity. Table 3.2 presents a comparison of the considered FAHP methods in terms of their advantages and disadvantages.

3.3 Literature Review on FAHP Applications in Supplier Selection

This section presents recently published papers on FAHP applications to the supplier selection problem.

Kahraman et al. (2003) used the FAHP method to select the best supplier firm providing the most satisfaction for the criteria determined. Yang and Chen (2006) proposed an integrated model by combining the analytic hierarchy

TABLE 3.2

The Advantages and Disadvantages of the Five Studied FAHP Methods

FAHP Method	Advantages	Disadvantages
Van Laarhoven and Pedrycz (1983)	• Evaluations of multiple decision makers can be modeled	• A solution to the linear equations may not always be found • The computational requirement is quite high • It allows only triangular fuzzy numbers • It derives fuzzy weights and requires defuzzification
Buckley (1985)	• It is easy to extend to the fuzzy case • Computational easiness • It guarantees a unique solution	• The computational requirement is quite high • It derives fuzzy weights and requires defuzzification
Chang (1996)	• The computational requirement is relatively low • Computational easiness • Similar to conventional AHP • It derives crisp weights	• It allows only triangular fuzzy numbers • It could lead to a wrong decision, because it may assign zero weights
Mikhailov (2000)	• The computational requirement is relatively low • It does not require complete fuzzy comparison matrices • It derives crisp weights	• It allows only triangular fuzzy numbers
Mikhailov (2003)	• The computational requirement is relatively low • It does not require complete fuzzy comparison matrices • It derives crisp weights	• It allows only triangular fuzzy numbers

process (AHP) and gray relational analysis (GRA) for considering the qualitative and quantitative criteria for supplier selection in an outsourcing manufacturing organization. Lu et al. (2007) presented a multiobjective decision-making process for GSCM to help the supply chain manager in measuring and evaluating suppliers' performance based on integration of the AHP method and fuzzy logic. Lee (2009) proposed an analytical approach to select suppliers under a fuzzy environment. An FAHP model, which incorporates the benefits, opportunities, costs, and risks (BOCR) concept, was constructed to evaluate various aspects of suppliers. Awasthi et al. (2010) proposed a fuzzy multicriteria model for evaluating environmental performance of suppliers using the fuzzy TOPSIS (Technique for Order of Preference by Similarity to Ideal Solution) method.

Bai and Sarkis (2010) developed a green supplier evaluation model considering economic, environmental, and social issues, and they used rough set theory to deal with the information vagueness. Kuo et al. (2010) developed a green supplier selection model, which is a hybrid method that integrates artificial neural network (ANN) and data envelopment analysis (DEA) called ANN-multiattribute decision analysis (ANN-MADA). ANN-MADA considers both practicality in traditional supplier selection criteria and environmental regulations. Sanayei et al. (2010) proposed a multicriteria decision-making (MCDM) method based on fuzzy set theory and the VIKOR (VlseKriterijumska Optimizacija I Kompromisno Resenje) method, to deal with both qualitative and quantitative criteria for supplier selection problems in the supply chain system. Zhu et al. (2010) presented the development of a methodology to evaluate suppliers using portfolio analysis based on the analytical network process (ANP) and environmental factors.

Büyüközkan and Çifçi (2011) proposed fuzzy ANP within a multiperson decision-making scheme under incomplete preference relations for sustainable supplier selection considering incomplete information. Büyüközkan and Çifçi (2012) examined GSCM and its capability dimensions to propose an evaluation framework for green suppliers. They proposed a novel hybrid analytic approach based on the fuzzy DEMATEL (The Decision Making Trial and Evaluation Laboratory), fuzzy ANP, and fuzzy TOPSIS methodologies to evaluate strategic decisions for GSCM. Paksoy et al. (2012) proposed an organizational strategy development in distribution channel management using FAHP and hierarchical fuzzy TOPSIS. Shaw et al. (2012) presented an integrated approach for selecting the appropriate supplier in the supply chain, addressing the carbon emission issue, using FAHP and fuzzy MOLP. Xiao et al. (2012) integrated the fuzzy cognitive map (FCM) and fuzzy soft set model for solving the supplier selection problem under operational risks. In this method, not only the dependent and feedback effect among criteria, but also the uncertainties in the decision-making process are considered.

Junior et al. (2013) presented a supplier selection decision method based on fuzzy inference that integrates both types of approaches: a noncompensatory rule for sorting in the qualification stages and a compensatory rule for ranking in the final selection. In the qualification stages, fuzzy inference combined with a fuzzy rule-based classification method is used to categorize suppliers.

Govindan et al. (2013) explored sustainable supply chain initiatives and examined the problem of identifying an effective model based on the triple bottom line (TBL) approach (economic, environmental, and social aspects) for supplier selection operations in supply chains. Qualitative performance evaluation was performed by using fuzzy numbers to find the criteria weights and then fuzzy TOPSIS was applied for ranking the suppliers.

Hsu et al. (2013) utilized the DEMATEL approach to recognize the influential criteria of carbon management in GSCM for improving the overall performance of suppliers. By considering the interrelationships among the criteria, DEMATEL was applied to deal with the importance and causal relationships among the evaluation criteria of supplier selection. Kannan et al. (2013) presented an integrated approach to determine the best green suppliers, according to the economic and environmental criteria. FAHP was used for calculating the weights of the criteria and fuzzy TOPSIS was applied for ranking the suppliers. The weights of the criteria and the ranks of suppliers were incorporated into the MOLP model to determine the optimal order quantity from each supplier while subjecting them to some resource constraints.

Shen et al. (2013) examined GSCM to propose a fuzzy multicriteria approach for the evaluation of green suppliers. Fuzzy set theory was applied to translate subjective human perceptions into a solid crisp value and linguistic preferences are combined through fuzzy TOPSIS to generate an overall performance score for each supplier. Tseng and Chiu (2013) aimed to capture linguistic preferences in the selection of an alternative supplier with environmental regulations. The study contributed to the GSCM literature by developing (1) valid and reliable GSCM criteria based on expert team and environmental and nonenvironmental literatures and (2) an approach to use GRA given the linguistic preferences.

Akman (2015) proposed a methodology using segmentation and evaluation approaches together by means of fuzzy c-means and VIKOR methods in order to determine which suppliers should be included in the green supplier development programs (GSDPs) and to help the manager to prepare plans for GSDPs. Kannan et al. (2014) proposed a framework, which was built on the criteria of GSCM practices, using fuzzy TOPSIS to select green suppliers for a Brazilian electronics company. In the study, a Spearman rank correlation coefficient and sensitivity analysis were performed. Dobos and Vörösmarty (2014) developed a supplier selection method based on the DEA-like composite indicator model. Fahimnia et al. (2015) presented a thorough bibliometric and network analysis of over 1000 published studies that provided insights not previously fully grasped or evaluated by other reviews on GSCM. Govindan et al. (2015) analyzed research in international scientific journals and conference proceedings from 1997 to 2011 that focus on green supplier selection. They proposed three research questions: (1) Which selection approaches are commonly applied? (2) What environmental and other selection criteria for green supplier management are popular? (3) What limitations exist?

Hashemi et al. (2015) proposed a comprehensive green supplier selection model taking into account both economic and environmental criteria. The ANP was used to deal with the interdependencies among the criteria and the improved GRA was applied to rank the suppliers. Igoulalene et al. (2015) presented two fuzzy hybrid approaches to the strategic supplier selection problem. The problem was formulated as a multistakeholder multicriteria (MSMC) decision-making problem. In the first hybrid approach, the fuzzy consensus-based possibility measure and fuzzy TOPSIS method were combined. In the second hybrid approach, the fuzzy consensus-based neat ordered weighted average (OWA) and goal programming model were combined.

Kannan et al. (2015) proposed fuzzy axiomatic design (FAD), which is an MCDM approach to select the best green supplier for a Singapore-based plastic manufacturing company. The environmental criteria were developed and the FAD method evaluated the requirements of both the manufacturer (design needs) and the supplier (functional needs). Because of the multiple criteria, a multiobjective optimization model of a fuzzy nature was developed. Kar (2015) presented the application of a hybrid approach, which integrates fuzzy set theory, AHP, and neural networks, for group decision support for the supplier selection problem. Discriminant analysis (DA) was used for the purpose of supplier base rationalization through which suppliers were mapped to highly suitable and less suitable supplier classes.

3.4 Proposed Supplier Selection Model

In this section, the proposed supplier selection model with green issues is introduced and a comparison of FAHP methods is presented. The framework of the solution process is given in Figure 3.2. First, an MOLP model is developed that includes green issues such as carbon emission, CSR, green packaging, green handling, and waste management policy. In the developed MOLP model, numerous objectives should be compromised simultaneously. To attain a compromise solution, we used an integrated approach based on FAHP methods.

The assumptions of the developed MOLP problem are as follows:

- Only one type of product can be purchased from different suppliers.
- Quantity discounts are not taken into consideration.
- We do not allow shortages for any of the suppliers.
- Demand is known and deterministic.

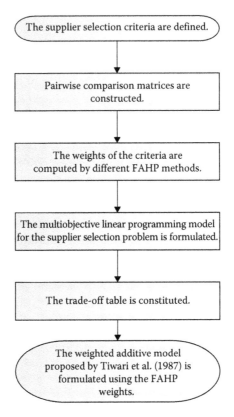

FIGURE 3.2
Framework of the proposed supplier selection model.

3.4.1 Formulation of Proposed Supplier Selection Problem

Indices

i	set of suppliers	$i = \{1, 2, ..., I\}$
j	set of objectives,	$j = \{1, 2, ..., J\}$
k	set of constraints,	$k = \{1, 2, ..., K\}$

Parameters

N The maximum number of available suppliers

D Demand for the planning period

C_i Unit purchasing cost of one unit of item from supplier i

L_i The percentage of items delivered late by supplier i

PCO_i The amount of CO_2 emitted during the production of one unit of item in supplier i

HCO_i The amount of CO_2 emitted during the handling of one unit of item in supplier i

E_i The amount of energy needed for producing one unit of item in supplier i

W_i The amount of waste generated during the production of one unit of item in supplier i

GP_i The green packaging level of supplier i

CSR_i The level of CSR of supplier i

Cap_i The capacity of supplier i

$Cap_i^{CO_2}$ The carbon emission capacity of supplier i

Decision variables

X_i The amount of order item from supplier i

Y_i If supplier i is selected 1; otherwise, 0

The MOLP model for the supplier selection problem with green issues is formulated as follows:

Objective functions

$$\text{Minimize } Z_1 = \sum_{i=1}^{I} C_i \cdot X_i \tag{3.27}$$

$$\text{Minimize } Z_2 = \sum_{i=1}^{I} L_i \cdot X_i \tag{3.28}$$

$$\text{Maximize } Z_3 = \sum_{i=1}^{I} CSR_i \cdot X_i \tag{3.29}$$

$$\text{Minimize } Z_4 = \sum_{i=1}^{I} PCO_i \cdot X_i \tag{3.30}$$

$$\text{Minimize } Z_5 = \sum_{i=1}^{I} HCO_i \cdot X_i \tag{3.31}$$

$$\text{Minimize } Z_6 = \sum_{i=1}^{I} E_i \cdot X_i \tag{3.32}$$

$$\text{Minimize} \, Z_7 = \sum_{i=1}^{I} W_i \cdot X_i \tag{3.33}$$

$$\text{Maximize} \, Z_8 = \sum_{i=1}^{I} GP_i \cdot X_i \tag{3.34}$$

The objective functions of the MOLP problem consist of eight parts, which are total purchasing cost, percentage of late delivery, CSR, carbon emissions during production, carbon emissions during handling, energy consumption, waste generation, and green packaging level.

The purchasing cost 3.27 refers to the price that manufacturers pay to buy a unit of item. The late delivery 3.28 represents the performance of the supplier during the delivery of goods. When manufacturers order an item from a supplier in a specific time period, the ordered goods must be delivered on time. The manufacturers expect the goods to be delivered within a reasonable time. If the supplier fails to do what it promised, the manufacturers can claim compensation or penalty. The CSR function 3.29 is the decision-making and implementation process that guides all company activities in the protection and promotion of international human rights, labor and environmental standards, and compliance with legal requirements within its operations and in its relations to the societies and communities where it operates (Emmett and Sood, 2010). The CO_2 emission contains the emissions during the production of one unit of item at the supplier 3.30. The CO_2 emission contains the emissions during the handling of one unit of an item at the supplier 3.31. Handling emissions include the emissions of the items during movement, protection, and storage. The energy function 3.32 represents the energy efficiency of the supplier. If energy consumption of a supplier is less than the others, the supplier is more efficient in the use of energy and materials. The waste function 3.33 indicates the waste generation of the supplier. Green suppliers generate less waste than the others and this function shows the greenness of the supplier. The green packaging function 3.34 aims to reduce the environmental impact of packaging, in which more recyclable materials such as papers and plastics can be used in the life cycle of the items.

Constraints

$$\sum_{i=1}^{I} X_i \geq D \tag{3.35}$$

$$X_i \leq Cap_i \tag{3.36}$$

$$\sum_{i=1}^{I} PCO_i \cdot X_i + \sum_{i=1}^{I} HCO_i \cdot X_i \leq Cap_i^{CO_2} \tag{3.37}$$

$$\sum_{i=1}^{I} Y_i \leq N \tag{3.38}$$

$$X_i \leq Cap_i \cdot Y_i \tag{3.39}$$

$$X_i \geq 0 \text{ and integer} \tag{3.40}$$

$$Y_i = \{0,1\} \tag{3.41}$$

Equation 3.35 ensures that the demands of all manufacturers are fully met; Equation 3.36 indicates that the total quantity of products sold from suppliers cannot exceed the capacity of those suppliers.

Equation 3.37 indicates that the CO_2 emissions generated from suppliers cannot exceed the CO_2 emission capacity of those suppliers. Equation 3.38 ensures that the total number of potential suppliers cannot exceed the maximum allowed suppliers. Equation 3.39 guarantees that if supplier i is selected for production, the amount of production should be less than the capacity of supplier i. Equation 3.40 enforces the nonnegativity restriction on the decision variables and finally Equation 3.41 represents the binary variable.

3.4.2 Solution Procedure of Proposed Supplier Selection Problem

In the solution process, an integrated approach using FAHP and MOLP is carried out to solve the proposed supplier selection problem. The FAHP methods are used to determine the weights of supplier selection criteria with green issues. These weights are multiplied with each membership function of the objectives and crisp formulation is performed by using the "weighted additive model (WAM)" proposed by Tiwari et al. (1987) as follows:

$$\mu_D(x) = \sum_{j=1}^{J} w_j \mu_{z_j}(x) + \sum_{k=1}^{K} \beta_k \mu_{g_k}(x) \tag{3.42}$$

$$\sum_{j=1}^{J} w_j + \sum_{k=1}^{K} \beta_k = 1, \ w_j, \beta_k \geq 0 \qquad (3.43)$$

In Equations 3.42 and 3.43, w_j and β_k indicate the relative weights of the fuzzy objective functions and fuzzy constraints, respectively.

The steps of the proposed algorithm can be summarized as follows:

Step 1: The criteria of the supplier selection problem are determined.

Step 2: A fuzzy pairwise comparison matrix of the criteria is constructed. Experts in the fields of purchasing management, product management, quality management, and production management evaluate the criteria using a nine-point scale given in Table 3.1.

Step 3: The weights of the criteria are computed by using the FAHP methods.

Step 4: The MOLP model for the problem is formulated.

Step 5: The trade-off table is constituted to obtain efficient extreme solutions for all objectives.

Step 6: The WAM is formulated by using the criteria weights and membership functions of the objectives as follows:

$$\text{Max} \sum_{j=1}^{J} w_j \mu_{z_j}(x)$$

s.t.

$$\lambda_j \leq \mu_{Z_j}(x), \ j = 1, 2, \ldots, J$$

$$g_k(x) \leq b_k, \ k = 1, 2, \ldots, K \qquad (3.44)$$

$$\sum_{j=1}^{J} w_j = 1, \ w_j \geq 0$$

$$\lambda_j \in [0, 1], \ j = 1, 2, \ldots, J$$

$$x_i \geq 0, \ i = 1, 2, \ldots, I$$

Step 7: The WAM in Equation 3.44 is solved and optimal distribution results are obtained.

3.4.3 A Numerical Example of the Proposed Supplier Selection Model with Green Issues

In this section, a numerical example is presented to apply the proposed solution procedure for the supplier selection problem with green issues. Hypothetical data are generated to determine the best supplier and the parameters of the supplier selection problem are given in Table 3.3. In the problem, the minimum demand of manufacturers is considered 500 units and the maximum number of available suppliers is taken as 7. It is assumed that there are 10 potential suppliers that produce a specific item and are evaluated with respect to the 10 criteria.

Here, we described the most important criteria in the supplier selection problem considering green issues: CSR is defined as the "actions that appear to further some social good, beyond the interests of the firm and that which is required by law." CSR activities can include green issues such as energy conservation, reducing emissions, material recycling, reducing packaging materials, and sourcing materials from suppliers to manufacturing facilities (McWilliams and Siegel, 2001, 2006; Sprinkle and Maines, 2010). Green handling is the process of carrying the goods from a specific location to another location using greener technologies. Energy consumption should be considered to understand the environmental protection level of suppliers. Waste management involves redesigning products or changing societal patterns of consumption, use, and waste generation to prevent the creation of waste and minimize the toxicity of waste that is produced (Davidson, 2011). Green packaging causes less damage to the environment than traditional types of packaging because packaging waste is one of the highest sources of environmental degradation (Emmett and Sood, 2010).

A committee consisting of a purchasing manager, product manager, quality manager, and production manager evaluated the green packaging level and CSR of suppliers using an importance scale between 1 and 10. The fuzzy pairwise comparison matrix evaluated by the committee using the linguistic variables in Table 3.2 is shown in Table 3.4. The weights of the criteria computed using FAHP methods are given in Table 3.5.

CSR, corporate social responsibility; FAHP, fuzzy analytic hierarchy process; FP, fuzzy programmimg; FPP, fuzzy preference programming.

According to the results of the FAHP methods given in Table 3.5, "cost" has the highest importance for supplier selection, with "late delivery," "CSR," "production emission," and "handling emission" coming after that.

The MOLP model of the supplier selection problem is formulated using Equations 3.27 through 3.41. The WAM is constituted by using criteria weights calculated from FAHP methods and the given in Equation 3.44. The trade-off table, including efficient extreme solutions for all objectives, is given in Table 3.6.

TABLE 3.3

The Parameters of the Supplier Selection Problem with Green Issues

Suppliers	Cost ($)	L. Delivery (%)	CSR (Level)	Em. Prod (kg)	Em. Hand (kg)	Energy (W)	Waste (g)	Packaging (Level)	Capacity (Unit)	CO₂ Cap. (kg)
1	35	0.195	10	5.301	5.379	22	15	9	367.000	478.427
2	24	0.226	1	4.296	5.080	26	50	5	684.000	158.282
3	37	0.849	7	3.859	2.059	24	58	1	755.000	723.025
4	16	0.368	9	4.542	5.283	8	47	4	810.000	511.458
5	39	0.278	3	1.189	1.940	27	38	9	653.000	1067.937
6	98	0.120	10	1.256	4.549	17	70	6	448.000	378.951
7	29	0.442	7	4.833	5.748	16	77	8	672.000	591.005
8	52	0.547	7	5.406	3.245	6	41	7	387.000	786.752
9	87	0.652	4	4.630	1.081	22	13	1	348.000	272.695
10	57	0.405	4	5.007	3.556	26	95	2	115.000	478.582

TABLE 3.4

The Fuzzy Pairwise Comparison Matrix for Supplier Selection Criteria

Criteria	Cost	L. Delivery	CSR	Em. Prod	Em. Hand	Energy	Waste	Packaging
Cost	(1, 1, 1)	(2, 3, 4)	(1, 2, 3)	(1, 2, 3)	(2, 3, 4)	(2, 3, 4)	(4,5,6)	(4, 5, 6)
L. Delivery	(1/4, 1/3, 1/2)	(1, 1, 1)	(4, 5, 6)	(4, 5, 6)	(2, 3, 4)	(2, 3, 4)	(1,2,3)	(1, 2, 3)
CSR	(1/3, 1/2, 1)	(1/6, 1/5, 1/4)	(1, 1, 1)	(1, 1, 1)	(2, 3, 4)	(2, 3, 4)	(1,2,3)	(1, 2, 3)
Em. Prod.	(1/3, 1/2, 1)	(1/6, 1/5, 1/4)	(1, 1, 1)	(1, 1, 1)	(2, 3, 4)	(2, 3, 4)	(2,3,4)	(1, 1, 1)
Em. Hand	(1/4, 1/3, 1/2)	(1/4, 1/3, 1/2)	(1/4, 1/3, 1/2)	(1/4, 1/3, 1/2)	(1, 1, 1)	(1, 1, 1)	(1,2,3)	(1, 1, 1)
Energy	(1/4, 1/3, 1/2)	(1/4, 1/3, 1/2)	(1/4, 1/3, 1/2)	(1/4, 1/3, 1/2)	(1/3, 1/2, 1)	(1, 1, 1)	(1,2,3)	(1, 2, 3)
Waste	(1/6, 1/5, 1/4)	(1/3, 1/2, 1)	(1/3, 1/2, 1)	(1/4, 1/3, 1/2)	(1/3, 1/2, 1)	(1/3, 1/2, 1)	(1,1,1)	(3, 4, 5)
Packaging	(1/6, 1/5, 1/4)	(1/3, 1/2, 1)	(1/3, 1/2, 1)	(1, 1, 1)	(1, 1, 1)	(1/3, 1/2, 1)	(1/5,1/4,1/3)	(1, 1, 1)

CSR, corporate social responsibility.

TABLE 3.5

The Weights of the Supplier Selection Criteria Using FAHP Methods

Methods Criteria	Van Laarhoven and Pedrycz	Buckley	Chang	Mikhailov—FP	Mikhailov—FPP $\alpha = 0$	$\alpha = 0.5$
Cost	0.3011	0.2760	0.3203	0.2045	0.3216	0.2596
Late delivery	0.2226	0.2119	0.2901	0.2727	0.4284	0.4807
CSR	0.1201	0.1234	0.1602	0.0909	0.1071	0.1081
Em. production	0.1129	0.1174	0.1496	0.0909	0.1071	0.1081
Em. handling	0.0661	0.0731	0.0210	0.0909	0	0
Energy	0.0630	0.0699	0.0204	0.0909	0	0
Waste	0.0629	0.0697	0.0380	0.0681	0	0
Packaging	0.0508	0.0582	0	0.0909	0.0357	0.0432

TABLE 3.6

The Trade-Off Table for the Supplier Selection Problem

Objective Functions	Z_1	Z_2	Z_3	Z_4	Z_5	Z_6	Z_7	Z_8
Min Z_1	**22,029.00**	221.04	3513.00	1800.69	1787.87	8907.00	24,759.00	3028.00
Min Z_2	23,395.00	**214.63**	3366.00	1794.64	1682.25	9415.00	25,821.00	2728.00
Max Z_3	22,617.00	221.76	**3573.00**	1808.22	1815.17	9009.00	25,179.00	3053.00
Min Z_4	24,698.00	238.90	3258.00	**1743.93**	1600.09	9021.00	24,761.00	2652.00
Min Z_5	27,469.00	246.96	3073.00	1831.79	**1550.16**	10,089.00	27,911.00	2617.00
Min Z_6	24,815.00	241.32	3294.00	1781.89	1611.84	**8832.00**	24,788.00	2634.00
Min Z_7	25,845.00	232.82	3286.00	1784.82	1585.51	9611.00	**23,269.00**	2895.00
Max Z_8	22,617.00	221.76	3273.00	1808.22	1815.17	9009.00	25,179.00	**3064.00**
The worst values	*27,469.00*	*246.96*	*3073.00*	*1831.79*	*1815.17*	*10,089.00*	*27,911.00*	*2617.00*

After determining the trade-off table, the membership functions of the each objective are obtained as follows:

$$\mu_1(x) = \begin{cases} 1, Z_1(x) \leq 22,029 \\ \dfrac{27,469 - Z_1(x)}{27,469 - 22,029}, & 22,029 \leq Z_1(x) \leq 27,469 \\ 0, Z_1(x) \geq 27,469 \end{cases} \qquad (3.45)$$

$$\mu_2(x) = \begin{cases} 1, Z_2(x) \le 214.63 \\ \dfrac{246.96 - Z_2(x)}{246.96 - 214.63}, & 214.63 \le Z_2(x) \le 246.96 \\ 0, Z_2(x) \ge 246.96 \end{cases} \tag{3.46}$$

$$\mu_3(x) = \begin{cases} 1, Z_3(x) \ge 3573 \\ \dfrac{Z_3(x) - 3073}{3573 - 3073}, & 3073 \le Z_3(x) \le 3573 \\ 0, Z_3(x) \le 3073 \end{cases} \tag{3.47}$$

$$\mu_4(x) = \begin{cases} 1, Z_4(x) \le 1743.93 \\ \dfrac{1831.79 - Z_4(x)}{1831.79 - 1743.93}, & 1743.93 \le Z_4(x) \le 1831.79 \\ 0, Z_4(x) \ge 1831.79 \end{cases} \tag{3.48}$$

$$\mu_5(x) = \begin{cases} 1, Z_5(x) \le 1550.16 \\ \dfrac{1815.17 - Z_5(x)}{1815.17 - 1550.16}, & 1550.16 \le Z_5(x) \le 1815.17 \\ 0, Z_5(x) \ge 1815.17 \end{cases} \tag{3.49}$$

$$\mu_6(x) = \begin{cases} 1, Z_6(x) \le 8832 \\ \dfrac{10,089 - Z_6(x)}{10,089 - 8832}, & 8832 \le Z_6(x) \le 10,089 \\ 0, Z_6(x) \ge 10,089 \end{cases} \tag{3.50}$$

$$\mu_7(x) = \begin{cases} 1, Z_7(x) \le 23,269 \\ \dfrac{27,911 - Z_7(x)}{27,911 - 23,269}, & 23,269 \le Z_7(x) \le 27,911 \\ 0, Z_7(x) \ge 27,911 \end{cases} \tag{3.51}$$

$$\mu_8(x) = \begin{cases} 1, Z_8(x) \geq 3064 \\ \dfrac{Z_8(x) - 2617}{3064 - 2617}, & 2617 \leq Z_8(x) \leq 3064 \\ 0, Z_8(x) \leq 2617 \end{cases} \quad (3.52)$$

The WAM is formed by using criteria weights obtained from the FAHP methods. For example, the WAM formulation formed by the criteria weights from the method of Van Laarhoven and Pedrycz (1983) is given as

$$\text{Max} \begin{pmatrix} 0.3011\lambda_1 + 0.2226\lambda_2 + 0.1201\lambda_3 + 0.1129\lambda_4 + \\ 0.0661\lambda_5 + 0.0630\lambda_6 + 0.0629\lambda_7 + 0.0508\lambda_8 \end{pmatrix} \quad (3.53)$$

s.t.

$$\lambda_1 \leq \mu_1(x) = \frac{27,469 - Z_1(x)}{27,469 - 22,029}$$

$$\lambda_2 \leq \mu_2(x) = \frac{246.96 - Z_2(x)}{246.96 - 214.63}$$

$$\lambda_3 \leq \mu_3(x) = \frac{Z_3(x) - 3073}{3573 - 3073}$$

$$\lambda_4 \leq \mu_4(x) = \frac{1831.79 - Z_4(x)}{1831.79 - 1743.93}$$

$$\lambda_5 \leq \mu_5(x) = \frac{1815.17 - Z_5(x)}{1815.17 - 1550.16}$$

$$\lambda_6 \leq \mu_6(x) = \frac{10,089 - Z_6(x)}{10,089 - 8832}$$

$$\lambda_7 \leq \mu_7(x) = \frac{27,911 - Z_7(x)}{27,911 - 23269}$$

$$\lambda_8 \leq \mu_8(x) = \frac{Z_8(x) - 2617}{3064 - 2617}$$

Equations $(3.9) - (3.15)$

$$\lambda_j \in [0,1], \ j = 1,2,...,8$$

The supplier selection problem in Equation 3.53 is solved by using the GAMS-CPLEX (The General Algebraic Modeling System) solver and its computation time is less than 5 CPU seconds. Similarly, other WAM formulations are constituted using other FAHP weights. The objective function values and

TABLE 3.7

The Objective Function Values of the Supplier Selection Problem

Method Objective		Van Laarhoven and Pedrycz		Buckley		Chang		Mikhailov—FP		Mikhailov—FPP			
										$\alpha = 0$		$\alpha = 0.5$	
i		λ_i	$Z_i(x)$	λ_i	$Z_i(x)$	λ_i	$Z_i(x)$	λ_i	$Z_i(x)$	λ_i	$Z_i(x)$	λ_i	$Z_i(x)$
1		0.933	23,684.00	0.933	22,395.00	0.949	22,305.00	0.923	22,395.00	0.908	22,395.00	0.933	23,546.00
2		0.937	236.66	0.957	216.66	0.881	218.47	0.917	216.66	0.837	216.66	0.957	216.66
3		0.916	3031.00	0.926	3531.00	0.916	3531.00	0.916	3531.00	0.916	3201.00	0.946	3531.00
4		0.532	1785.07	0.522	1785.07	0.637	1775.79	0.532	1785.07	0.532	1785.07	0.512	1785.07
5		0.047	1762.81	0.092	1802.81	0.073	1795.70	0.147	1802.81	0	1752.11	0	1802.81
6		0.974	8815.00	0.856	8865.00	0.888	8973.00	0.874	8865.00	0	8865.00	0	8865.00
7		0.664	23,711.00	0.660	24,831.00	0.642	24,933.00	0.694	24,831.00	0	24,831.00	0	23,803.00
8		0.987	3048.00	0.987	3058.00	0	3022.00	0.967	3058.00	0.987	3058.00	0.997	3098.00

FP, fuzzy programming; FPP, fuzzy preference programming.

TABLE 3.8

The Optimal Distribution of the Supplier Selection Problem with Green Issues

Method Variable	Van Laarhoven and Pedrycz		Buckley		Chang		Mikhailov— FP		Mikhailov—FPP			
									$\alpha = 0$		$\alpha = 0.5$	
i	X_i	Y_i	X_i	Y_i	X_i	Y_i	X_i	Y_i	X_i	Y_i	X_i	Y_i
1	44	1	44	1	40	1	24	1	44	1	44	1
2	0	0	15	1	0	0	20	1	0	0	20	1
3	94	1	79	1	100	1	84	1	94	1	74	1
4	52	1	50	1	52	1	52	1	52	1	52	1
5	100	1	80	1	100	1	100	1	100	1	78	1
6	65	1	53	1	60	1	65	1	65	1	65	1
7	55	1	67	1	55	1	60	1	55	1	65	1
8	90	1	82	1	84	1	85	1	90	1	82	1
9	0	0	30	1	9	1	0	0	0	0	0	0
10	0	0	0	0	0	0	10	1	0	0	20	1

FP, fuzzy programming; FPP, fuzzy preference programming.

the optimal distributions of the decision variables for the supplier selection problem are given in Tables 3.7 and 3.8, respectively.

According to the different FAHP weights, similar results are obtained. When the proposed model is solved using Chang's weights, the total purchasing cost is calculated as $22,305 and that value is better than other solutions.

It is observed that the maximum number of available suppliers is obtained by the FAHP methods of Buckley's geometric mean, Mikhailov's fuzzy prioritization, and Mikhailov's fuzzy preference programming with $\alpha = 0.5$. The results show that all the methods can be used in MOLP models to handle the conflicting criteria.

3.5 Conclusions

GSCM has played an important role in allowing companies to achieve higher service levels while meeting customer expectations and environmental regulations. Strategically, manufacturers should work with suppliers to produce goods while dealing with increasing environmental issues. Environmental issues represent an opportunity for purchasing to further influence SCM (Sarkis, 2006). In this chapter, the importance of the evaluation of supplier performance is handled for different environmental issues, such as carbon emissions, CSR, green packaging, green handling, and waste management policy. Determination of the appropriate suppliers is a critical problem for companies. Many previous studies of supplier selection problems have been considered as MCDM problems. We employed an integrated approach to find available suppliers. First, we proposed an MOLP model that includes different green issues. Then, several FAHP methods were used to determine the criteria weights of the problem. Finally, the WAM is constituted and the optimum solution is found. A case study is implemented to illustrate the applicability of the developed model. This chapter is the first study to apply an integrated method using FAHP and MOLP to a supplier selection problem with green issues.

For future research, we suggest applying heuristic/metaheuristic techniques and other MCDM methods, such as TOPSIS, VIKOR, ARAS (Additive Ratio Assessment), to supplier selection problems.

References

Akman, G. (2015), Evaluating suppliers to include green supplier development programs via fuzzy c-means and VIKOR methods, *Computers & Industrial Engineering*, 86, 69–82.

Awasthi, A., Chauhan, S. S., Goyal, S. K. (2010), A fuzzy multi criteria approach for evaluating environmental performance of suppliers, *International Journal of Production Economics*, 126(2), 370–378.

Bai, C., Sarkis, J. (2010), Green supplier development: Analytical evaluation using rough set theory, *Journal of Cleaner Production*, 18(12), 1200–1210.

Bellman, R. E., Zadeh, L. A. (1970), Decision-making in a fuzzy environment, *Management Science*, 17(4), 141–164.

Bozbura, F. T., Beskese, A. (2007), Prioritization of organizational capital measurement indicators using fuzzy AHP, *International Journal of Approximate Reasoning*, 44, 124–147.

Buckley, J. J. (1985), Fuzzy hierarchical analysis, *Fuzzy Sets and Systems*, 17(3), 233–247.

Büyüközkan, G., Çifçi, G. (2011), A novel fuzzy multi-criteria decision framework for sustainable supplier selection with incomplete information, *Computers in Industry*, 62(2), 164–174.

Büyüközkan, G., Çifçi, G. (2012), A novel hybrid MCDM approach based on fuzzy DEMATEL, fuzzy ANP and fuzzy TOPSIS to evaluate green suppliers, *Expert Systems with Applications*, 39, 3000–3011.

Büyüközkan, G., Kahraman, C., Ruan, D. (2004), A fuzzy multi-criteria decision approach for software development strategy selection, *International Journal of General Systems*, 33(2–3), 259–280.

Chang, D. Y. (1996), Applications of the extent analysis method on fuzzy AHP, *European Journal of Operational Research*, 95(3), 649–655.

Davidson, G. (2011), Waste Management Practices: Literature Review, Dalhousie University—Office of Sustainability.

Dobos, I., Vörösmarty, G. (2014), Green supplier selection and evaluation using DEA-type composite indicators, *International Journal of Production Economics*, 157, 273–278.

Emmett, S., Sood, V. (2010), *Green Supply Chains: An Action Manifesto*, New York, NY: John Wiley & Sons, ISBN: 978-0-470-68941-7.

Fahimnia, B., Sarkis, J., Davarzani, H. (2015), Green supply chain management: A review and bibliometric analysis, *International Journal of Production Economics*, 162, 101–114.

Govindan, K., Khodaverdi, R., Jafarian, A. (2013), A fuzzy multi criteria approach for measuring sustainability performance of a supplier based on triple bottom line approach, *Journal of Cleaner Production*, 47, 345–354.

Govindan, G., Rajendran, S., Sarkis, J., Murugesan, P. (2015), Multi criteria decision making approaches for green supplier evaluation and selection: A literature review, *Journal of Cleaner Production*, 98, 66–83.

Hashemi, S. H., Karimi, A., Tavana, M. (2015), An integrated green supplier selection approach with analytic network process and improved Grey relational analysis, *International Journal of Production Economics*, 159, 178–191.

Hsu, C.-W., Kuo, T.-C., Chen, S.-H., Hu, A. H. (2013), Using DEMATEL to develop a carbon management model of supplier selection in green supply chain management, *Journal of Cleaner Production*, 56, 164–172

Igoulalene, I., Benyoucef, L., Tiwari, M. K. (2015), Novel fuzzy hybrid multi-criteria group decision making approaches for the strategic supplier selection problem, *Expert Systems with Applications*, 42, 3342–3356.

Javanbarg, M. B., Scawthorn, C., Kiyano, J., Shahbodaghkhan, B. (2012), Fuzzy AHP-based multicriteria decision making systems using particle swarm optimization, *Expert Systems With Applications*, 39(1), 960–966.

Junior, F. R. L., Osiro, L., Carpinetti, L. C. R. (2013), A fuzzy inference and categorization approach for supplier selection using compensatory and non-compensatory decision rules, *Applied Soft Computing*, 13, 4133–4147.

Kahraman, C., Cebeci, U., Ulukan, Z. (2003), Multi-criteria supplier selection using fuzzy AHP, *Logistics Information Management*, 16(6), 382–394.

Kannan, D., de Sousa Jabbour, A. B. L., Jabbour, C. J. C. (2014), Selecting green suppliers based on GSCM practices: Using fuzzy TOPSIS applied to a Brazilian electronics company, *European Journal of Operational Research*, 233, 432–447.

Kannan, D., Govindan, K., Rajendran, S. (2015), Fuzzy axiomatic design approach based green supplier selection: A case study from Singapore, *Journal of Cleaner Production*, 96, 194–208.

Kannan, D., Khodaverdi, R., Olfat, L., Jafarian, A., Diabat, A. (2013), Integrated fuzzy multi criteria decision making method and multiobjective programming approach for supplier selection and order allocation in a green supply chain, *Journal of Cleaner Production*, 47, 355–367.

Kar, A. K. (2015), A hybrid group decision support system for supplier selection using analytic hierarchy process, fuzzy set theory and neural network, *Journal of Computational Science*, 6, 23–33.

Kuo, R. J., Wang, Y. C., Tien, F. C. (2010), Integration of artificial neural network and MADA methods for green supplier selection, *Journal of Cleaner Production*, 18(12), 1161–1170.

Lee, A. H. I. (2009), A fuzzy supplier selection model with the consideration of benefits, opportunities, costs and risks, *Expert Systems with Applications*, 36, 2879–2893.

Lu, Y. Y.,Wu, C. H., Kuo, T. C. (2007), Environmental principles applicable to green supplier evaluation by using multi-objective decision analysis, *International Journal of Production Research*, 45(18–19), 4317–4331.

McWilliams, A., Siegel, D. (2001), Corporate social responsibility: A theory of the firm perspective, *Academy of Management Review*, 26(1), 117–127.

McWilliams, A., Siegel, D. (2006), Corporate social responsibility: Strategic implications, *Journal of Management Studies*, 43(1), 1–18.

Mikhailov, L. (2000), A fuzzy programming method for deriving priorities in the analytic hierarchy process, *Journal of the Operational Research Society*, 51(3), 341–349.

Mikhailov, L. (2002), Fuzzy analytical approach to partnership selection in formation of virtual enterprises, *Omega*, 30, 393–401.

Mikhailov, L. (2003), Deriving priorities from fuzzy pairwise comparison judgements, *Fuzzy Sets and Systems*, 134, 365–385.

Mikhailov, L., Tsvetinov, P. (2004), Evaluation of services using a fuzzy analytic hierarchy process, *Applied Soft Computing*, 5(1), 23–33.

Paksoy, T., Yapici Pehlivan, N., Kahraman, C. (2012) Organizational strategy development in distribution channel management using fuzzy AHP and hierarchical fuzzy TOPSIS, *Expert Systems with Applications*, 39(3), 2822–2841.

Saaty, T. L. (1980), *The Analytic Hierarchy Process*, New York, NY: McGraw-Hill.

Sanayei, A., Mousavi, S. F., Yazdankhah, A. (2010), Group decision making process for supplier selection with VIKOR under fuzzy environment, *Expert Systems with Applications*, 37, 24–30.

Sarkis, J. (2006), *Greening the Supply Chain*, London: Springer-Verlag. doi: 10.1007/1-84628-299-3.

Shaw, K., Shankar, R., Yadav, S. S., Thakur, L. S. (2012), Supplier selection using fuzzy AHP and fuzzy multi-objective linear programming for developing low carbon supply chain, *Expert Systems with Applications*, 39, 8182–8192.

Shen, L., Olfat, L., Govindan, K., Khodaverdi, R., Diabat, A. (2013), A fuzzy multi criteria approach for evaluating green supplier's performance in green supply chain with linguistic preferences, *Resources, Conservation and Recycling*, 74, 170–179.

Sprinkle, G. B., Maines, L. A., (2010), The benefits and costs of corporate social responsibility, *Business Horizons*, 53, 445–453.

Srivastava, S. K. (2007), Green supply-chain management: A state-of-the-art literature review, *International Journal of Management Reviews*, 9(1), 53–80.

Tiwari, R. N., Dharmahr, S., Rao, J. R. (1987), Fuzzy goal programming—An additive model, *Fuzzy Sets and Systems*, 24(1), 27–34.

Tseng, M.-L., Chiu, A. S. F. (2013), Evaluating firm's green supply chain management in linguistic preferences, *Journal of Cleaner Production*, 40, 22–31.

Tzeng, G. H., Huang, J. J. (2011), *Multiple Attribute Decision Making: Methods and Applications*, Boca Raton, FL: Chapman and Hall/CRC.

Van Laarhoven, P. J. M., Pedrycz, W. (1983), A fuzzy extension of Saaty's priority theory, *Fuzzy Sets and Systems*, 11(1–3), 229–241.

Xiao, Z., Chen, W. Li, L. (2012), An integrated FCM and fuzzy soft set for supplier selection problem based on risk evaluation, *Applied Mathematical Modelling*, 36, 1444–1454.

Yang, C.-C., Chen, B.-S. (2006), Supplier selection using combined analytical hierarchy process and gray relational analysis, *Journal of Manufacturing Technology Management*, 17(7), 926–941.

Zhu, Q., Dou, Y., Sarkis, J. (2010), A portfolio-based analysis for green supplier management using the analytical network process, *Supply Chain Management: An International Journal*, 15(4), 306–319.

Zimmermann, H. J. (1978), Fuzzy programming and linear programming with several objective functions, *Fuzzy Sets and Systems*, 1(1), 45–55.

4

Data Mining Group Decision-Making with FAHP: An Application in Supplier Evaluation and Segmentation

Mohammad Hasan Aghdaie

CONTENTS

4.1 Introduction

There has been an increasing emphasis on supply chain management (SCM) by both practitioners and academics during the last decade and suppliers play leading roles in today's SCM. Every supplier-related decision can be considered an important component of production and logistics plans for many enterprises. That is why a substantial portion of literature on SCM is related to supplier evaluation and selection by using a wide variety of multiple attribute decision-making (MADM) methods based on considering diverse criteria in different industries (Chai et al. 2013; Govindan et al. 2015; Dweiri et al. 2016; Sodenkamp et al. 2016; Tavana et al. 2016).

77

To face the dynamic and unpredictable market in a fierce competitive environment, every organization needs to create a strong relationship with its own suppliers. The better a firm's relation with the supplier, the more successful it will be. Therefore, creating a strong relationship with a supplier is beyond just selection. Thus, thinking strategically about firms' suppliers and maybe avoiding "one single strategy" for all suppliers can be a mandatory approach for many companies (Dyer et al. 1998).

Supplier segmentation is one of the most important parts of SCM and rationally happens after supplier selection. Supplier segmentation is the process of dividing a set of suppliers into a distinct subset of suppliers, where any subset can be treated differently.

Today with the huge amount of data and more complicated environment, managers need to apply more sophisticated tools for their decisions. Data mining (DM) is a discipline that can extract the hidden and unknown information from a cornucopia of data. Among the most frequently used tools of DM, such as concept description (characterization and discrimination), association, classification, clustering, and prediction, cluster analysis is widely used as a segmentation technique in many everyday problems. This technique is a convenient method for the identification and definition of segments.

The decision-making process usually deals with alternatives in the presence of different and often conflicting criteria. This kind of decision-making in operations research is MADM. In 1980, Thomas L. Saaty developed a novel method, namely the analytic hierarchy process (AHP), which very soon became one of the dominant decision-making tools (Saaty, 1980). Decision makers have widely recognized that most decisions made in real-world situations have vague and unclear constrains because of their complexity. Thus, many decision-making problems cannot be exactly defined or precisely represented in a crisp value (Bellman and Zadeh, 1970). To deal with these problems, Zadeh (1965) suggested employing fuzzy set theory as a modeling tool for complex systems that are controllable by humans but are hard to define exactly. The fuzzy analytic hierarchy process (FAHP) method is one of the famous MADM methods for complicated situations when the decision maker has to evaluate based on linguistic terms. In most real-life situations, the decision-making process needs to be conducted by a group of experts or panels. One of the advantages of FAHP is its capability of being applied by a group of experts as a group decision-making (GDM) tool, namely the fuzzy group analytical hierarchical process (FGAHP).

Today, business deals with more complexity, a fast-changing environment, and a huge amount of data so using more integrated decision-making tools is unavoidable. Integration of DM tools with operations research methods is recognized as a new synergy by Meisel and Mattfeld (2010) and in the field of MADM, a brand new synergy is combining the MADM and DM approaches (Aghdaie et al. 2014). The aim of this chapter is to propose a novel integrated FGAHP, simple additive weighting (SAW), and two-stage cluster analysis as a DM tool for supplier evaluation and segmentation. A case study in the fast moving consumer goods (FMCG) industry was conducted to show the applicability of the model in a real-world situation.

The outline of this chapter is as follows. Section 4.2 is the literature review for supplier segmentation. Section 4.3 outlines the proposed methodology combining FGAHP, SAW, and DM. Section 4.4 comprises an in-depth case study to show the applicability of the proposed model and its results. Finally, a conclusion and future research directions are provided in Section 4.5.

4.2 Literature Review

This section focuses on a literature review for supplier segmentation. In marketing, especially industrial market segmentation, occasionally the focus is on the demand-side of the market and its purpose is to group customers into different classes based on their similarities. When enterprises cooperate with potential suppliers, it is hard to manage all suppliers individually, so segmenting the supply side of the market can be a useful approach as well. Furthermore, literature relatively neglected the supply side of the market (Erevelles and Stevenson, 2006). Supplier segmentation is business-to-business (B2B) segmentation and it rationally happens after supplier evaluation and selection. In addition, it is a step between supplier selection and supplier relationship management (Rezaei and Ortt, 2013a). As mentioned earlier, supplier segmentation is studied by a few pieces of research. For example, Parasuraman (1980) recognized the significance of supplier segmentation and concentrated on the supply side of the market. As a pioneer, he defined a stepwise procedure for vendor selection with four steps: (1) identification of key features of vender segments, (2) identification of critical vendor characteristics, (3) selection of relevant dimensions for vendor segmentation, and (4) identification of vendor segments. Another pioneer is Kraljic (1983) who introduced the first comprehensive portfolio model for the firm's purchased product based on two criteria: (1) the profit impact and (2) supply risk. His model has two values, "low" and "high," and it divides supply into four classes: (1) noncritical items, (2) leverage items, (3) bottleneck items, and (4) strategic items. Olsen and Ellram (1997) introduced a portfolio model with two dimensions: (1) difficulty of managing the purchase situation and (2) strategic importance of the purchase. Dyer et al. (1998) studied supplier–automaker relationships in the United States, Japan, and Korea. According to Dyer et al. (1998), strategic segmentation is critical and studied at a durable arm's length and with strategic partnership. Bensaou (1999) used supplier's specific investments and buyer's specific investment criteria. Kaufman et al. (2000) applied technology and collaboration for supplier segmentation. Van

Weele (2000) applied profit impact and supply risk based on a portfolio model. The supplier's commitment and the commodity's importance were used as two factors by Svensson (2004) for buyer relation segmentation. Hallikas et al. (2005) applied supply/buyer dependency risk. Rezaei and Ortt (2012) reviewed the literature of supplier segmentation approaches and proposed a model with two dimensions, capabilities and willingness, in a conceptual framework. In addition, they provided a comprehensive criteria list for supplier segmentation as well. According to Rezaei and Ortt (2012), supplier capability is defined as a "complex bundles of skills and accumulated knowledge, exercised through organizational processes that enable firms to co-ordinate activities and make use of their assets in different business functions that are important for a buyer." Willingness is defined by Rezaei and Ortt (2012) as "confidence, commitment and motivation to engage in a (long-term) relationship with a buyer." A year later, Rezaei and Ortt (2013a) used FAHP and fuzzy preference relations for supplier segmentation. In the same year, the fuzzy rule was applied by Rezaei and Ortt (2013b) to segmentation as well.

In this section, the most important supplier segmentation approaches were introduced and explained. Two portfolio matrix (2 × 2 matrix) analyses are frequently used by researchers for segmentation. Although reaching one complete definition of supplier segmentation is not an easy task, according to Rezaei and Ortt (2013a) it is defined as "the identification of the capabilities and willingness of suppliers by a particular buyer in order for the buyer to engage in a strategic and effective partnership with the suppliers with regard to a set of evolving business functions and activities in the supply chain management."

4.3 Proposed Methodology

This section presents a novel hybrid model that uses FGAHP, SAW, and two-stage cluster analysis for segmentation. The first part of the proposed methodology describes FGAHP as a GDM tool for the evaluation process. The second part explains a SAW method for scoring suppliers. The third part describes a two-stage cluster analysis to segment suppliers. Finally, the last part explains the proposed framework for the supplier evaluation and segmentation process.

4.3.1 Fuzzy Group Analytic Hierarchy Process

AHP was developed in 1980 and it has been commonly used in various decision situations. The strengths of AHP are (1) viewing a complex and ill-structured problem as a simple hierarchical structure with alternatives and decision attributes, (2) having the capability of identifying the relative weight of the factors and the total values of each alternative weight in multiple attribute problems based on using a series of pairwise comparisons, and (3) calculating the consistency index (CI) of pairwise comparisons, which is the ratio of the DM's inconsistency.

Since the real world is full of ambiguity and vagueness, orthodox MADM cannot handle problems with imprecise information. In addition, incorporation of imprecise information and vagueness is an unavoidable requirement of fine decision-making models. MADM techniques need to tolerate vagueness; however, conventional MADM, like traditional AHP, does not contain uncertainty in its pairwise comparisons (Yu, 2002). To deal with this kind of problem, Zadeh (1965) proposed using fuzzy set theory for complicated systems that are hard to define. Thus, fuzzy multiple attribute decision-making (FMADM) was developed. As mentioned earlier, in decision-making, there is a fuzzy concept in comparisons as well. That is why there are many FAHP methods that are proposed by different authors in the literature (De Grann, 1980; Buckley, 1985; Chang, 1996; Cheng, 1997; Deng, 1999; Leung and Cao, 2000; Mikhailov, 2004; Van Laarhoven and Pedrycz, 1983). Furthermore, it is easier to understand and can effectively handle both qualitative and quantitative data (Liao, 2011). In this study, the FAHP approach that was introduced by Chang (1996) was preferred. In his approach, for the pairwise comparison scale, triangular fuzzy numbers (TFNs) can be used. This is simpler and needs less complicated calculations compared with other proposed FAHP methods.

Let $X = \{x_1, x_2, x_3, ..., x_n\}$ be an object set and $U = \{u_1, u_2, u_3, ..., u_m\}$ be a goal set. According to the method of Chang's (1996) extent analysis, each object is taken and extent analysis for each goal, g_i, is performed, with m extent analysis values for each object, given as $M_{gi}^1, M_{gi}^2, ..., M_{gi}^m, i = 1, 2, ..., n$, where all the $M_{gi}^j (j = 1, 2, ..., m)$ are TFNs, representing the performance of the object x_i with regard to each goal u_j.

The steps of the Chang FAHP are given below:

Step 1: The value of fuzzy synthetic extent with respect to the ith object is defined as

$$S_i = \sum_{j=1}^{m} M_{gi}^j \otimes \left[\sum_{i=1}^{n} \sum_{j=1}^{m} M_{gi}^j \right]^{-1} \tag{4.1}$$

To obtain $\sum_{j=1}^{m} M_{gi}^j$, perform the fuzzy addition operation of m extent analysis values for a particular matrix such that operation m extent analysis values for a particular matrix such that

$$\sum_{j=1}^{m} M_{gi}^{j} = \left(\sum_{j=1}^{m} l_j, \sum_{j=1}^{m} m_j, \sum_{j=1}^{m} u_j \right)$$ (4.2)

and obtain $\left[\sum_{i=1}^{n} \sum_{j=1}^{m} M_{gi}^{j} \right]^{-1}$, perform the fuzzy addition operation of

$M_{gi}^{j} (j = 1, 2, ..., m)$ values and such that

$$\sum_{i=1}^{n} \sum_{j=1}^{m} M_{gi}^{j} = \left(\sum_{i=1}^{n} l_i, \sum_{i=1}^{n} m_i, \sum_{i=1}^{n} u_i \right)$$ (4.3)

and then compute the inverse of the vector in Equation 4.3 such that

$$\left[\sum_{i=1}^{n} \sum_{j=1}^{m} M_{gi}^{j} \right]^{-1} = \left(\frac{1}{\sum_{i=1}^{n} u_i}, \frac{1}{\sum_{i=1}^{n} m_i}, \frac{1}{\sum_{i=1}^{n} l_i} \right)$$ (4.4)

Step 2: The degree of possibility of $M_2(l_2, m_2, u_2) \geq M_1(l_1, m_1, u_1)$ is defined as

$$V(M_2 \geq M_1) = \sup_{y \geq x}[\min(\mu_{M_1}(x), \mu_{M_2}(y))]$$ (4.5)

and can be equivalently expressed as follows:

$$V(M_2 \geq M_1) = \text{hgt}(M_1 \cap M_2) = \mu_{M_2}(d) = \begin{cases} 1, \text{if } (m_2 \geq m_1) \\ 0, \text{if } (l_1 \geq u_2) \\ \dfrac{l_1 - u_2}{(m_2 - u_2) - (m_1 - l_1)}, \text{otherwise} \end{cases}$$ (4.6)

where d is the ordinate of the highest intersection point D between μ_{M_1} and μ_{M_2} (see Figure 4.1).

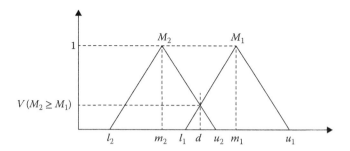

FIGURE 4.1
Intersection point "*d*" between two fuzzy numbers M_1 and M_2.

To compare M_1 and M_2, the values of both $V(M_1 \geq M_2)$ and $V(M_2 \geq M_1)$ are required.

Step 3: The degree possibility of a convex fuzzy number to be greater than k convex fuzzy numbers $M_i(i = 1, 2, ..., k)$ can be defined by Equation 4.7.

$$V(M \geq M_1, M_2, ..., M_K) = V[M \geq M_1] \text{ and } V[M \geq M_2] \text{ and...and } V[M \geq M_k]$$
$$= \min(V[M \geq M_i], i = 1, 2, ..., k) \tag{4.7}$$

Assume that

$$d'(A_i) = \min(S_i \geq S_k) \tag{4.8}$$

for $k = 1, 2, ..., n$; $k \neq i$. Then, the weight vector is given in Equation 4.10:

$$W' = (d'(A_1), d'(A_2), ..., d'(A_n))^T \tag{4.9}$$

where $A_i(i = 1, 2, ..., n)$ has n elements.

Step 4: The normalized weight vectors are defined as

$$W = (d(A_1), d(A_2), ..., d(A_n))^T \tag{4.10}$$

where W is a nonfuzzy number.

In our approach, we increased Chang's (1996) FAHP scale into nine TFNs. The fuzzy pairwise scale (linguistic variables for the evaluation process) is defined in Table 4.1.

TABLE 4.1

Linguistic Variables for Fuzzy Pairwise Scale

Linguistic Scale	Triangular Fuzzy Scale	Triangular Fuzzy Reciprocal Scale
Equally important	$\tilde{1} = (1,1,1)$	$\tilde{1}^{-1} = (1,1,1)$
Judgment values between equally and moderately	$\tilde{2} = (1,2,3)$	$\tilde{2}^{-1} = \left(\frac{1}{3}, \frac{1}{2}, 1\right)$
Moderately more important	$\tilde{3} = (2,3,4)$	$\tilde{3}^{-1} = \left(\frac{1}{4}, \frac{1}{3}, \frac{1}{2}\right)$
Judgment values between moderately and strongly	$\tilde{4} = (3,4,5)$	$\tilde{4}^{-1} = \left(\frac{1}{5}, \frac{1}{4}, \frac{1}{3}\right)$
Strongly more important	$\tilde{5} = (4,5,6)$	$\tilde{5}^{-1} = \left(\frac{1}{6}, \frac{1}{5}, \frac{1}{4}\right)$
Judgment values between strongly and very strongly	$\tilde{6} = (5,6,7)$	$\tilde{6}^{-1} = \left(\frac{1}{7}, \frac{1}{6}, \frac{1}{5}\right)$
Very strongly more important	$\tilde{7} = (6,7,8)$	$\tilde{7}^{-1} = \left(\frac{1}{8}, \frac{1}{7}, \frac{1}{6}\right)$
Judgment values between very strongly and extremely	$\tilde{8} = (7,8,9)$	$\tilde{8}^{-1} = \left(\frac{1}{9}, \frac{1}{8}, \frac{1}{7}\right)$
Extremely more important	$\tilde{9} = (8,9,9)$	$\tilde{9}^{-1} = \left(\frac{1}{9}, \frac{1}{9}, \frac{1}{8}\right)$
If factor i has one of the above numbers assigned to it when compared to factor j, then j has the reciprocal value when compared with i		Reciprocals of above $M_1^{-1} \approx (1/u_1, 1/m_1, 1/l_1)$

For aggregation of group decisions, geometric mean operations were used. Geometric mean operations were applied in many studies for GDM (Davies, 1994; Buckley, 1985).

$$l_{ij} = \left(\prod_{k=1}^{k} l_{ijk}\right)^{1/k}, \quad m_{ij} = \left(\prod_{k=1}^{k} m_{ijk}\right)^{1/k}, \quad u_{ij} = \left(\prod_{k=1}^{k} u_{ijk}\right)^{1/k} \tag{4.11}$$

For sensitivity analysis of pairwise comparisons, Saaty (1977) proposed a CI, which is related to the eigenvalue method:

$$CI = \frac{\lambda_{max} - n}{n-1} \tag{4.12}$$

where n = dimension of the matrix and λ_{max} = maximal eigenvalue.

TABLE 4.2

Average Random Index for Corresponding Matrix Size

Size of matrix (n)	1	2	3	4	5	6	7	8	9	10	11
Random index (RI)	0	0	0.52	0.89	1.11	1.25	1.35	1.40	1.45	1.49	1.51

Source: Saaty, T. L., *Journal of Mathematical Psychology*, 15, 234–281, 1977.

The consistency ratio (CR), the ratio of CI and random index (RI), is calculated using

$$CR = CI / RI \qquad (4.13)$$

Based on the average CI of 500 randomly filled matrixes, less than 0.1 is an acceptable value for CR (Ishizaka and Labib, 2011). Table 4.2 shows the RI with respect to different size matrices.

CR is used to monitor the validity of the pairwise comparisons. As the CR index is crisp, a defuzzification method (in this study a fuzzy means approach) was applied to defuzzify the TFNs. For more information on defuzzification methods, see Leekwijck and Kerre (1999), Hussu (1995), Yager and Filev (1993), de Oliveira (1995), Rondeau et al. (1997), and others. If CR is larger than 0.1, the group of experts is required to revise their pairwise comparisons until the CR is within the acceptable range.

4.3.2 SAW Method

Churchman and Ackoff (1954) introduced and applied the SAW method to deal with a portfolio selection problem. SAW is one of the simplest MADM methods and the best alternative can be selected according to the following equation:

$$A^* = \left\{ u_i(x) \middle| \max_i u_i(x) \middle| i = 1, 2, \dots, n \right\} \qquad (4.14)$$

Also

$$u_i(x) = \sum_i^n w_j r_{ij}(x) \qquad (4.15)$$

where $u_i(x)$ represents the utility of the ith alternative and $i = 1, 2, \dots, n$; w_j indicates the weight of the jth criterion; and $r_{ij}(x)$ is the normalized preferred rating of the ith alternative with respect to the jth criterion. It is assumed that all the criteria are independent. For calculating the normalized preferred ratings $r_{ij}(x)$ of the ith alternative with respect to the jth criterion, there are two possible forms:

First form

For the benefit criterion, $r_{ij}(x) = x_{ij}/x_j^*$, where $x_j^* = \max_i x_{ij}$ or let x_j^* be the desired level, and it is clear $0 \le r_{ij}(x) \le 1$.

For the cost criterion, $r_{ij}(x) = (1/x_{ij})/(1/x_j^*) = (\max_i x_j^*)/(x_{ij})$ or let x_j^* be the desired level.

Second form

For the benefit criterion, $r_{ij} = (x_{ij} - x_j^*)/(x_j^* - x_j^-)$, where $x_j^* = \max_i x_{ij}$ and $x_j^- = \min_i x_{ij}$ or let x_j^* be the desired level and x_j^- be the worst level.

For the cost criterion, and $r_{ij} = (x_j^- - x_{ij})/(x_j^- - x_j^*)$.

Therefore, the synthesized performance is

$$p_i = \sum_j^m w_j r_{ij} \tag{4.16}$$

where p_i is the synthesizing performance value of the ith alternative; w_j represents the weights of the jth criterion; and r_{ij} is the normalized preferred ratings of the ith alternative with respect to the jth criterion.

4.3.3 Data Mining

With the huge amount of data available and with the increasing comprehension that necessitates deeper analysis in the current business environment, DM has been applied in many fields. DM is the process of extracting valid, novel, useful, and understandable information (knowledge) from a large amount of data. According to Han and Kamber (2006), some of the functions of DM are association, classification, clustering, and prediction.

Cluster analysis is one of the most frequently used DM techniques for segmentation in marketing literature and in practice (Wedel and Kamakura, 2001). Especially, it is used frequently for customer segmentation (Chiu and Tavella, 2008; Dillon et al. 1993). In addition, it is an easy way to identify and define segments (Hong, 2012). Hierarchical and nonhierarchical clustering methods are two famous classes of clustering. In 1967, MacQueen introduced a clustering algorithm, K-means, which soon became one of the most famous algorithms (Hung and Tsai, 2008). Although it was proposed 50 years ago, it is still one of the most frequently used clustering algorithms (Anil, 2010). In addition, it is simple, easy, quick, and efficient. Its drawback, however, is that the number of clusters should be supplied as an input parameter. To tackle this problem, Punj and Steward (1983) proposed using Ward's method (1967) as a new integration, namely two-stage cluster analysis. Ward's method (1967) is one of the most popular hierarchical clustering methods (Mingoti and Lima, 2006). It can be used as a tool to find the optimum number of clusters. In addition, Ward's minimum variance has worked well in earlier studies (Hair et al. 1995). Some of the recent applications of two-stage cluster analysis with Ward's method and K-means are Kazemzadeh et al. (2009), Aghdaie and Tafreshi (2012), Aghdaie et al. (2013), and others.

4.3.4 The Supplier Segmentation Model

This section elaborates on the supplier segmentation model. To this end, the four-phase conceptual framework is proposed by integration of MADM and DM (see Figure 4.2). The following paragraphs will explain every phase of the model.

The first phase (model preparation) has five steps. In the first step, a decision-making team (a group of experts) was selected to carry out supplier segmentation. Then, according to an in-depth literature review, a comprehensive criteria list with the most overarching factors was created. Next, a panel of experts studied and accepted the criteria list. The Delphi method was used to build a consensus among the panel in this stage. After that, the hierarchical structure of the problem was created. Every criterion was classified into the cost criterion group (the smaller the better) or the benefit criterion group (the larger the better). In addition, criteria were classified into a qualitative or quantitative group. For evaluating the qualitative criteria, TFNs were used. After constructing the problem structure, suppliers were identified and evaluated based on their past performance and experts' personal evaluations. The decision matrix was created next.

In the second phase, FGAHP was applied by the group of experts to evaluate and weight the criteria. In the next step of this phase, the sensitivity analysis of the experts' pairwise comparisons was checked by the CR index.

In the third phase, the SAW approach was used to score the suppliers. For the defuzzification method, a fuzzy mean approach was applied. After SAW calculations, the scores of suppliers were ready for cluster analysis.

In the DM phase, two-stage cluster analysis (Punj and Stewart, 1983) that is based on the integration of the K-means algorithm (MacQueen, 1967) and Ward's (1963) method was applied to cluster suppliers according to their scores. Schwarz's Bayesian information criterion (BIC) and an average silhouette measure were used to assess partitioning. For interpretation, every cluster was explained and some points for managing them were proposed.

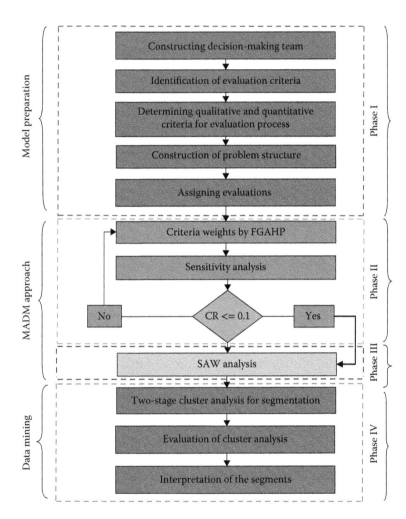

FIGURE 4.2
The proposed model for supplier segmentation and evaluation.

4.4 Illustrative Case Study

4.4.1 Company Profile

XYZ Company is the biggest producer of processed food and dairy products in Iran. Dairy goods, processed meat, ice cream, and nondairy beverages like juices and alcohol-free beers are the major products of the company. It has more than 15,000 employees, with annual turnover of $1 billion and 550 million liters of annual dairy production. In Iran, XYZ products are purchased by more than 140,000 supermarkets, 20,000 restaurants and fast food establishments, 1000 hotels, and 600 chain stores. The company has five main categories of suppliers. They provide equipment, packages, flavor, and essence, milk and vegetables, and fruits. All of these items are delivered to the plant and then processed. At the end, the final products are distributed in the market through a direct-store delivery method.

As the company is very big, managing a large number of suppliers is a very challenging task. Thus, the supplier segmentation project was defined at the highest level of management. Dairy sales composed 80% of the total sales of the XYZ Company so a supplier segmentation project was defined for dairy suppliers at first. The Iran dairy market demands high quality and low price products, which have to be delivered at the right time. In addition, the market size has multiplied recently in terms of customers. In this dynamic environment, managing suppliers is a critical task. To deal with this issue, FMCG companies should segment their suppliers fruitfully.

After realizing the criticality of defining the project in the company, a project team with seven experts was selected for carrying out the project. The number seven is suggested by Saaty and Özdemir (2015) to obtain valid and consistent judgments when using the AHP. The group consisted of two females and five males. The characteristics of the group of experts are shown in Table 4.3.

To make a useful criteria list, we asked the experts to check the whole list and select the most suitable criteria. The basic accepted criteria in the model are extracted from the reviewed literature and experts' ideas (see Table 4.4). The last column of Table 4.4 shows the related literature source for selected subcriteria as well.

Next, the problem was constructed in a hierarchical structure as follows (see Figure 4.3). The four-level hierarchy of the problem included the goal at the first level, two criteria/dimensions at the second level, 12 subcriteria at the third level, and finally suppliers at the fourth level as alternatives.

In this study, we used Rezaei and Ortt (2012) for dividing supplier evaluation into two main categories: (1) willingness and (2) potentials. Among all subcriteria, two criteria, P_1 (cost/price) and P_3 (rank of the supplier in related industry) are cost criteria (a minimum of cost criterion is desirable); the others are benefit criteria. This kind of classification is absolutely essential for SAW analysis in scoring the suppliers. Furthermore, two criteria, W_1 (duration relationship) and P_3 (rank of the supplier in related industry), were considered quantitative and the others were qualitative criteria. For the evaluation of the qualitative criteria, a TFN scale (see Table 4.1) was used. For aggregation of group decisions, geometric mean operations of TFNs were applied. A fuzzy mean approach was used as a defuzzification method to make crisp numbers.

4.4.2 Results and Discussion

This section focuses on the numerical results. In this study, a hybrid FGAHP, SAW, and two-stage cluster analysis were applied to segment suppliers. The

TABLE 4.3

The Characteristics of the Seven Decision-Making Experts

Code	Gender	Age	Education Level	Experience (Years)	Job Title	Job Responsibility
(D1)	Male	57	Master's in management	>29	Manager of the dairy factory	In charge of the most important policies and decisions of the factory.
(D2)	Male	50	Master's in industrial engineering	>25	Project manager and supply chain manager	Managing the engineering team, supply chain, suppliers, and new projects.
(D3)	Male	53	MBA in operations research	>20	Manager of production planning	In charge of all production planning decisions.
(D4)	Male	48	MBA in marketing	>15	Import and export manager	Responsible for all abroad relations.
(D5)	Female	45	PhD in business administration	>17	Marketing and sales manager	Responsible for R&D, new products, marketing research, and pricing decisions.
(D6)	Female	42	MSc in industrial management	>13	Manager of supply and procurement	Responsible for supplying all necessities of the production process.
(D7)	Male	39	MBA in strategy	>12	Strategic manager	Responsible for strategic decision-making.

proposed model was used in a large-sized FMCG company in Iran. The outline of this section is as follows. FGAHP results are explained in Section 4.4.2.1. The obtained SAW results are explained in Section 4.4.2.2. Finally, clustering results with interpretations are explained in Section 4.4.2.3.

4.4.2.1 FGAHP Results

This section focuses on the results obtained from FGAHP. The four-phase model is explained in Figure 4.2. In the first part of the second phase, FGAHP was used to weight the dimensions and subcriteria by building the fuzzy pairwise comparison matrixes for criteria and subcriteria shown in Tables 4.5 through 4.7, respectively. We asked experts to assign pairwise evaluations. Linguistic variables for a fuzzy scale (see Table 4.1) were used by a

TABLE 4.4

Factors Taken from the Review of the Related Literature and Experts' Ideas That Are Relevant to Supplier Evaluation and Segmentation

Code	Criteria	Subcriteria	Related Literature Source
W	**Willingness**		
W_1		Duration of relationship	Hoyt and Huq (2000)
W_2		Commitment to quality and continuous improvement in product and process	Kannan and Tan (2002), Svensson (2004), Urgal-González and García-Vázquez (2007)
W_3		Ethical standards	Kannan and Tan (2002)
W_4		Willingness to share information, ideas, technology, and cost savings	Kannan and Tan (2002)
W_5		Relationship closeness	Choi and Hartley (1996), Kaufman et al. (2000), Chan (2003)
W_6		Willingness to invest in specific equipment	Urgal-González and García-Vázquez (2007)
P	**Potentials**		
P_1		Price/cost	Dickson (1966), Weber et al. (1991), Day (1994), Swift (1995), Choi and Hartley (1996), Kannan and Tan (2002), Chan (2003), Rezaei and Davoodi (2011)
P_2		Quality	Dickson (1966), Weber et al. (1991), Tan et al. (2002), Chan (2003), Rezaei and Davoodi (2011)
P_3		Rank, reputation, and position in industry	Dickson (1966), Weber et al. (1991), Swift (1995), Choi and Hartley (1996)
P_4		Flexibility	Alimardani et al. (2013)
P_5		Production, manufacturing/transformation facilities, and capacity	Dickson (1966), Weber et al. (1991), Day (2010)
P_6		Technology	Dickson (1966), Weber et al. (1991), Swift (1995), Choi and Hartley (1996), Chan (2003)

group of experts for pairwise evaluations. Then for aggregation of seven comparison matrixes, a geometric mean was applied (see Equation 4.11). Each cell of the mentioned tables (see Tables 4.5 through 4.7) is the result of the geometric mean of seven fuzzy pairwise comparisons (TFNs) of the experts. When all fuzzy aggregated comparison matrixes were constructed, an FAHP procedure based on Chang's (1996) extent analysis was applied. By using Equations 4.1 through 4.10, the relative weights of subcriteria were calculated. The resulting six scores of the willingness and potentials are presented in Figure 4.4.

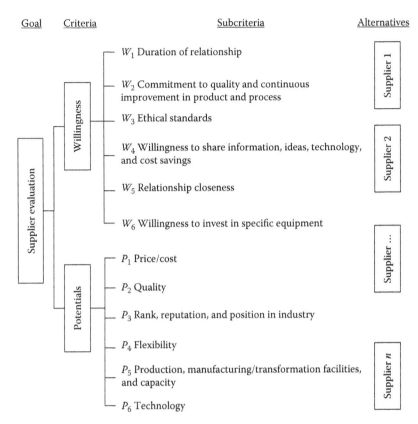

FIGURE 4.3
Decision hierarchy for supplier evaluation.

TABLE 4.5

Fuzzy Aggregated Pairwise Comparison Matrix for the Criteria

	W	P	Weights
W	(1, 1, 1)	(0.95, 1.23, 1.63)	0.62
P	(0.62, 0.82, 1.05)	(1, 1, 1)	0.38

In the next step, the CR index of each matrix was checked to evaluate the consistency using Equations 4.12 and 4.13. According to the experts' evaluations, the CI indexes are 0.064 and 0.078, which is within the acceptable range (see Figure 4.4).

4.4.2.2 SAW Analysis Results

After calculating the criteria and subcriteria weights, SAW was used to calculate the score of each supplier. In doing this, the relative weights of

TABLE 4.6

Fuzzy Aggregated Pairwise Comparison Matrix for Willingness Subcriteria

	W_1	W_2	W_3	W_4	W_5	W_6
W_1	(1, 1, 1)	(0.36, 0.49, 0.82)	(0.14, 0.17, 0.20)	(0.25, 0.34, 0.53)	(0.23, 0.31, 0.46)	(1.09, 1.63, 2.28)
W_2	(1.22, 2.03, 2.78)	(1, 1, 1)	(0.18, 0.22, 0.28)	(0.22, 0.28, 0.40)	(0.25, 0.33, 0.52)	(2.00, 3.05, 4.07)
W_3	(4.98, 6.03, 6.96)	(3.59, 4.62, 5.64)	(1, 1, 1)	(2.42, 3.48, 4.50)	(0.85, 1.19, 1.60)	(3.52, 4.70, 5.79)
W_4	(1.87, 2.97, 4.02)	(2.16, 3.25, 4.30)	(0.22, 0.29, 0.41)	(1, 1, 1)	(0.33, 0.43, 0.67)	(2.26, 2.69, 3.00)
W_5	(2.19, 3.26, 4.29)	(1.92, 2.99, 4.03)	(0.62, 0.84, 1.17)	(1.49, 2.32, 3.07)	(1, 1, 1)	(3.77, 4.82, 5.85)
W_6	(0.44, 0.62, 0.92)	(0.25, 0.33, 0.50)	(0.17, 0.21, 0.28)	(0.28, 0.37, 0.52)	(0.17, 0.21, 0.27)	(1, 1, 1)

TABLE 4.7

Fuzzy Aggregated Pairwise Comparison Matrix for Potentials Subcriteria

	P_1	P_2	P_3	P_4	P_5	P_6
P_1	(1, 1, 1)	(0.30, 0.36, 0.48)	(0.31, 0.41, 0.59)	(1.35, 2.38, 3.39)	(1.15, 1.79, 2.52)	(4.19, 5.23, 6.26)
P_2	(1.95, 2.78, 3.52)	(1, 1, 1)	(0.53, 0.67, 1.00)	(3.17, 4.23, 5.27)	(2.38, 3.47, 4.52)	(1.81, 2.56, 3.28)
P_3	(1.45, 2.21, 3.01)	(1.00, 1.49, 1.87)	(1, 1, 1)	(3.52, 4.57, 5.59)	(2.52, 3.54, 4.54)	(4.74, 5.76, 6.77)
P_4	(0.29, 0.42, 0.74)	(0.19, 0.24, 0.32)	(0.18, 0.22, 0.28)	(1, 1, 1)	(0.33, 0.44, 0.67)	(1.29, 2.12, 2.87)
P_5	(0.40, 0.56, 0.87)	(0.22, 0.29, 0.42)	(0.21, 0.28, 0.40)	(1.49, 2.28, 3.02)	(1, 1, 1)	(1.84, 2.29, 3.93)
P_6	(0.16, 0.19, 0.24)	(0.33, 0.42, 0.67)	(0.15, 0.17, 0.21)	(0.35, 0.47, 0.77)	(0.25, 0.37, 0.54)	(1, 1, 1)

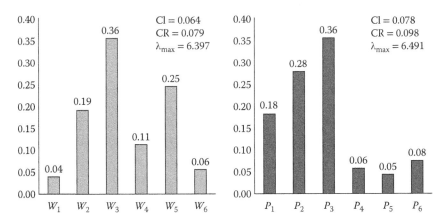

FIGURE 4.4
Defuzzified weight of willingness criteria and potentials criteria.

subcriteria were calculated. The weights are presented in the second row of Table 4.8. Then, experts used fuzzy numbers (see Table 4.1) for evaluations of qualitative subcriteria. Among all subcriteria, two criteria W_1 (duration of relationship) and P_3 (rank of the supplier in related industry) were considered as quantitative and the others were regarded as qualitative criteria. Also, two criteria, P_1 and P_3, were cost criteria (a minimum of cost criterion is desirable); the others were benefit criteria. For aggregation of expert evaluations, geometric mean operations were applied to TFNs (see Equation 4.11). The initial aggregated defuzzified decision matrix is presented in Table 4.8. Each cell of Table 4.8 shows the aggregated fuzzy mean of TFNs.

The weighted normalized decision matrix is shown in Table 4.9. Then, SAW analysis was used to calculate the score of every supplier. Equations 4.14 through 4.16 were applied for SAW analysis. The last column of Table 4.9 shows the score of each supplier. The final scores of suppliers were applied as an input for cluster analysis.

4.4.2.3 Clustering Results and Interpretations

The purpose of using cluster analysis was to partition suppliers into different groups and deal with each cluster suitably. The crisp output of SAW analysis was applied as the clustering input. In the first step of two-stage clustering, Ward's method was applied to calculate the optimal number of clusters. The initial seed points were taken from the derived cluster centers, which were derived from Ward's method. Then, the K-means method was used to segment them. Four clusters were derived as an optimal number by using the algorithm. The quality of the clustering is good with an average silhouette coefficient of 0.80, which is greater than 0.50. If an average silhouette coefficient is greater than 0.5, then this indicates reasonable clustering, whereas a coefficient less than 0.2 denotes a problematic solution (Tsiptsis and Chorianopoulos, 2009).

TABLE 4.8

The Initial Aggregated Defuzzified Decision-Making Matrix

Supplier No.	W_1	W_2	W_3	W_4	W_5	W_6	P_1	P_2	P_3	P_4	P_5	P_6
Subcriteria relative weights	0.024	0.0119	0.221	0.070	0.153	0.035	0.069	0.105	0.135	0.022	0.017	0.028
1	8.22	8.15	7.41	8.18	7.46	7.96	8.03	2.37	2.94	7.64	7.9	8.12
2	3.45	3.26	3.66	7.64	5.28	5.54	5.54	4.47	3.80	4.71	7.20	3.23
3	3.67	4.05	4.82	3.54	4.56	3.94	3.94	5.31	4.76	3.50	3.72	6.37
4	3.34	5.95	2.12	4.11	2.30	3.60	3.60	4.42	6.24	5.36	4.57	4.22
⋮	⋮	⋮	⋮	⋮	⋮	⋮	⋮	⋮	⋮	⋮	⋮	⋮
98	8.67	7.81	8.31	8.28	8.21	8.06	8.06	3.00	3.20	7.55	7.35	7.96
99	5.44	4.56	6.23	5.34	3.70	2.53	2.53	5.24	3.17	6.32	4.24	4.79
100	4.77	6.49	3.77	4.66	5.55	4.01	4.01	4.32	4.75	4.57	3.04	4.48
101	8.37	7.50	7.64	7.64	7.90	7.58	7.94	2.17	2.69	7.88	8.06	7.79

TABLE 4.9

The Weighted Normalized Decision-Making Matrix

Supplier No.	W_1	W_2	W_3	W_4	W_5	W_6	P_1	P_2	P_3	P_4	P_5	P_6	Score
1	0.02	0.12	0.22	0.07	0.15	0.03	0.02	0.03	0.011	0.02	0.02	0.03	0.85
2	0.01	0.06	0.13	0.07	0.11	0.03	0.02	0.07	0.08	0.01	0.02	0.01	0.61
3	0.01	0.07	0.17	0.03	0.10	0.02	0.03	0.09	0.06	0.01	0.01	0.02	0.61
4	0.01	0.09	0.06	0.04	0.05	0.02	0.04	0.07	0.05	0.02	0.01	0.02	0.46
⋮	⋮	⋮	⋮	⋮	⋮	⋮	⋮	⋮	⋮	⋮	⋮	⋮	⋮
98	0.02	0.12	0.22	0.07	0.15	0.03	0.02	0.05	0.10	0.02	0.02	0.03	0.86
99	0.01	0.07	0.18	0.05	0.07	0.01	0.05	0.08	0.09	0.02	0.01	0.02	0.66
100	0.01	0.10	0.11	0.04	0.10	0.02	0.03	0.06	0.05	0.01	0.01	0.02	0.57
101	0.02	0.12	0.22	0.07	0.15	0.03	0.02	0.03	0.13	0.02	0.02	0.03	0.87

TABLE 4.10

Average SAW Scores of the Segments

	Segment 1, $n = (8)$	Segment 2, $n = (43)$	Segment 3, $n = (27)$	Segment 4, $n = (23)$	Mean, $n = (101)$
Average SAW scores	0.85	0.58	0.63	0.47	0.59

SAW, simple additive weighting.

Most computations were done by Microsoft Excel 2010 while Predictive Analytics SoftWare (PASW) statistics 18 was applied for two-step cluster analysis. The smallest cluster size is 8% and the largest one is 42%. The ratio of sizes is 5.38. Table 4.10 provides a basis for comparison between mean values of the segments on suppliers' scores. Also, the average SAW scores on profiles for every segment are presented in Table 4.10. As a consequence, the four segments had 8 (8%), 43 (42%), 27 (27%), and 23 (23%) suppliers, respectively. The averages of the SAW scores in each segment are presented in Table 4.10.

The main interpretations of the segments can be defined as follows:

Segment 1: This segment consisted of eight suppliers with the highest scores. The mean SAW score of the segment is 0.85 (SD = 0.035). These suppliers are superstars, as they have high potentials and a high willingness level. Regarding willingness, for instance, XYZ Company can take advantage of willingness to invest in specific equipment or share information and ideas. Considering potentials, they are reputable and loyal suppliers with high-quality products and the most competitive prices. It is a small but strategic segment. The firm should increase the willingness by providing new strategic planes in order to profit as much as possible from working with them. Also, keeping a sustainable relationship with them is obligatory.

Segment 2: These suppliers are typical suppliers. The mean SAW score of the segment is 0.58 (SD = 0.014). They are suppliers with a mediocre willingness and good potentials. The company can take advantage of them by concentrating on their potentials. From the experts' perspective, the company ought to find the reasons behind the low levels of willingness. For example, it may be a good idea to share the risk and reward with this kind of supplier to motivate them for more collaboration.

Segment 3: These suppliers have high willingness with average potentials. The mean SAW score of the segment is 0.63 (SD = 0.013). They are golden suppliers as they are located under the superstars. One of the possible approaches to deal with them may be that the company aids them to boost their potentials. In addition, according to experts' interviews with the suppliers, they were ready to participate in new technologies, and transform their facilities and capacity. Also, they were going to spend money on management standard systems in order to improve their quality. Since they need more money for improving their potentials, the firm can eliminate some suppliers from Segment 4 and expand its buying interactions with these suppliers.

Segment 4: The worst suppliers located in this segment. The mean SAW score of this segment is 0.47 (SD = 0.022). Not only do they have low potentials, but also they have low willingness to cooperate. They are dormant suppliers. Generally, it is advisable to replace them. Most of them (35 out of 43) are long-standing suppliers and according to the experts' point of view, replacing them may be a suitable strategy; even when the demand is high it may be possible to supply using other suppliers. For the eight very new suppliers in this segment, the experts' idea was to work for another year with them. Maybe after some months, these new suppliers can improve their potentials and willingness.

4.5 Conclusion and Future Research Directions

Suppliers are one of the cruxes of SCM and companies become dependent on their suppliers. They are different so organizations should ponder strategically about their suppliers' relationships and avoid having one single strategy. Segmentation is one of the practical approaches to deal with heterogeneity in firm–supplier connections. Since there are hosts of possible multiple criteria for evaluation and segmentation of suppliers, consequently segmentation can be viewed as a multiple criteria decision-making problem.

To put it in a nutshell, the major contributions of this chapter are as follows:

First, this chapter suggested a novel hybrid FMADM and DM for supplier evaluation and segmentation. Second, it proposed FGAHP for weight calculation as a GDM tool. Third, it suggested SAW analysis for scoring suppliers. Fourth, the use of a two-stage clustering approach for segmentation of suppliers is another contribution. Fifth, the chapter suggested a supplier segmentation framework by reviewing the literature in depth and applying the concepts to a FMCG company.

To extend the current direction of this chapter, it is proposed to use other FAHP procedures and compare the results. In addition, using other MADM methods is recommended. Another possible further research direction can be using other clustering methods. For aggregation of group decisions, geometric mean operations were used. Therefore, further study can apply other aggregation methods and compare the results. According to the results of this study, experts chose critical factors that were found to influence the segmentation process. This proves the role of experts in the decision-making process. To extend the current direction of this chapter, it is highly recommended to create new studies applying other criteria for segmentation. Furthermore, the integrated model proposed an analytical framework for dealing with problems in conflicting management situations. Therefore, further research can apply this method as an adaptable approach to other segmentation situations or other industries.

Acknowledgments

The author expresses his gratitude to the Solico Group (a multinational food company) for the financial support of this project. He also thanks the respectful editors for their valuable comments.

References

Aghdaie, M. H., Hashemkhani Zolfani, S., Zavadskas, E. K. (2013). A hybrid approach for market segmentation and market segment evaluation and selection: An integration of data mining and MADM. *Transformations in Business & Economics*, 12(29B), pp. 415–436.

Aghdaie, M. H., Hashemkhani Zolfani, S., Zavadskas, E. K. (2014). Synergies of data mining and multiple attribute decision making. *The 2nd International Scientific Conference Contemporary Issues in Business, Management and Education*, Vilnius, Lithuania, 201(14), pp. 767–776.

Aghdaie, M. H., Tafreshi, P. F. (2012). Using two-stage clustering and conjoint analysis for benefit segmentation of Iranian laptop buyers. *The 3rd International Conference on Emergency Management and Management Sciences*, Beijing, China, pp. 816–819.

Alimardani, M., Hashemkhani Zolfani, S., Aghdaie, M. H., Tamosaitiene, J. (2013). A novel hybrid SWARA and VIKOR methodology for supplier selection in an agile environment. *Technological and Economic Development of Economy*, 19(3), pp. 533–548.

Anil, K. J. (2010). Data clustering: 50 years beyond K-means. *Pattern Recognition Letters*, 31(8), pp. 651–666.

Bellman, R. E., Zadeh, L. A. (1970). Decision-making in a fuzzy environment. *Management Science*, 17(4), pp. 141–164.

Bensaou, B. M. (1999). Portfolios of buyer–supplier relationships. *Sloan Management Review*, 40(4), pp. 35–44.

Berger, P., Gerstenfeld, A., Zeng, A. Z. (2004). How many suppliers are best? A decision-analysis approach. *Omega*, 32(1), pp. 9–15.

Buckley, J. J. (1985). Fuzzy hierarchical analysis. *Fuzzy Sets and Systems*, 17(3), pp. 233–247.

Chai, J., Liu, J. N. K., Ngai, E. W. T. (2013). Application of decision-making techniques in supplier selection: A systematic review of literature. *Expert Systems with Applications*, 40(10), pp. 3872–3885.

Chan, F. T. S. (2003). Interactive selection model for supplier selection process: An analytic hierarchy process approach. *International Journal of Production Research*, 41(15), pp. 3549–3579.

Chang, D. Y. (1996). Applications of the extent analysis method on fuzzy AHP. *European Journal of Operational Research*, 95(3), pp. 649–655.

Cheng, C. H. (1997). Evaluating naval tactical missile systems by fuzzy AHP based on the grade value of membership function. *European Journal of Operational Research*, 96(2), pp. 343–350.

Chiu, S., Tavella, D. (2008). *Data Mining and Market Intelligence for Optimal Marketing Returns*. Oxford, United Kingdom: Butterworth-Heinemann.

Choi, T. Y., Hartley, J. L. (1996). An exploration of supplier selection practices across the supply chain. *Journal of Operations Management*, 14(4), pp. 333–343.

Churchman, C. W., Ackoff, R. L. (1954). An approximate measure of value. *Journal of Operations Research Society of America*, 2(1), pp. 172–187.

Davies, M. A. P. (1994). A multicriteria decision model application for managing group decisions. *Journal of the Operational Research Society*, 45(1), pp. 47–58.

Day, G. S. (2010). The capabilities of market-driven organisations. *Journal of Marketing*, 58(4), pp. 37–52.

Day, M., Magnan, G. M., Moeller, M. M. (2010). Evaluating the bases of supplier segmentation: A review and taxonomy. *Industrial Marketing Management*, 39(4), pp. 625–639.

De Grann, J. G. (1980). Extension to the multiple criteria analysis method of T.L. Satty. *A Report for National Institute for Water Supply*. Voorburg, the Netherlands.

Deng, H. (1999). Multicriteria analysis with fuzzy pairwise comparison. *International Journal of Approximate Reasoning*, 21(3), pp. 215–231.

de Oliveira, J. (1995). A set-theoretical defuzzification method. *Fuzzy Sets and Systems*, 76(1), pp. 63–71.

Dickson, G. W. (1966). An analysis of vendor selection systems and decisions. *Journal of Purchasing*, 2(1), pp. 5–17.

Dillon, W. R., Kumar, A., Borrero, M. S. (1993). Capturing individual differences in paired comparisons: An extended BTL model incorporating descriptor variables. *Journal of Marketing Research*, 30(1), pp. 42–51.

Dweiri, F., Kumar, S., Khan, S. A., Jain, V. (2016). Designing an integrated AHP based decision support system for supplier selection in automotive industry. *Expert Systems with Applications*, 62(2), pp. 273–283.

Dyer, J. H., Cho, D. S., Chu, W. (1998). Strategic supplier segmentation: The next 'best practice' in supply chain management. *California Management Review*, 40(2), pp. 57–77.

Erevelles, S., Stevenson, T. H. (2006). Enhancing the business-to-business supply chain: Insights from partitioning the supply-side. *Industrial Marketing Management*, 35(4), pp. 481–492.

Govindan, K., Rajendran, S., Sarkis, J., Murugesan, P. (2015). Multi criteria decision making approaches for green supplier evaluation and selection: A literature review. *Journal of Cleaner Production*, 98(1), pp. 66–83.

Hair, J. F., Anderson, R. E., Tatham, R. L., Black, W. C. (1995). *Multivariate Data Analysis*, 4th edition. Upper Saddle River, NJ: Prentice Hall.

Hallikas, J., Puumalainen, K., Vesterinen, T., Virolainen, V.-M. (2005). Risk based classification of supplier relationships. *Journal of Purchasing & Supply Management*, 11(2–3), pp. 72–82.

Han, J., Kamber, M. (2006). *Data Mining: Concepts and Techniques*, 2nd edition. Massachusetts, US: Morgan Kaufmann Publisher.

Hong, C. H. (2012). Using the Taguchi method for effective market segmentation. *Expert Systems with Applications*, 39(5), pp. 5451–5459.

Hoyt, J., Huq, F. (2000). From arms-length to collaborative relationships in the supply chain: An evolutionary process. *International Journal of Physical Distribution & Logistics Management*, 30(9), pp. 750–764.

Hung, C., Tsai, C. (2008). Market segmentation based on hierarchical self-organizing map for markets of multimedia on demand. *Expert Systems with Applications*, 34(1), pp. 780–787.

Hussu, A. (1995). Fuzzy control and defuzzification. *Mechatronics*, 5(5), pp. 513–526.

Ishizaka, A., Labib, A. (2011). Review of the main developments in the analytic hierarchy process. *Expert Systems with Applications*, 38(11), pp. 14336–14345.

Kannan, V. R., Tan, K. C. (2002). Supplier selection and assessment: Their impact on business performance. *Journal of Supply Chain Management*, 38(4), pp. 11–21.

Kaufman, A., Wood, C. H., Theyel, G. (2000). Collaboration and technology linkages: A strategic supplier typology. *Strategic Management Journal*, 21(6), pp. 649–663.

Kazemzadeh, R. B., Behzadian, M., Aghdasi, M., Albadvi, A. (2009). Integration of marketing research techniques into house of quality and product family design. *International Journal of Advance Manufacturing Technology*, 41(9–10), pp. 1019–1033.

Kraljic, P. (1983). Purchasing must become supply management. *Harvard Business Review*, 61(5), pp. 109–117.

Leekwijck, W. V., Kerre, E. E. (1999). Defuzzification: Criteria and classification. *Fuzzy Sets and Systems*, 108(2), pp. 159–178.

Leung, L. C., Cao, D. (2000). On consistency and ranking of alternatives in fuzzy AHP. *European Journal of Operational Research*, 124(1), pp. 102–113.

Liao, C-N. (2011). Fuzzy analytic hierarchy process and multi-segment goal programming applied to new product segmented under price strategy. *Computers and Industrial Engineering*, 61(3), pp. 831–841.

MacQueen, J. B. (1967). Some methods for classification and analysis of multivariate observations. *Proceedings of the Fifth Berkeley Symposium on Mathematical Statistics and Probability*, 1, pp. 281–297, Berkeley, CA.

Meisel, S., Mattfeld, D. (2010). Synergies of operations research and data mining. *European Journal of Operational Research*, 206(1), pp. 1–10.

Mikhailov, L. (2004). A fuzzy approach to deriving priorities from interval pairwise comparison judgments. *European Journal of Operational Research*, 159(3), pp. 687–704.

Mingoti, S. A., Lima, J. O. (2006). Comparing SOM neural network with fuzzy c-means, K-means and traditional hierarchical clustering algorithms. *European Journal of Operational Research*, 174(3), pp. 1742–1759.

Olsen, R. F., Ellram, L. M. (1997). A portfolio approach to supplier relationships. *Industrial Marketing Management*, 26(2), pp. 101–113.

Parasuraman, A. (1980). Vendor segmentation: An additional level of market segmentation. *Industrial Marketing Management*, 9(1), pp. 59–62.

Punj, G., Stewart, D. W. (1983). Cluster analysis in marketing research: Review and suggestions for application. *Journal of Marketing Research*, 20(2), pp. 134–148.

Rezaei, J., Davoodi, M. (2011). Multi-objective models for lot-sizing with supplier selection. *International Journal of Production Economics*, 130(1), pp. 77–86.

Rezaei, J., Ortt, R. (2012). A multi-variable approach to supplier segmentation. *International Journal of Production Research*, 50(16), pp. 4593–4611.

Rezaei, J., Ortt, R. (2013a). Multi-criteria supplier segmentation using a fuzzy preference relations based AHP. *European Journal of Operational Research*, 225(1), pp. 75–84.

Rezaei, J., Ortt, R. (2013b). Supplier segmentation using fuzzy logic. *Industrial Marketing Management*, 42(4), pp. 507–517.

Rondeau, L., Ruelas, R., Levrat, L., Lamotte, M. (1997). A defuzzification method respecting the fuzzification. *Fuzzy Sets and Systems*, 86(3), pp. 311–320.

Saaty, T. L. (1977). A scaling method for priorities in hierarchical structures. *Journal of Mathematical Psychology*, 15(3), pp. 234–281.

Saaty, T. L. (1980). *The Analytic Hierarchy Process: Planning, Priority Setting, Resource Allocation*. New York, NY: McGraw-Hill.

Saaty, T. L., Özdemir, M. S. (2015). How many judges should there be in a group? *Annals of Data Science*, 1(3–4), pp. 359–368.

Sodenkamp, M. A., Tavana, M., Di Caprio, D. (2016). Modeling synergies in multi-criteria supplier selection and order allocation: An application to commodity trading. *European Journal of Operational Research*, 254(3), pp. 859–874.

Svensson, G. (2004). Supplier segmentation in the automotive industry: A dyadic approach of a managerial model. *International Journal of Physical Distribution & Logistics Management*, 34(1), pp. 12–38.

Swift, C. O. (1995). Preferences for single sourcing and supplier selection criteria. *Journal of Business Research*, 32(2), pp. 105–111.

Tan, K. C., Lyman, S. B., Wisner, J. D. (2002). Supply chain management: A strategic perspective. *International Journal of Operations & Production Management*, 22(6), pp. 614–631.

Tavana, M., Fallahpour, A., Di Caprio, D., Santos-Arteaga, F. J. (2016). A hybrid intelligent fuzzy predictive model with simulation for supplier evaluation and selection. *Expert Systems with Applications*, 61(1), pp. 129–144.

Tsiptsis, K., Chorianopoulos, A. (2009). *Data Mining Techniques in CRM: Inside Customer Segmentation*. Chichester, UK: John Wiley & Sons.

Urgal-González, B., García-Vázquez, J. M. (2007). The strategic influence of structural manufacturing decisions. *International Journal of Operations & Production Management*, 27(6), pp. 605–626.

Van Laarhoven, P. J. M., Pedrycz, W. (1983). A fuzzy extension of Saaty's priority theory. *Fuzzy Sets and Systems*, 11(1–3), pp. 229–241.

Van Weele, A. J. (2000). *Purchasing and Supply Chain Management*. London: Business Press, Thomson Learning.

Ward, J. (1963). Hierarchical grouping to optimize an objective function. *Journal of the American Statistical Association*, 58(301), pp. 236–244.

Weber, C. A., Current, J. R., Benton, W. C. (1991). Vendor selection criteria and methods. *European Journal of Operational Research*, 50(1), pp. 2–18.

Wedel, M., Kamakura, W. A. (2001). *Market Segmentation Conceptual and Methodological Foundations*. Dordrecht, The Netherlands: Kluwer Academic Publishers.

Yager, R. R., Filev, D. (1993). On the issue of defuzzification and selection based on fuzzy set. *Fuzzy Sets and Systems*, 55(3), pp. 255–271.

Yu, C. S. (2002). A GP-AHP method for solving group decision-making fuzzy AHP problems. *Computers and Operations Research*, 29(14), pp. 1969–2001.

Zadeh, L. (1965). Fuzzy sets. *Information and Control*, 8(3), pp. 338–353.

5

Group Decision-Making under Uncertainty: FAHP Using Intuitionistic and Hesitant Fuzzy Sets

Cengiz Kahraman and Fatih Tüysüz

CONTENTS

5.1 Introduction

The decision-making process can be described as defining the decision goals, gathering the related criteria and possible alternatives, evaluating the alternatives, and selecting an alternative or ranking the alternatives. Multicriteria decision-making (MCDM) methods are an important set of tools for addressing challenging decisions since they enable the decision maker to handle uncertainty, complexity, and conflicting objectives. In MCDM, the predetermined alternatives are ranked according to the decision makers' evaluation of multiple criteria. MCDM problems usually require both quantitative and qualitative factors to be evaluated. The main problem for qualitative criteria is that the values for such criteria are often imprecisely defined for the decision makers. In the evaluation of the qualitative criteria, the desired value and the weight of importance for the criteria are usually defined in linguistic terms. In crisp or conventional methods, it is difficult to express the character and significance of criteria exactly or clearly (Chou et al., 2008). The conventional approaches tend to be less effective in dealing with the imprecision or vagueness of the linguistic assessment (Kahraman et al., 2003; Tolga et al., 2013; Kaya and Kahraman, 2011). Due to this reason, the use of fuzzy set theory in MCDM for evaluating various factors and alternatives seems more convenient by allowing decision makers to express their ideas more adequately (Zadeh, 1965; Beskese et al., 2004; Kulak et al., 2010).

Uncertainty is unavoidable in the decision-making process. Klir and Yuan (1995) describe uncertainty and categorize it as vagueness and ambiguity. Vagueness or imprecision refers to a lack of definite or sharp distinction, and ambiguity is due to unclear distinction of various alternatives. The analytic hierarchy process (AHP) procedure comprises both vagueness and ambiguity in assigning pairwise comparisons and evaluating alternatives (Sadiq and Tesfamariam, 2009).

AHP is one of the most widely used MCDM methods since it enables handling of both tangibles and intangibles and also has simple mathematical calculations. The AHP method uses crisp or objective mathematics to represent the subjective or personal preferences of an individual or a group in MCDM (Saaty and Vargas, 2001). One of the important drawbacks of classical or crisp AHP is that it cannot adequately handle the inherent uncertainty and vagueness (Xu and Liao, 2014). In classical AHP, the pairwise comparisons are performed by using crisp numbers within the 1–9 scale. In real-life problems, the decision makers may be unable to assign crisp evaluation values to the comparison judgments due to limited knowledge, the subjectivity of the qualitative evaluation criteria, or the variations of individual judgments in group decision-making (Leung and Cao, 2000). Since decision makers' evaluations in pairwise comparisons contain uncertainty, this may cause decision makers to feel more confident in providing fuzzy judgment than crisp comparisons (Wang and Chen, 2008). In order to overcome this, fuzzy set theory is the most convenient and most common approach. The fuzzy extensions of AHP are usually in the form of fuzzifying the pairwise comparisons by using fuzzy numbers. Fuzzy AHP (FAHP) is an extension of crisp AHP in which fuzzy sets are incorporated with the pairwise comparisons to model the uncertainty in human judgment and preference. There are different FAHP methods proposed by various authors. These methods are systematic approaches to the alternative selection and justification problem using the concepts of fuzzy set theory and hierarchical structure analysis. The methods differ from each other mainly in deriving priorities from fuzzy pairwise comparison matrices.

The first fuzzy extension of FAHP was introduced by Van Laarhoven and Pedrycz (1983). They used triangular fuzzy numbers and Lootsma's logarithmic least squares method for deriving fuzzy weights and fuzzy performance scores of alternatives. Buckley (1985) used trapezoidal fuzzy numbers and derived the fuzzy weights and fuzzy performance scores by using the geometric mean method. Boender et al. (1989) modified Van Laarhoven and Pedrycz's (1983) method and proposed a more robust approach to the normalization of the local priorities. Chang (1996) proposed an extent analysis method for deriving priorities from comparison matrices whose elements are defined as triangular fuzzy numbers. The drawback of this method is that it is possible to obtain the value of zero for initial weights or local priorities for some elements of the decision structure (Wang et al., 2008). Such a computed zero local priority may cause some of the information not to be considered in the calculations (Li et al., 2008). Cheng (1997) proposed a FAHP method based on both probability and possibility measures in which performance scores are represented by membership functions and the aggregate weights are calculated by using entropy concepts.

Mikhailov (2003) proposed a fuzzy extension of AHP, which obtains crisp priorities based on an α-cut of fuzzy numbers using a technique called fuzzy preference programming (FPP). The main drawback of this approach is that each comparison matrix must be constructed as an individual FPP model, which increases the complexity of the solving procedure (Yu and Cheng, 2007).

To overcome this disadvantage, Yu and Cheng (2007) proposed a multiple objective programming approach to obtain all local priorities for crisp or interval judgments at one time in which the priorities of all comparison matrices can be obtained simultaneously.

In addition to the abovementioned ordinary fuzzy sets extensions, there some other generalizations and extensions of fuzzy sets for better modeling uncertainty have recently been developed. The motivation behind these studies is to overcome some problems and drawbacks of ordinary fuzzy sets. When more than one source of vagueness appears simultaneously, ordinary fuzzy sets may encounter problems with modeling the situation and dealing with imprecise information (Rodriguez et al., 2012). These generalizations are type 2 fuzzy sets, intuitionistic fuzzy sets (IFSs), and hesitant fuzzy sets (HFSs) (Rodriguez et al., 2012). The common underlying idea behind these extensions set is that assigning the membership degree of an element to a fixed set is often not clear (Torra, 2010). In ordinary fuzzy sets, the membership degree of the element in a universe takes a single value between 0 and 1. These single values cannot represent adequately the lack of knowledge and ordinary fuzzy sets are not capable of co-opting the lack of knowledge with membership degrees (Wang and Zhang, 2013). Ordinary fuzzy sets and type 2 fuzzy sets are limited since they are based on the elicitation of single and simple terms (Rodríguez et al., 2012). When the decision makers are hesitant and need to use rich and more complex linguistic variables to express their qualitative cognition, these two approaches become inadequate (Liao et al., 2015).

Therefore, IFSs, introduced by Atanassov (1986), which are an extension of Zadeh's fuzzy sets, can come into prominence in better simulating the human decision-making process and activities which require human expertise and knowledge that are inevitably imprecise or not totally reliable (Wang and Zhang, 2013; Xu and Liao, 2014).

Xu and Liao (2014) compared IFSs with fuzzy sets used in FAHP and revealed the advantages of IFSs over fuzzy sets. IFSs can represent membership degree, nonmembership degree, and hesitancy degree by the three grades of the membership function, respectively. It can be seen as a particular case of type 2 fuzzy sets, which are an extension of ordinary fuzzy sets and are defined as fuzzy sets whose membership values themselves are also fuzzy sets. However, triangular fuzzy numbers and the trapezoidal fuzzy numbers do not have this property. Triangular fuzzy numbers, trapezoidal fuzzy numbers, and interval-valued fuzzy numbers can only be used to depict the fuzziness of agreement but cannot reflect the disagreement of the decision maker. The ignorance or indeterminacy, which is associated to the lack of information, can be represented as the hesitancy function in IFSs, which is one of the most important characteristics of IFSs. Describing the opinions from three sides as membership degree, nonmembership degree, and indeterminacy degree or hesitancy degree ensures more comprehensive preference information.

Another important extension of fuzzy sets is HFSs, introduced by Torra (2010), which allows the membership degrees to have a set of possible values between 0 and 1 (Torra and Narukawa, 2009). The main idea behind the usage of HFSs in decision-making is that people are usually hesitant or irresolute for one thing or another, which makes it difficult to come to a final agreement. In other words, when more than one decision maker assigns the membership degree of an element x to a set A, the difficulty of establishing a common membership degree arises. This is not because of a margin of error (IFS) or some possibility distribution values (type 2 fuzzy sets), but because of having a set of possible values (Xu and Xia, 2011).

In this chapter, due to the reasons mentioned above and the recent developments in the literature, AHP based on IFSs and HFSs, which enables us to better represent the hesitancy and uncertainty of decision makers' cognition, is studied. The rest of this chapter is organized as follows. In Section 5.2, some basic concepts related to IFSs are presented. In Section 5.3, the literature related to intuitionistic FAHP methods is given. In Section 5.4, a new intuitionistic FAHP method is explained in detail, and a numerical example is given in Section 5.5. Section 5.6 presents the basic concepts of HFSs. In Section 5.7, some recent developments related to HFS-based AHP methods are given. Section 5.8 presents a HFS-based AHP method in detail, and a numerical application of the method is given in Section 5.9. Finally, the conclusions are presented.

5.2 Intuitionistic Fuzzy Sets

A generalization of fuzzy sets was proposed by Atanassov (1983) as IFSs, which incorporate the degree of hesitation. Hesitation is calculated by "1-membership degree–nonmembership degree." The concept of an IFS can be viewed as an alternative approach in the case where the available information is not sufficient to define the impreciseness using ordinary fuzzy sets.

In this section, some basic concepts related to fuzzy sets and IFSs, which are used for modeling uncertainty, are introduced.

Definition 1: A fuzzy set A in the universe of discourse $X = \{x_1, x_2, \ldots, x_n\}$ is defined as

$$A = \{x, \mu_x \mid x \in X\} \tag{5.1}$$

where μ_x is the membership function defined as $\mu_x(x): X \rightarrow [0, 1]$, which indicates the membership degree of the element x to the set A.

Atanassov's IFSs (1986) are an extension of the fuzzy sets (Zadeh, 1965) that take into account the membership value as well as the nonmembership value for describing any x in X such that the sum of membership and nonmembership is less than or equal to 1. In the following, we define the basic definitions of an IFS, which are mainly taken from Atanassov (1986, 1999, 2012).

Definition 2: Let $X \neq \varnothing$ be a given set. An IFS in X is an object A given by

$$\tilde{A} = \{x, \mu_{\tilde{A}}(x), v_{\tilde{A}}(x); x \in X\} \tag{5.2}$$

where the membership function $\mu_{\tilde{A}}: X \rightarrow [0,1]$ and the nonmembership $v_{\tilde{A}}: X \rightarrow [0,1]$ satisfy the condition

$$0 \leq \mu_{\tilde{A}}(x) + v_{\tilde{A}}(x) \leq 1 \tag{5.3}$$

for every $x \in X$. X defines the possible range of values for a variable x, and A is an IFS defined over X using membership and nonmembership functions $\in [0,1]$. If the above expression reduces to $\mu_{\tilde{A}}(x) + v_{\tilde{A}}(x) = 1$, then an IFS becomes an ordinary fuzzy set.

Definition 3: For an IFS, the degree of hesitation or nondeterminancy or nonspecificity $\pi_{\tilde{A}}(x)$ of the element x in IFS A is defined as follows:

$$\pi_{\tilde{A}}(x) = 1 - \mu_{\tilde{A}}(x) - v_{\tilde{A}}(x) \; ; \; \pi_{\tilde{A}}(x): X \rightarrow [0,1] \tag{5.4}$$

Therefore, for ordinary fuzzy sets the degree of hesitation $\pi_{\tilde{A}}(x) = 0$.

Definition 4: An intuitionistic fuzzy number (IFN) $\tilde{A}$ is defined as follows:

1. An intuitionistic fuzzy subset of the real line
2. Normal, that is, there is a $x_0 \in \mathbb{R}$ such that $\mu_{\tilde{A}}(x_0) = 1$ $\left(so \; v_{\tilde{A}}(x_0) = 0\right)$

3. A convex set for the membership function $\mu_{\tilde{A}}(x)$, that is,

$$\mu_{\tilde{A}}(\lambda x_1 + (1-\lambda)x_2) \geq \min(\mu_{\tilde{A}}(x_1), \mu_{\tilde{A}}(x_2)) \quad \forall x_1, x_2 \in \mathbb{R}, \lambda \in [0,1]$$

4. A concave set for the nonmembership function $v_{\tilde{A}}(x)$, that is,

$$v_{\tilde{A}}(\lambda x_1 + (1-\lambda)x_2) \leq \max(v_{\tilde{A}}(x_1), v_{\tilde{A}}(x_2)) \quad \forall x_1, x_2 \in \mathbb{R}, \lambda \in [0,1]$$

Definition 5: The α-cut of an IFN is defined as the set

$$\tilde{A}_\alpha = \{x \in X | \mu_{\tilde{A}}(x) \geq \alpha, v_{\tilde{A}}(x) \leq 1-\alpha\} \tag{5.5}$$

Definition 6: A triangular intuitionistic fuzzy number (TIFN) $\tilde{A}$ is a subset of an IFS in $\mathbb{R}$, with the following membership function and nonmembership function:

$$\mu_{\tilde{A}}(x) = \begin{cases} \dfrac{x - a_1}{a_2 - a_1}, & \text{for } a_1 \leq x \leq a_2 \\[2mm] \dfrac{a_3 - x}{a_3 - a_2}, & \text{for } a_2 \leq x \leq a_3 \\[2mm] 0, & \text{otherwise} \end{cases} \tag{5.6}$$

and

$$v_{\tilde{A}}(x) = \begin{cases} \dfrac{a_2 - x}{a_2 - a_1'}, & \text{for } a_1' \leq x \leq a_2 \\[2mm] \dfrac{x - a_2}{a_3' - a_2}, & \text{for } a_2 \leq x \leq a_3' \\[2mm] 1, & \text{otherwise} \end{cases} \tag{5.7}$$

where $a_1' \leq a_1 \leq a_2 \leq a_3 \leq a_3'$, $0 \leq \mu_{\tilde{A}}(x) + v_{\tilde{A}}(x) \leq 1$ and TIFN (Figure 5.1) is denoted by $\tilde{A}_{TIFN} = (a_1, a_2, a_3; a_1', a_2, a_3')$.

Definition 7: A trapezoidal intuitionistic fuzzy number (TrIFNs) $\tilde{A}$ is a subset of an IFS in R with membership function and nonmembership function as follows:

$$\mu_{\tilde{A}}(x) = \begin{cases} \dfrac{x - a_1}{a_2 - a_1}, & \text{for } a_1 \leq x \leq a_2 \\[2mm] 1, & \text{for } a_2 \leq x \leq a_3 \\[2mm] \dfrac{a_4 - x}{a_4 - a_3}, & \text{for } a_3 \leq x \leq a_4 \\[2mm] 0, & \text{otherwise} \end{cases} \tag{5.8}$$

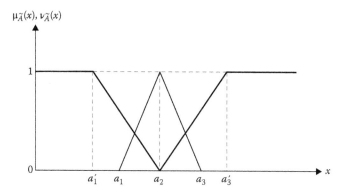

$\mu_{\tilde{A}}(x),\, \nu_{\tilde{A}}(x)$

FIGURE 5.1
Membership and nonmembership functions of triangular intuitionistic fuzzy number.

and

$$
v_{\tilde{A}}(x) = \begin{cases}
\dfrac{a_2 - x}{a_2 - a_1'}, & \text{for } a_1' \leq x \leq a_2 \\[2mm]
0, & \text{for } a_2 \leq x \leq a_3 \\[2mm]
\dfrac{x - a_3}{a_4' - a_3}, & \text{for } a_3 \leq x \leq a_4' \\[2mm]
1, & \text{otherwise}
\end{cases}
\tag{5.9}
$$

where $a_1' \leq a_1 \leq a_2 \leq a_3 \leq a_4 \leq a_4'$, $0 \leq \mu_{\tilde{A}}(x) + v_{\tilde{A}}(x) \leq 1$ and TrIFN (Figure 5.2) is denoted by $\tilde{A}_{\text{TrIFN}} = (a_1, a_2, a_3, a_4; a_1', a_2, a_3, a_4')$.

We now give the arithmetic operations for TrIFNs, which are also valid for TIFNs.

Addition: If $\tilde{A}_{\text{TrIFN}} = (a_1, a_2, a_3, a_4; a_1', a_2, a_3, a_4')$ and $\tilde{B}_{\text{TrIFN}} = (b_1, b_2, b_3, b_4; b_1', b_2, b_3, b_4')$ are two TrIFNs, then $\tilde{C} = \tilde{A} \oplus \tilde{B}$ is also a TrIFN and

$$
\tilde{C} = \left(a_1 + b_1,\, a_2 + b_2,\, a_3 + b_3,\, a_4 + b_4; a_1' + b_1',\, a_2 + b_2,\, a_3 + b_3,\, a_4' + b_4'\right) \tag{5.10}
$$

Multiplication: If $\tilde{A}_{\text{TrIFN}} = (a_1, a_2, a_3, a_4; a_1', a_2, a_3, a_4')$ and $\tilde{B}_{\text{TrIFN}} = (b_1, b_2, b_3, b_4; b_1', b_2, b_3, b_4')$ are two TrIFNs, then $\tilde{C} = \tilde{A} \otimes \tilde{B}$ is also a TrIFN and

$$
\tilde{C} \cong \left(a_1 b_1,\, a_2 b_2,\, a_3 b_3,\, a_4 b_4; a_1' b_1',\, a_2 b_2,\, a_3 b_3,\, a_4' b_4'\right) \tag{5.11}
$$

Multiplication by a real number λ where $\lambda > 0$:

$$
\lambda \tilde{A} = \left(\left[1 - \left(1 - \mu_1^-\right)^{\lambda}, 1 - \left(1 - \mu_1^+\right)^{\lambda} \right], \left[\left(v_1^-\right)^{\lambda}, \left(v_1^+\right)^{\lambda} \right] \right) \tag{5.12}
$$

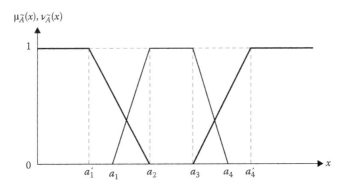

FIGURE 5.2
Membership and nonmembership functions of trapezoidal intuitionistic fuzzy number.

Subtraction: Subtraction is defined as follows (Atanassov, 2012):
If $\upsilon_B(x) > 0$ and $\upsilon_A(x)\pi_B(x) \le \pi_A(x)\upsilon_B(x)$,

$$\tilde{A} \ominus \tilde{B} = \left\{ x, \frac{\mu_A(x) - \mu_B(x)}{1 - \mu_B(x)}, \left. \frac{\upsilon_A(x)}{\upsilon_B(x)} \right| x \in X \right\} \qquad (5.13)$$

where $\pi_A(x)$ and $\pi_B(x)$ are the unknown degrees (hesitancy degrees) of the IFNs A and B, respectively.

For simplicity, the following linear operation for TrIFNs (Kumar and Hussain, 2014) can also be used:

If $\tilde{A}_{\text{TrIFN}} = \left(a_1, a_2, a_3, a_4; a_1', a_2, a_3, a_4' \right)$ and $\tilde{B}_{\text{TrIFN}} = \left(b_1, b_2, b_3, b_4; b_1', b_2, b_3, b_4' \right)$ are two TrIFNs, then $\tilde{C} = \tilde{A} \ominus \tilde{B}$ is also a TrIFN and

$$\tilde{C} = \left(a_1 - b_4,\ a_2 - b_3,\ a_3 - b_2,\ a_4 - b_1; a_1' - b_4',\ a_2 - b_3,\ a_3 - b_2,\ a_4' - b_1' \right) \quad (5.14)$$

Division: The division operation is defined as follows (Atanassov, 2012):
If $\mu_B(x) > 0$ and $\mu_A(x)\pi_B(x) \le \pi_A(x)\mu_B(x)$,

$$\tilde{A} \oslash \tilde{B}\ \tilde{A} \oslash \tilde{B} = \left\{ x, \frac{\mu_A(x)}{\mu_B(x)}, \left. \frac{\upsilon_A(x) - \upsilon_B(x)}{1 - \upsilon_B(x)} \right| x \in X \right\} \qquad (5.15)$$

For simplicity, the following linear operation for TrIFNs (Mahapatra and Roy, 2013) can be used:

If $\tilde{A}_{\text{TrIFN}} = \left(a_1, a_2, a_3, a_4; a_1', a_2, a_3, a_4' \right)$ and $\tilde{B}_{\text{TrIFN}} = \left(b_1, b_2, b_3, b_4; b_1', b_2, b_3, b_4' \right)$ are two TrIFNs, then $\tilde{C} = \tilde{A} \oslash \tilde{B}$ is also a TrIFN:

$$\tilde{C} = \left(a_1/b_4,\ a_2/b_3,\ a_3/b_2,\ a_4/b_1; a_1'/b_4',\ a_2/b_3,\ a_3/b_2,\ a_4'/b_1' \right) \qquad (5.16)$$

Definition 8: Let $D \subseteq [0,1]$ be the set of all closed subintervals of the interval and let X be a universe of discourse. An interval-valued intuitionistic fuzzy set (IVIFS) in $\tilde{A}$ over X is an object having the form

$$\tilde{A} = \left\{ < x, \mu_{\tilde{A}}(x), v_{\tilde{A}}(x) > \big| x \in X \right\} \tag{5.17}$$

where $\mu_{\tilde{A}} \to D \subseteq [0,1]$, $v_{\tilde{A}}(x) \to D \subseteq [0,1]$ with the condition $0 \leq \sup\mu_{\tilde{A}}(x) + \sup v_{\tilde{A}}(x) \leq 1$, $\forall x \in X$.

The intervals $\mu_{\tilde{A}}(x)$ and $v_{\tilde{A}}(x)$ denote the membership function and the nonmembership function of the element x to the set $\tilde{A}$, respectively. Thus, for each $x \in X$, $\mu_{\tilde{A}}(x)$ and $v_{\tilde{A}}(x)$ are closed intervals and their starting and ending points are denoted by $\mu_{\tilde{A}}^{-}(x)$, $\mu_{\tilde{A}}^{+}(x)$, $v_{\tilde{A}}^{-}(x)$, and $v_{\tilde{A}}^{+}(x)$, respectively. IVIFS $\tilde{A}$ is then denoted by

$$\tilde{A} = \left\{ < x, \left[\mu_{\tilde{A}}^{-}(x),\ \mu_{\tilde{A}}^{+}(x) \right], \left[v_{\tilde{A}}^{-}(x),\ v_{\tilde{A}}^{+}(x) \right] > \big| x \in X \right\} \tag{5.18}$$

where $0 \leq \mu_{\tilde{A}}^{+}(x) + v_{\tilde{A}}^{+}(x) \leq 1$, $\mu_{\tilde{A}}^{-}(x) \geq 0$, $v_{\tilde{A}}^{-}(x) \geq 0$.

For each element x, the unknown degree (hesitancy degree) of an IVIFS of $x \in X$ in $\tilde{A}$ is defined as follows:

$$\pi_{\tilde{A}}(x) = 1 - \mu_{\tilde{A}}(x) - v_{\tilde{A}}(x) = \left(\left[1 - \mu_{\tilde{A}}^{+}(x) - v_{\tilde{A}}^{+}(x) \right],\ 1 - \mu_{\tilde{A}}^{-}(x) - v_{\tilde{A}}^{-}(x) \right) \tag{5.19}$$

For convenience, let $\mu_{\tilde{A}}(x) = \left[\mu_{\tilde{A}}^{-}(x),\ \mu_{\tilde{A}}^{+}(x) \right] = \left[\mu_{\tilde{A}}^{-},\ \mu_{\tilde{A}}^{+} \right]$, $v_{\tilde{A}}(x) = \left[v_{\tilde{A}}^{-}(x),\ v_{\tilde{A}}^{+}(x) \right] = \left[v_{\tilde{A}}^{-},\ v_{\tilde{A}}^{+} \right]$, so $\tilde{A} = \left(\left[\mu_{\tilde{A}}^{-},\ \mu_{\tilde{A}}^{+} \right], \left[v_{\tilde{A}}^{-},\ v_{\tilde{A}}^{+} \right] \right)$.

Some arithmetic operations with interval-valued intuitionistic fuzzy numbers (IVIFNs) and $\lambda \geq 0$ are given in the following:

Let $\tilde{A} = \left(\left[\mu_{\tilde{A}}^{-}, \mu_{\tilde{A}}^{+} \right], \left[v_{\tilde{A}}^{-}, v_{\tilde{A}}^{+} \right] \right)$ and $\tilde{B} = \left(\left[\mu_{\tilde{B}}^{-}, \mu_{\tilde{B}}^{+} \right], \left[v_{\tilde{B}}^{-}, v_{\tilde{B}}^{+} \right] \right)$ be two IVIFNs. Then,

$$\tilde{A} \oplus \tilde{B} = \left(\left[\mu_{\tilde{A}}^{-} + \mu_{\tilde{B}}^{-} - \mu_{\tilde{A}}^{-}\mu_{\tilde{B}}^{-},\ \mu_{\tilde{A}}^{+} + \mu_{\tilde{B}}^{+} - \mu_{\tilde{A}}^{+}\mu_{\tilde{B}}^{+} \right], \left[v_{\tilde{A}}^{-}v_{\tilde{B}}^{-}, v_{\tilde{A}}^{+}v_{\tilde{B}}^{+} \right] \right) \tag{5.20}$$

$$\tilde{A} \otimes \tilde{B} = \left(\left[\mu_{\tilde{A}}^{-}\mu_{2}^{-}, \mu_{\tilde{A}}^{+}\mu_{2}^{+} \right], \left[v_{\tilde{A}}^{-} + v_{2}^{-} - v_{\tilde{A}}^{-}v_{2}^{-}, v_{\tilde{A}}^{+} + v_{2}^{+} - v_{\tilde{A}}^{+}v_{2}^{+} \right] \right) \tag{5.21}$$

Using the extension principle, the arithmetic operations for IVIFNs can be obtained by the general equation given in Equation 5.22 (Li, 2011):

$$\tilde{A} \circledast \tilde{B} = \left\{ \begin{array}{l} < z, \left[\max_{z=x^{*}y} \min\left\{ \mu_{\tilde{A}}^{-}(x), \mu_{\tilde{B}}^{-}(y) \right\}, \max_{z=x^{*}y} \min\left\{ \mu_{\tilde{A}}^{+}(x), \mu_{\tilde{B}}^{+}(y) \right\} \right], \\ \left[\min_{z=x^{*}y} \max\left\{ v_{\tilde{A}}^{-}(x), v_{\tilde{B}}^{-}(y) \right\}, \min_{z=x^{*}y} \max\left\{ v_{\tilde{A}}^{+}(x), v_{\tilde{B}}^{+}(y) \right\} \right] > \big| (x,y) \in X \times Y \end{array} \right\} \tag{5.22}$$

where the symbol "*" stands for one of the algebraic operations. As an example, the subtraction operation for IVIFNs can be defined as in Equation 5.23:

$$\tilde{A} \ominus \tilde{B} = \left\{ \begin{array}{l} < z, \left[\max_{z=x-y} \min \left\{ \mu_{\tilde{A}}^-(x), \mu_{\tilde{B}}^-(y) \right\}, \max_{z=x-y} \min \left\{ \mu_{\tilde{A}}^+(x), \mu_{\tilde{B}}^+(y) \right\} \right], \\ \left[\min_{z=x-y} \max \left\{ v_{\tilde{A}}^-(x), v_{\tilde{B}}^-(y) \right\}, \min_{z=x-y} \max \left\{ v_{\tilde{A}}^+(x), v_{\tilde{B}}^+(y) \right\} \right] > \left| (x,y) \in X \times Y \right. \end{array} \right\} \quad (5.23)$$

5.3 AHP Using IFSs

Before we present an AHP method based on IFSs, we give a brief review of the research that integrates IFSs with the AHP methods. Sadiq and Tesfamariam (2009) applied the concept of the IFS to AHP, which is called intuitionistic fuzzy-AHP (IF-AHP), to handle both vagueness and ambiguity-related uncertainties in the environmental decision-making process. The authors used IFSs in pairwise comparisons and calculated the intuitionistic fuzzy weights. The proposed IF-AHP methodology was demonstrated using fixed vagueness and ambiguity (nonspecificity) in establishing pairwise comparisons. They also offered to use different values of fuzzification factors and degrees of belief for different pairwise comparisons.

Abdullah et al. (2009) applied IFSs to an AHP method called IF-AHP to quantify vagueness uncertainties in AHP using IFSs for decision-making problems. The authors constructed several linear programming models to generate optimal weights for attributes. The concept of the IFS in AHP was introduced through pairwise comparisons. A ranking order was obtained via an IF-AHP evaluation process by using positive and negative components of IFSs in the concept of AHP.

Wang et al. (2011) proposed an IF-AHP method in which the decision information was represented by intuitionistic fuzzy values. They synthesized eigenvectors of the intuitionistic fuzzy comparison matrix.

Zhang et al. (2011) discussed the approximation of IFSs to fuzzy sets based on the relationships between IFSs and fuzzy sets. They also showed that an intuitionistic fuzzy complementary judgment matrix approximates to the fuzzy complementary judgment matrix and proved that the consistency adjustment of both is equal. Based on these results, the authors proposed an AHP method based on an intuitionistic judgment matrix.

Feng et al. (2012) proposed a Dempster–Shafer-AHP (DS-AHP) method combined with intuitionistic fuzzy information. The expected utility was used to transform a decision matrix with intuitionistic fuzzy information obtained from the assessments. Then, a nonlinear optimization model was used to combine the attributes.

Wu et al. (2013) proposed an interval-valued intuitionistic fuzzy-AHP (IVIF-AHP) method for MCDM problems. They developed a score function for IVIFNs and some new concepts such as an antisymmetric interval matrix, transfer interval matrix, and approximate optimal transfer matrix. The importance of their score function is that it can obtain the interval-valued fuzzy preference relations by converting them into interval preference relations. The approximate optimal transfer matrix presented in the study guarantees that it is a consistent interval multiplicative preference relation, which can be easily derived by the traditional method.

Kaur (2014) proposed a TIFN-based approach for the vendor selection problem using AHP. The crisp data of the vendors were represented in the form of TrIFNs. By applying AHP, which involves decomposition, pairwise comparison, and deriving priorities for the various levels of the hierarchy, an overall crisp priority was obtained for ranking the best vendor. In this approach, fuzzification was done for intuitionistic fuzzy pairwise comparison between criteria. Then, an IFS of weights was calculated using an intuitionistic fuzzy judgment matrix. Intuitionistic defuzzification was used to convert the final IF-AHP score into a crisp value for ranking of alternatives.

Xu and Liao (2014) proposed a new way to check the consistency of an intuitionistic preference relation and then introduced an automatic procedure to repair the inconsistent one. Their proposed method improves the inconsistent intuitionistic preference relation without the participation of the decision maker, and thus, it can save much time and show some advantages over the AHP and the FAHP. They also developed a novel normalizing rank summation method to derive the priority vector of an intuitionistic preference relation, on which the priorities of the hierarchy in the IF-AHP are derived. The procedure of the IF-AHP is given in detail, and an example concerning global supplier development is used to demonstrate the results.

Abdullah and Najib (2016a) proposed a new preference scale in the framework of the IVIF-AHP. The comparison matrix judgment is expressed in IVIFNs with the degree of hesitation. The proposed new preference scale concurrently considers the membership function, the nonmembership function, and the degree of hesitation of IVIFNs. To define the weight entropy of the aggregated matrix of IVIFNs, a modified interval-valued intuitionistic fuzzy weighted averaging was proposed, by considering the interval number of the hesitation degree. Three MCDM problems were used to test the proposed method. A comparison of the results was also presented to check the feasibility of the proposed method. It was shown that the ranking order of the proposed method is slightly different from that of the other two methods because of the inclusion of the hesitation degree in defining the preference scale.

Abdullah and Najib (2016b) proposed a new method of IF-AHP to deal with the uncertainty in decision-making. The new IF-AHP was applied to establish a preference in the sustainable energy planning decision-making problem. Three decision makers attached to Malaysian government agencies were interviewed to provide linguistic judgment prior to analyzing with the new IF-AHP.

Dutta and Guha (2015) proposed a preference programming-based weight determination method using IF-AHP in which decision makers express their pairwise comparisons by using generalized triangular intuitionistic fuzzy numbers (GTIFNs). The main property of their approach is that it does not require additional aggregation, normalization, and ranking procedures of IFNs for deriving priorities.

Keshavarzfard and Makui (2015) presented an intuitionistic fuzzy multiple attribute decision-making (MADM) approach for modeling and solving AHP problems with a small amount of relationship among various criteria. IF-AHP is used to evaluate the weighting for each criterion and then the intuitionistic fuzzy Decision-Making Trial and Evaluation Laboratory (DEMATEL) method is applied to establish contextual relationships among the criteria.

Onar et al. (2015) proposed an IVIF-AHP approach for the overall performance measurement of wind energy technology alternatives. A sensitivity analysis was also conducted to assess the robustness of the results obtained from the model and the comparative results showed that the proposed method produced a consistent ranking among the alternative technologies.

Fahmi et al. (2015) a applied a FAHP method based on IVIFs for selection of people applying for a position at a university. Bali et al. (2015) developed an integrated dynamic intuitionistic fuzzy MADM approach for a personnel promotion problem. Their model integrates AHP and dynamic evaluation by an intuitionistic fuzzy operator for personnel promotion. AHP was used to determine the weight of attributes based on the decision maker's opinions, and the dynamic operator was used to aggregate the evaluations of candidates for different years.

Tavana et al. (2016) presented a hybrid method that combines SWOT (Strengths, Weaknesses, Opportunities, and Threats) analysis with an IF-AHP model to rank the criteria and subcriteria characterizing reverse logistics outsourcing decision-making. They used an extension of the FPP model in IF-AHP to quantify the qualitative results obtained from the SWOT analysis.

Kahraman et al. (2016) proposed an IVIF-AHP and technique for order performance by similarity to ideal solution (TOPSIS)-based methodology for the evaluation of outsource manufacturers. IVIF-AHP was used to calculate the criteria weights, whereas fuzzy TOPSIS was used for prioritizing the alternative outsource manufacturers. They also gave an application to show the applicability of the proposed model.

Ren et al. (2016) proposed a method named group intuitionistic multiplicative (GIM)-AHP for solving group decision-making problems. Their study includes adjusting the individual intuitionistic multiplicative preference relations (IMPRs), aggregating the individual IMPRs, and deriving the priorities from the overall IMPR. They presented performance assessments of the hydropower stations in the southwest of China to demonstrate the effectiveness and applicability of the GIM-AHP.

Deepika and Kannan (2016) proposed an IF-AHP method and applied it for a global supplier selection problem. They also developed a method for checking the consistency and a normalized rank summation method for deriving the priority weights/vectors of a preference relation.

These IFS-based AHP methods differ from each other in terms of the fuzzification approach deriving the priorities and also the computational complexity. When applied to the same problem, these approaches may give different results. Studies related to the application of IFSs to both AHP and other MCDM methods are still in progress and it can be said that some new developments are still required.

5.4 Algorithmic Steps of the AHP Method Based on IVIFSs

AHP has been the most widely used MCDM method since its first introduction in the 1980s (Saaty, 1980). One of the most important reasons for that is that the mathematical calculations and logic behind it are very easy to understand. To be consistent with this, in this section, we present an AHP method utilizing IVIFSs, which is mainly based on Wu et al. (2013) with a new IVIF linguistic scale. The method proposes a fuzzy approach for MCDM problems with interval-valued intuitionistic fuzzy preference relations (IVIFPRs). Let $\tilde{R} = \left(\tilde{r}_{ij}\right)_{n \times n} = \left(\left[\mu^-_{ij}, \mu^+_{ij}\right], \left[v^-_{ij}, v^+_{ij}\right]\right)_{n \times n}$ be an interval-valued intuitionistic judgment matrix as follows:

$$\tilde{R} = \begin{bmatrix} \left(\left[\mu^-_{11}, \mu^+_{11}\right], \left[v^-_{11}, v^+_{11}\right]\right) & \cdots & \left(\left[\mu^-_{1n}, \mu^+_{1n}\right], \left[v^-_{1n}, v^+_{1n}\right]\right) \\ \vdots & \ddots & \vdots \\ \left(\left[\mu^-_{n1}, \mu^+_{n1}\right], \left[v^-_{n1}, v^+_{n1}\right]\right) & \cdots & \left(\left[\mu^-_{nn}, \mu^+_{nn}\right], \left[v^-_{nn}, v^+_{nn}\right]\right) \end{bmatrix} \quad (5.24)$$

The reciprocal value of $\left(\left[\mu^-_{ij}, \mu^+_{ij}\right], \left[v^-_{ij}, v^+_{ij}\right]\right)$ in $\tilde{R}$ is $\left(\left[v^-_{ji}, v^+_{ji}\right], \left[\mu^-_{ji}, \mu^+_{ji}\right]\right)$. For instance, the reciprocal value of $\left(\left[\mu^-_{n1}, \mu^+_{n1}\right], \left[v^-_{n1}, v^+_{n1}\right]\right)$ is $\left(\left[v^-_{1n}, v^+_{1n}\right], \left[\mu^-_{1n}, \mu^+_{1n}\right]\right)$.

The score judgment matrix of $\tilde{R}$ is represented by the matrix $\tilde{S} = \left(\tilde{s}_{ij}\right)_{n \times n} = \left[\mu^-_{ij} - v^+_{ij}, \mu^+_{ij} - v^-_{ij}\right]$ is as follows:

$$\tilde{S} = \begin{bmatrix} \left[\mu^-_{11} - v^+_{11}, \mu^+_{11} - v^-_{11}\right] & \cdots & \left[\mu^-_{1n} - v^+_{1n}, \mu^+_{1n} - v^-_{1n}\right] \\ \vdots & \ddots & \vdots \\ \left[\mu^-_{n1} - v^+_{n1}, \mu^+_{n1} - v^-_{n1}\right] & \cdots & \left[\mu^-_{nn} - v^+_{nn}, \mu^+_{nn} - v^-_{nn}\right] \end{bmatrix} \quad (5.25)$$

The interval multiplicative matrix $\tilde{A} = \left(\tilde{a}_{ij}\right)_{n \times n} = \left[10^{\left(\mu_{\overline{ij}} - v_{ij}^{+}\right)}, 10^{\left(\mu_{ij}^{+} - v_{\overline{ij}}\right)}\right]$ is given as follows:

$$\tilde{A} = \begin{bmatrix} \left[10^{\left(\mu_{\overline{11}} - v_{11}^{+}\right)}, 10^{\left(\mu_{1n}^{+} - v_{\overline{1n}}\right)}\right] & \cdots & \left[10^{\left(\mu_{\overline{ij}} - v_{ij}^{+}\right)}, 10^{\left(\mu_{ij}^{+} - v_{\overline{ij}}\right)}\right] \\ \vdots & \ddots & \vdots \\ \left[10^{\left(\mu_{\overline{n1}} - v_{n1}^{+}\right)}, 10^{\left(\mu_{n1}^{+} - v_{\overline{n1}}\right)}\right] & \cdots & \left[10^{\left(\mu_{\overline{nn}} - v_{nn}^{+}\right)}, 10^{\left(\mu_{nn}^{+} - v_{\overline{nn}}\right)}\right] \end{bmatrix} \tag{5.26}$$

The steps of the fuzzy MCDM method are summarized in the following:

Step 1: Assign the interval-valued intuitionistic preferences for each pair of alternatives and obtain the interval-valued intuitionistic preference relation $\tilde{R} = \left(\tilde{r}_{ij}\right)_{n \times n}$.

Step 2: Calculate the score judgment matrix $\tilde{S} = \left(\tilde{s}_{ij}\right)_{n \times n}$ and the interval multiplicative matrix $\tilde{A} = \left(\tilde{a}_{ij}\right)_{n \times n}$.

Step 3: Determine the priority vector of the interval multiplicative matrix $\tilde{A} = \left(\tilde{a}_{ij}\right)_{n \times n}$ by calculating the $\tilde{w}_i$ interval for each criterion using Equation 5.27:

$$\tilde{w}_i = \left[\frac{\sum_{j=1}^{n} \tilde{a}_{ij}^{-}}{\sum_{i=1}^{n} \sum_{j=1}^{n} \tilde{a}_{ij}^{+}}, \frac{\sum_{j=1}^{n} \tilde{a}_{ij}^{+}}{\sum_{i=1}^{n} \sum_{j=1}^{n} \tilde{a}_{ij}^{-}}\right] \tag{5.27}$$

Step 4: Construct the possibility degree matrix $P = \left(p_{ij}\right)_{n \times n}$ by comparing the obtained weights in Step 3. To do this, use Equation 5.28:

$$P\left(w_i \geq w_j\right) = \frac{\min\left\{L_{w_i} + L_{w_j}, \max\left(w_i^{+} - w_j^{-}, 0\right)\right\}}{L_{w_i} + L_{w_j}} \tag{5.28}$$

where $L_{w_i} = w_i^{+} - w_i^{-}$ and $L_{w_j} = w_j^{+} - w_j^{-}$ and $p_{ij} \geq 0$, $p_{ij} + p_{ji} = 1$, $p_{ii} = 1/2$.

Step 5: Prioritize the $P = \left(p_{ij}\right)_{n \times n}$ by using Equation 5.29:

$$w_i = \frac{1}{n}\left[\sum_{j=1}^{n} p_{ij} + \frac{n}{2} - 1\right] \tag{5.29}$$

Step 6: Normalize the weights vector obtained in Step 5.

TABLE 5.1

Linguistic Scale and the Corresponding IVIFSs

Linguistic Terms	Membership and Nonmembership Values
Absolutely low (AL)	([0.10, 0.25] , [0.65, 0.75])
Very low (VL)	([0.15, 0.30] , [0.60, 0.70])
Low (L)	([0.20, 0.35] , [0.55, 0.65])
Medium low (ML)	([0.25, 0.4] , [0.50, 0.60])
Approximately equal (AE)	([0.45, 0.55] , [0.30, 0.45])
Medium high (MH)	([0.50, 0.60] , [0.25, 0.40])
High (H)	([0.55,0.65] , [0.20, 0.35])
Very high (VH)	([0.60,0.70] , [0.15,0.30])
Absolutely high (AH)	([0.65,0.75] , [0.10,0.25])

The scale given in Table 5.1 is used to assign the interval-valued intuitionistic fuzzy preferences.

IVIFSs, interval-valued intuitionistic fuzzy sets.

Exactly equal (EE) = ([0.5, 0.5], [0.5, 0.5]).

5.5 Application of IVIF-AHP to Personnel Selection

The presented IVIF-AHP method will now be applied to a personnel selection problem. Assume that there are three candidates (A1, A2, A3) for a job position in a company. These candidates are evaluated according to four criteria which are experience (C1), education (C2), technical skills (C3), and communication skills (C4).

Step 1: Linguistic pairwise comparisons of the criteria with respect to goal and the alternatives with respect to criteria based on the linguistic scale (Table 5.1) are given in Tables 5.2 through 5.6.

The corresponding interval-valued intuitionistic values of the above pairwise comparisons are given in Tables 5.7 through 5.11.

Step 2: The score judgment matrix $\tilde{S}$ and the interval multiplicative matrix $\tilde{A}$ are obtained by applying Equations 5.25 and 5.26, respectively. As an example, by using the pairwise values from Table 5.7, the score judgment matrix given in Table 5.12 is obtained.

The interval multiplicative matrixes $\tilde{A}$ for the values given in Table 5.12 are obtained as given in Table 5.13.

Step 3: The $\tilde{w}_i$ interval values for each criterion and each alternative are calculated to determine the priority vector of the interval multiplicative matrix $\tilde{A}$ by using Equation 5.27. Table 5.14 gives the $\tilde{w}_i$ values for the values given in Table 5.13.

TABLE 5.2

Linguistic Pairwise Comparisons of Criteria with Respect to Goal

Goal	C1—Experience	C2—Education	C3—Technical Skills	C4—Communication Skills
C1	Exactly equal (EE)	High (H)	Medium high (MH)	Very high (VH)
C2	–	EE	Low (L)	Very low (VL)
C3	–	–	EE	Medium high (MH)
C4	–	–	–	EE

TABLE 5.3

Linguistic Pairwise Comparisons of Alternatives with Respect to C1

C1—Experience	A1	A2	A3
A1	Exactly equal (EE)	Medium high (MH)	Approximately equal (AE)
A2	–	EE	Medium low (ML)
A3	–	–	EE

TABLE 5.4

Linguistic Pairwise Comparisons of Alternatives with Respect to C2

C2—Education	A1	A2	A3
A1	Exactly equal (EE)	Very low (VL)	High (H)
A2	–	EE	Very high (VH)
A3	–	–	EE

TABLE 5.5

Linguistic Pairwise Comparisons of Alternatives with Respect to C3

C3—Technical Skills	A1	A2	A3
A1	Exactly equal (EE)	Very low (VL)	Medium low (ML)
A2	–	EE	Approximately equal (AE)
A3	–	–	EE

TABLE 5.6

Linguistic Pairwise Comparisons of Alternatives with Respect to C4

C4—Communication Skills	A1	A2	A3
A1	Exactly equal (EE)	Very high (VH)	Medium high (MH)
A2	–	EE	Low (L)
A3	–	–	EE

TABLE 5.7

Interval-Valued Intuitionistic Values of Pairwise Comparisons of Criteria with Respect to Goal

Goal	C1—Experience				C2—Education				C3—Technical Skills				C4—Communication Skills			
C1	0.50	0.50	0.50	0.50	0.55	0.65	0.20	0.35	0.50	0.60	0.25	0.40	0.60	0.70	0.15	0.30
C2	0.20	0.35	0.55	0.65	0.50	0.50	0.50	0.50	0.20	0.35	0.55	0.65	0.15	0.30	0.60	0.70
C3	0.25	0.40	0.50	0.60	0.55	0.65	0.20	0.35	0.50	0.50	0.50	0.50	0.50	0.60	0.25	0.40
C4	0.15	0.30	0.60	0.70	0.60	0.70	0.15	0.30	0.25	0.40	0.50	0.60	0.50	0.50	0.50	0.50

TABLE 5.8

Interval-Valued Intuitionistic Values of Pairwise Comparisons of Alternatives with Respect to C1

C1—Experience	A1				A2				A3			
A1	0.50	0.50	0.50	0.50	0.50	0.60	0.25	0.40	0.45	0.55	0.30	0.45
A2	0.25	0.40	0.50	0.60	0.50	0.50	0.50	0.50	0.25	0.40	0.50	0.60
A3	0.30	0.45	0.45	0.55	0.50	0.60	0.25	0.40	0.50	0.50	0.50	0.50

TABLE 5.9

Interval-Valued Intuitionistic Values of Pairwise Comparisons of Alternatives with Respect to C2

C2—Education	A1				A2				A3			
A1	0.50	0.50	0.50	0.50	0.15	0.30	0.60	0.70	0.55	0.65	0.20	0.35
A2	0.60	0.70	0.15	0.30	0.50	0.50	0.50	0.50	0.60	0.70	0.15	0.30
A3	0.20	0.35	0.55	0.65	0.15	0.30	0.60	0.70	0.50	0.50	0.50	0.50

TABLE 5.10

Interval-Valued Intuitionistic Values of Pairwise Comparisons of Alternatives with Respect to C3

C3—Technical Skills	A1				A2				A3			
A1	0.50	0.50	0.50	0.50	0.15	0.30	0.60	0.70	0.25	0.40	0.50	0.60
A2	0.60	0.70	0.15	0.30	0.50	0.50	0.50	0.50	0.45	0.55	0.30	0.45
A3	0.50	0.60	0.25	0.40	0.30	0.45	0.45	0.55	0.50	0.50	0.50	0.50

TABLE 5.11

Interval-Valued Intuitionistic Values of Pairwise Comparisons of Alternatives with Respect to C4

C4—Communication Skills	A1						A2						A3					
A1	0.50	0.50	0.50	0.50	0.60	0.70	0.15	0.30	0.50	0.60	0.25	0.40
A2	0.15	0.30	0.60	0.70	0.50	0.50	0.50	0.50	0.20	0.35	0.55	0.65
A3	0.25	0.40	0.50	0.60	0.55	0.65	0.20	0.35	0.50	0.50	0.50	0.50

TABLE 5.12

The Score Judgment Matrix $\tilde{S}$ for the Pairwise Judgments Given in Table 5.7

Goal	C1—Experience		C2—Education		C3—Technical Skills		C4—Communication Skills	
C1	0.00	0.00	0.20	0.45	0.10	0.35	0.30	0.55
C2	−0.45	−0.20	0.00	0.00	−0.45	−0.20	−0.55	−0.30
C3	−0.35	−0.10	0.20	0.45	0.00	0.00	0.10	0.35
C4	−0.55	−0.30	0.30	0.55	−0.35	−0.10	0.00	0.00

TABLE 5.13

The Interval Multiplicative Matrix $\tilde{A}$ for the Score Judgment Matrix Given in Table 5.12

Goal	C1—Experience		C2—Education		C3—Technical Skills		C4—Communication Skills	
C1	1.000	1.000	1.585	2.818	1.259	2.239	1.995	3.548
C2	0.355	0.631	1.000	1.000	0.355	0.631	0.282	0.501
C3	0.447	0.794	1.585	2.818	1.000	1.000	1.259	2.239
C4	0.282	0.501	1.995	3.548	0.447	0.794	1.000	1.000

TABLE 5.14

The $\tilde{w}_i$ Interval Values for the Values Given in Table 5.12

C1	[0.233; 0.606]
C2	[0.079; 0.174]
C3	[0.171; 0.432]
C4	[0.149; 0.369]

TABLE 5.15

The Possibility Degree Matrix P for the Values Given in Table 5.14

	C1	C2	C3	C4
C1	0.500	1.000	0.686	0.771
C2	0.000	0.500	0.009	0.082
C3	0.314	0.991	0.500	0.590
C4	0.229	0.918	0.410	0.500

TABLE 5.16

The Normalized Weights of the Criteria

Criteria	Normalized Weight (w)
C1	w
C2	0.133
C3	0.283
C4	0.255

TABLE 5.17

The Normalized Weights of the Alternatives with Respect to Criteria

	A1	A2	A3
W_{c1}	0.418	0.214	0.368
W_{c2}	0.345	0.476	0.179
W_{c3}	0.187	0.456	0.357
W_{c4}	0.454	0.173	0.373

TABLE 5.18

The Priority Weights and the Ranking of the Alternatives

Alternatives	Priority Weight	Rank
A1	0.352	1
A2	0.307	3
A3	0.341	2

Step 4: By using Equation 5.28, the weights obtained in the previous step (Table 5.14) are compared. Then the possibility degree matrix P is obtained as given in Table 5.15.

Step 5: By using Equation 5.29, P is defuzzified and in the last step they are normalized and the normalized weights are obtained. The normalized weight vector for the goal is obtained as in Table 5.16.

The same calculations are also made for the pairwise calculations given in Tables 5.8 through 5.11. The weight vectors of the alternatives with respect to criteria can be easily obtained as given in Table 5.17.

The ranking of the alternatives based on the overall results are given in Table 5.18. According to the results given in Table 5.18, A1 is selected as the best alternative.

5.6 Hesitant Fuzzy Sets

HFSs, developed by Torra (2010), are the extensions of regular fuzzy sets, which handle the situations where a set of values are possible for the membership of a single element. In other words, an HFS allows the membership degree to have a set of possible values between 0 and 1 (Torra and Narukawa,

2009). Torra and Narukawa (2009) state the difficulty of determining the membership value of an element on a set and specify that an HFS can be used in cases where uncertainty on the possible membership values is limited, such as when a group of experts may not agree on the membership of an element and discuss whether it is 0.3 or 0.5. In such cases, an HFS can represent the situation and instead of using an aggregation operator to get a single value, it is useful to deal with all the possible values (Torra, 2009). Torra and Narukawa (2009) and Torra (2010) investigated the relationship between HFSs and other generalizations of fuzzy sets. They showed that the envelope of an HF is an IFS and also showed that the operations that are proposed are consistent with those of IFSs when applied to the envelope of an HFS. Xia and Xu (2011) gave an intensive study on hesitant fuzzy information aggregation techniques and their application in decision-making. They developed some hesitant fuzzy operational rules based on the interconnection between the HFS and the IFS. To aggregate the hesitant fuzzy information, they proposed a series of operators under various situations and discussed the relationships among them.

Since people may have hesitancy in providing their preferences, HFS can be effectively used to represent these in different levels of the decision-making process. Some basic concepts related to HFSs, which are taken from Torra and Narukawa (2009) and Torra (2010), are as follows.

Definition 9: Let X be a fixed set. An HFS on X is given in terms of a function that when applied to X returns a subset of [0, 1]. The mathematical expression for an HFS is as follows:

$$E = \left\{ < x, h_E(x) > \mid x \in X \right\} \tag{5.30}$$

where $h_E(x)$ is a set of some values in [0, 1], denoting the possible membership degrees of the element $x \in X$ to the set E. Xu and Xia (2011) call $h = h_E(x)$ a hesitant fuzzy element (HFE).

Definition 10: Let h, h_1, and h_2 be three HFEs. Then basic operations on these elements are defined as follows:

$$h^-(x) = \min h(x) \tag{5.31}$$

$$h^+(x) = \max h(x) \tag{5.32}$$

where $h^-(x)$ and $h^+(x)$ are the lower and upper bounds of h, respectively.

$$h^c = \cup_{\gamma \in h} \left\{ 1 - \gamma \right\} \tag{5.33}$$

where h^c is the complement of h.

$$h^\lambda = \cup_{\gamma \in h} \left\{ \gamma^\lambda \right\} \tag{5.34}$$

$$\lambda h = \cup_{\gamma \in h} \left\{ 1 - (1 - \gamma)^\lambda \right\} \tag{5.35}$$

$$h_1 \cup h_2 = \cup_{\gamma_1 \in h_1, \gamma_2 \in h_2} \max\{\gamma_1, \gamma_2\} \tag{5.36}$$

$$h_1 \cap h_2 = \cup_{\gamma_1 \in h_1, \gamma_2 \in h_2} \min\{\gamma_1, \gamma_2\} \tag{5.37}$$

$$h_1 \oplus h_2 = \cup_{\gamma_1 \in h_1, \gamma_2 \in h_2} \{\gamma_1 + \gamma_2 - \gamma_1\gamma_2\} \tag{5.38}$$

$$h_1 \oplus h_2 = \cup_{\gamma_1 \in h_1, \gamma_2 \in h_2} \{\gamma_1\gamma_2\} \tag{5.39}$$

Definition 11: Given an HFE h, the envelope of h, $A_{env(h)}$, is an IFS, which is defined as follows:

$$A_{env}(h) = \{x, \mu(x), v(x)\} \tag{5.40}$$

where

$$\mu(x) = h^-(x) \tag{5.41}$$

$$v(x) = 1 - h^+(x) \tag{5.42}$$

The relationship between HFEs and intuitionistic fuzzy values (IFVs) is as follows:

$$A_{env}(h_c) = \left(A_{env}(h)\right)^c \tag{5.43}$$

$$A_{env}(h_1 \cup h_2) = A_{env}(h_1) \cup A_{env}(h_2) \tag{5.44}$$

$$A_{env}(h_1 \cap h_2) = A_{env}(h_1) \cap A_{env}(h_2) \tag{5.45}$$

5.7 Hesitant FAHP

HFS extensions of the AHP method in literature are limited as compared to IFS extensions. Rodriguez et al. (2012) presented a different approach using the name hesitant fuzzy linguistic term set (HFLTS) to deal with the situations when ordinary fuzzy linguistic approaches that aim to use a single linguistic term are incapable of handling the hesitation of decision makers. The importance of this approach is that it provides a linguistic and computational basis to increase the richness of linguistic elicitation based on the fuzzy linguistic approach and the use of context-free grammars by using comparative terms.

Rodríguez et al. (2013) proposed a new group decision model based on HFLTSs in order to enhance the elicitation of flexible and rich linguistic expressions. The drawback of this model is that it considers only a single criterion and cannot be used for complex MCDM problems. This model is based on m experts evaluating n alternatives on a single criterion. Later, Zhu and Xu (2014) proposed a methodology called AHP-hesitant group decision-making (AHP-HGDM) in which hesitant judgments that each include several possible values are used to indicate the original judgments provided by the decision makers. They introduced hesitant multiplicative preference relations (HMPRs) to collect the hesitant judgments, and then they developed a hesitant multiplicative programming method (HMPM) as a new prioritization method to derive ratio-scale priorities from HMPRs. The practicality and effectiveness of the proposed method were also illustrated by an example.

Mousavi et al. (2014) proposed a method called hesitant fuzzy-AHP (HF-AHP) in which decision makers' evaluations for comparison matrices are expressed by linguistic variables and then the DMs' judgments are aggregated by utilizing the hesitant fuzzy geometric operator. They also gave a real-world example to verify the proposed method.

Yavuz et al. (2015) extended HFLTS to multicriteria evaluation that considers hesitancy of the experts in defining membership degrees or functions. In this model, linguistic term sets are used together with context-free grammar such as "at most medium importance," "between low and high importance," etc. This model can handle a complex multicriteria problem with a hierarchical structure and use a fuzzy representation for comparative linguistic expressions based on a fuzzy envelope for HFLTS.

Öztayşi et al. (2015) developed a hesitant FAHP method involving multiexperts' linguistic evaluations aggregated by an ordered weighted averaging (OWA) operator. The developed method was successfully applied to a multicriteria supplier selection problem.

Zhu et al. (2016) proposed a hesitant AHP method with new concepts by using a new stochastic prioritization method. They defined two indices, which are an expected geometric consistency index to check the consistency degrees of individual hesitant comparison matrices and an expected geometric consensus index to check the consensus degrees of multiple hesitant comparison matrices.

Onar et al. (2016) proposed a new hesitant fuzzy quality function deployment (QFD) approach for selection of computer workstations. QFD was used to define design requirements of the computer workstations. They used hesitant FAHP based on HFLTS to determine the weights of criteria and used hesitant fuzzy TOPSIS to select the most suitable alternative.

Zhou and Xu (2016) introduced the hesitant fuzzy preference format and defined the hesitant fuzzy continuous preference term. Based on this approach, they presented a model framework of the asymmetric hesitant fuzzy sigmoid preference relation (AHSPR) in the AHP. The authors indicated that their model requires complex calculations and a more simplified calculation process needs to be developed.

Tüysüz and Şimşek (2016) applied an HFLTS-based AHP method as an HFS extension of AHP for assessing and prioritizing the factors used in the performance evaluation of the branches of a cargo company operating in Turkey, and Tüysüz and Çelikbilek (2016) used an HFLTS-based AHP approach for determining the importance of the factors used in the evaluation of renewable energy resources. Due to the flexibility of the model in defining linguistic terms, an HFLTS-based AHP method is presented in detail in the next section.

5.8 HFLTS-Based AHP Method

An HFLTS is able to mathematically represent and solve decision-making problems with multiple linguistic assessments and enhance the elicitation of flexible and rich linguistic expressions. Before we explain and give the algorithmic steps of the HFLTS-based AHP method, some basic concepts related to HFLTSs, which are taken from Rodriguez et al. (2012) will be given.

Definition 12: An HFLTS H_s is an ordered finite subset of consecutive linguistic terms of a linguistic term set S which can be shown as $S = \{s_0, s_1, \ldots, s_g\}$.

Definition 13: Assume that E_{G_H} is a function that converts linguistic expressions into an HFLTS, H_S. Let G_H be a context-free grammar that uses the linguistic term set S. Let S_{ll} be the expression domain generated by G_H. This relation can be shown as $E_{G_H} : S_{ll} \rightarrow H_S$.

Using the following transformations, comparative linguistic expressions are converted into an HFLTS:

$$E_{G_H}(s_i) = \{s_i | s_i \in S\} \tag{5.46}$$

$$E_{G_H}(\text{at most } s_i) = \{s_j | s_j \in S \text{ and } s_j \leq s_i\} \tag{5.47}$$

$$E_{G_H}(\text{lower than } s_i) = \{s_j | s_j \in S \text{ and } s_j < s_i\} \tag{5.48}$$

$$E_{G_H}(\text{at least } s_i) = \{s_j | s_j \in S \text{ and } s_j \geq s_i\} \tag{5.49}$$

$$E_{G_H}(\text{greater than } s_i) = \{s_j | s_j \in S \text{ and } s_j > s_i\} \tag{5.50}$$

$$E_{G_H}(\text{between } s_i \text{ and } s_j) = \{s_k | s_k \in S \text{ and } s_i \leq s_k \leq s_j\} \tag{5.51}$$

Definition 14: The envelope of an HFLTS is represented by $\mathrm{env}(H_S)$, and it is a linguistic interval whose limits are obtained by its maximum value and minimum value:

$$\mathrm{env}(H_S) = \left[H_{S^-}, H_{S^+} \right], \quad H_{S^-} \le H_{S^+} \tag{5.52}$$

where

$$H_{S^-} = \min(s_i) = s_j, \quad s_i \in H_S \text{ and } s_i \ge s_j \forall i \tag{5.53}$$

$$H_{S^+} = \max(s_i) = s_j, \quad s_i \in H_S \text{ and } s_i \le s_j \forall i \tag{5.54}$$

The algorithmic steps of the HFLTS-based AHP method are as follows:

Step 1: Define the semantics and syntax of the linguistic term set S and the context-free grammar G_H, where $G_H = \{V_N, V_T, I, P\}$

$$V_N = \{\text{primary term, composite term, unary relation, binary relation, conjunction}\}$$

$$V_T = \{\text{lower than, greater than, at least, at most, between, and}, s_0, s_1, \ldots, s_g\}$$

$$I \in V_N$$

The production rules can be obtained by Equation 5.55:

$$P = \begin{cases} I ::= \text{primary term}|\text{composite term}, \text{ composite term} ::= \text{unary} \\ \text{relation primary term}|\text{binary relation primary term} \\ \text{conjunction primary term, primary term} ::= s_0|s_1|\ldots|s_g, \\ \text{unary relation} ::= \text{lower than}|\text{greater than}|\text{at least}| \\ \text{at most, binary relation} ::= \text{between, conjunction} ::= \text{and} \end{cases} \tag{5.55}$$

Step 2: Gather the pairwise comparisons from the experts. In the domain of group decision-making, m decision makers ($E = \{e_1, e_2, \ldots, e_m\}$) try to select the best alternative among n alternatives ($X = \{x_1, x_2, \ldots, x_n\}$) where $m > 1$ and $n > 1$. In this case, a matrix composed of preference relations ($p^k[0,1]$) are formed as given in Equation 5.56:

$$p^k = \begin{pmatrix} p_{11}^k & \cdots & p_{1m}^k \\ \vdots & \ddots & \vdots \\ p_{n1}^k & \cdots & p_{nm}^k \end{pmatrix} \tag{5.56}$$

where p_{ij}^k shows the degree of preference of the alternative x_i over x_j according to expert e_k. In this step, the preference matrix is constructed for the criteria.

Step 3: Transform the preference relations into an HFLTS by using the transformation function E_{G_H}. For each HFLTS obtain an envelope $\left[p_{ij}^{k-}, p_{ij}^{k+} \right]$.

Step 4: Obtain the pessimistic and optimistic collective preference relations (P_C^- and P_C^+). Compute the pessimistic and optimistic collective preference for each alternative using two-tuple sets. The two-tuple set associated with S is defined as $S = S \times [0.5, 0.5]$. The function $\Delta : [0, g] \to S$ is given in Equation 5.57:

$$\Delta(\beta) = (s_i, \alpha) \text{ with } \begin{cases} i = \text{round}(\beta) \\ \alpha = \beta - i \end{cases} \tag{5.57}$$

where round assigns to β the integer number $i \in \{0, 1, \ldots, g\}$ closest to β and $\Delta^{-1} : S \to [0, g]$ is defined as shown in Equation 5.58:

$$\Delta^{-1}(s_i, \alpha) = i + \alpha \tag{5.58}$$

Step 5: Build a vector of intervals $V^R = \left(p_1^R, p_2^R, \ldots, p_n^R \right)$ of collective preferences for the alternatives $p_i^R = \left[p_i^-, p_i^+ \right]$.

Step 6: Calculate the midpoints of the intervals and normalize the results to find the weights.

5.9 Application of HFLTS-Based AHP Method to Personnel Selection

The presented HFLTS-based AHP method is applied to the same personnel selection problem given in Section 5.5. A group of three experts (E1, E2, E3) evaluates three candidates (A1, A2, A3) according to four criteria, which are experience (C1), education (C2), technical skills (C3), and communication skills (C4). These three experts evaluate the decision-making criteria as well as the three candidates using their expertise and experience.

Step 1: First, the semantics and syntax of the linguistic term set S and the context-free grammar G_H are defined. S can be defined as follows:

$$S = \left\{ \begin{array}{l} \text{absolutely low (n), very low (vl), low (l), medium (m),} \\ \text{high (h), very high (vh), absolutely high (ah)} \end{array} \right\}$$

Step 2: The pairwise comparisons of experts, which represent the preference relations (p^ks) as given in Equation 5.56, are collected. Tables 5.19 through 5.23 present the linguistic pairwise evaluations of the experts.

Step 3: The preference relations are transformed into an HFLTS using the transformation function E_{G_H}. Then, the envelope of each HFLTS or the HFLTS intervals is obtained. For the sake of keeping the study

TABLE 5.19

Pairwise Evaluations of the Criteria with Respect to Goal

E1's Linguistic Pairwise Evaluations				
Goal	C1—Experience	C2—Education	C3—Technical Skills	C4—Communication Skills
C1	–	is (h)	at most (m)	at least (vh)
C2	is (l)	–	between (l and m)	at most (vl)
C3	at least (h)	between (m and h)	–	is (m)
C4	at most (vl)	at least (vh)	is (m)	–

E2's Linguistic Pairwise Evaluations				
Goal	C1—Experience	C2—Education	C3—Technical Skills	C4—Communication Skills
C1	–	is (h)	is (m)	is (vh)
C2	is (l)	–	is (l)	is (vl)
C3	is (m)	is (h)	–	is (m)
C4	is (vl)	is (vh)	is (m)	–

E3's Linguistic Pairwise Evaluations				
Goal	C1—Experience	C2—Education	C3—Technical Skills	C4—Communication Skills
C1	–	at least (h)	between (l and m)	at least (vh)
C2	at most (l)	–	at most (m)	at most (vl)
C3	between (m and h)	at least (h)	–	between (m and h)
C4	at most (vl)	at least (vh)	between (l and m)	–

TABLE 5.20

Pairwise Evaluations of the Alternatives with Respect to C1

E1's Linguistic Pairwise Evaluations		
C1—Experience	**A1**	**A2**	**A3**
A1	–	is (h)	at most (l)
A2	is (l)	–	between (m and h)
A3	at least (h)	between (l and m)	–

E2's Linguistic Pairwise Evaluations		
C1—Experience	**A1**	**A2**	**A3**
A1	–	between (l and m)	at most (h)
A2	between (m and h)	–	at most (l)
A3	at least (l)	at least (h)	–

E3's Linguistic Pairwise Evaluations		
C1—Experience	**A1**	**A2**	**A3**
A1	–	between (l and m)	between (l and m)
A2	between (m and h)	–	at most (l)
A3	between (m and h)	at least (h)	–

TABLE 5.21

Pairwise Evaluations of the Alternatives with Respect to C2

E1's Linguistic Pairwise Evaluations		
C2—Education	**A1**	**A2**	**A3**
A1	–	is (vl)	at most (h)
A2	is (vh)	–	is (ah)
A3	at least (l)	is (n)	–

E2's Linguistic Pairwise Evaluations		
C2—Education	**A1**	**A2**	**A3**
A1	–	between (l and m)	between (l and m)
A2	between (m and h)	–	at most (l)
A3	between (m and h)	at least (h)	–

E3's Linguistic Pairwise Evaluations		
C2—Education	**A1**	**A2**	**A3**
A1	–	at most (l)	between (vl and l)
A2	at least (h)	–	at most (vl)
A3	between (vh and ah)	at least (h)	–

TABLE 5.22

Pairwise Evaluations of the Alternatives with Respect to C3

E1's Linguistic Pairwise Evaluations			
C3—Technical Skills	A1	A2	A3
A1	–	at most (l)	at most (vl)
A2	at least (h)	–	at most (vl)
A3	at least (vh)	at least (vh)	–

E2's Linguistic Pairwise Evaluations			
C3—Technical Skills	A1	A2	A3
A1	–	is (h)	between (m and h)
A2	is (l)	–	at most (vl)
A3	between (l and m)	at least (vh)	–

E3's Linguistic Pairwise Evaluations			
C3—Technical Skills	A1	A2	A3
A1	–	at least (h)	between (vl and l)
A2	at most (l)	–	at most (vl)
A3	between (vh and ah)	at least (vh)	–

TABLE 5.23

Pairwise Evaluations of the Alternatives with Respect to C4

E1's Linguistic Pairwise Evaluations			
C4—Communication Skills	A1	A2	A3
A1	–	between (vh and ah)	is (m)
A2	between (vl and l)	–	at most (l)
A3	is (m)	at least (h)	–

E2's Linguistic Pairwise Evaluations			
C4—Communication Skills	A1	A2	A3
A1	–	is (h)	between (vl and l)
A2	is (l)	–	at most (vl)
A3	between (vh and ah)	at least (vh)	–

E3's Linguistic Pairwise Evaluations			
C4—Communication Skills	A1	A2	A3
A1	–	between (vh and ah)	between (vl and l)
A2	between (vl and l)	–	at most (l)
A3	between (vh and ah)	at least (h)	–

TABLE 5.24

Obtained Envelopes for the HFLTS Given in Table 5.17

HFLTS Intervals for E1's Evaluations

Goal	C1—Experience	C2—Education	C3—Technical Skills	C4—Communication Skills
C1	–	[h, h]	[n, m]	[vh, ah]
C2	[l, l]	–	[l, m]	[n, vl]
C3	[h, ah]	[m, h]	–	[m, m]
C4	[n, vl]	[vh, ah]	[m, m]	–

HFLTS Intervals for E2's Evaluations

Goal	C1—Experience	C2—Education	C3—Technical Skills	C4—Communication Skills
C1	–	is (h)	is (m)	is (vh)
C2	[l, l]	–	is (l)	is (vl)
C3	[m, m]	[h, h]	–	is (m)
C4	[vl, vl]	[vh, vh]	[m, m]	–

HFLTS Intervals for E3's Evaluations

Goal	C1—Experience	C2—Education	C3—Technical Skills	C4—Communication Skills
C1	–	[h, ah]	[l, m]	[vh, ah]
C2	[n, l]	–	[n, m]	[n, vl]
C3	[m, h]	[h, ah]	–	[m, h]
C4	[n, vl]	[vh, ah]	[l, m]	–

short and making the methodology clearer, the rest of the calculations will be given for only the evaluations of criteria with respect to goal (Table 5.19). For the evaluations of alternatives with respect to criteria, the same calculations can be easily performed. Table 5.24 presents the obtained envelopes for the values given in Table 5.19.

Step 4: The pessimistic and optimistic collective preferences (P_C^- and P_C^+) are calculated using two-tuple operations. Before these calculations, the scale given in Table 5.25 is assigned to the linguistic terms.

For example, the pessimistic collective preference value for C1 with respect to C2 is calculated as follows based on the values given in Table 5.24:

$$P_{C_{12}}^- = \Delta\left(\frac{1}{3}\left(\Delta^{-1}(h, 4) + \Delta^{-1}(h, 4) + \Delta^{-1}(h, 4)\right)\right)$$

$$= \Delta\left(\frac{1}{3}(4 + 4 + 4)\right) = \Delta(4) = (h, -.0)$$

TABLE 5.25

The Scale for Linguistic Terms

Absolutely low (n)	0
Very low (vl)	1
Low (l)	2
Medium (m)	3
High (h)	4
Very high (vh)	5
Absolutely high (ah)	6

TABLE 5.26

Pessimistic and Optimistic Collective Preferences for the Criteria

Pessimistic Collective Preferences (P_C^-)

Goal	C1—Experience	C2—Education	C3—Technical Skills	C4—Communication Skills
C1	–	(h, –0)	(l, –.33)	(vh, –0)
C2	(vl, .33)	–	(vl, .33)	(n, .33)
C3	(m, .33)	(h, –.33)	–	(m, –0)
C4	(n, .33)	(vh, –0)	(m, –.33)	–

Optimistic Collective Preferences (P_C^+)

Goal	C1—Experience	C2—Education	C3—Technical Skills	C4—Communication Skills
C1	–	(vh, –.33)	(m, –.0)	(ah, –.33)
C2	(l, –0)	–	(m, –.33)	(vl, –.33)
C3	(h, .33)	(vh, –.33)	–	(m, .33)
C4	(vl, –0)	(ah, –.33)	(m, –.0)	–

Similarly, the optimistic collective preference value for C1 with respect to C2 is calculated as follows based on the values given in Table 5.24:

$$P_{C_{12}}^+ = \Delta\left(\frac{1}{3}\left(\Delta^{-1}(h,\, 4) + \Delta^{-1}(h,\, 4) + \Delta^{-1}(ah,\, 6)\right)\right)$$

$$= \Delta\left(\frac{1}{3}(4 + 4 + 6)\right) = \Delta(4.67) = (vh,\, -.33)$$

Table 5.26 gives the pessimistic and optimistic collective preferences for the values given in Table 5.24.

Step 5: The linguistic intervals are converted to interval utilities. Finally, in **Step 6**, midpoints of interval utilities are obtained and then weights are obtained by normalizing those midpoints.

Table 5.27 gives linguistic intervals of the criteria, interval utilities associated with them, midpoints, and obtained weights of all four criteria.

When Steps 3–6 are applied for the evaluations of alternatives with respect to each criterion (for the values given in Tables 5.20 through 5.23), the weights of the alternatives can be easily obtained as given in Table 5.28.

A ranking of the alternatives based on the overall results is given in Table 5.29. According to the results given in Table 5.29, A3 is selected as the best alternative.

TABLE 5.27

Weights of the Criteria with Respect to Goal

Criteria	Linguistic Intervals	Interval Utilities	Midpoints	Weights
C1—Experience	[(h, −.44),(h, .44)]	[3.56, 4.44]	4.00	0.33
C2—Education	[(vl, −.0),(l, −.11)]	[1.00, 1.89]	1.44	0.12
C3—Technical skills	[(m, .33),(h, .11)]	[3.33, 4.11]	3.72	0.31
C4—Communication skills	[(m, −.33),(m, .22)]	[2.67, 3.22]	2.94	0.24

TABLE 5.28

Weights of the Alternatives with Respect to Criteria

C1—Experience	Linguistic Intervals	Interval Utilities	Midpoints	Weights
A1	[(l, −.33),(m, .17)]	[1.67, 3.17]	2.42	0.27
A2	[(l, −.17),(m, −.0)]	[1.83, 3.00]	2.42	0.27
A3	[(m, .17),(vh, .17)]	[3.17, 5.17]	4.17	0.46

C2—Education	Linguistic Intervals	Interval Utilities	Midpoints	Weights
A1	[(vl, −0),(m, −.17)]	[1.00, 2.83]	1.92	0.21
A2	[(m, −0),(h, −.0)]	[3.00, 4.00]	3.50	0.38
A3	[(m, -0),(vh, -.33)]	[3.00, 4.67]	3.83	0.41

C3—Technical Skills	Linguistic Intervals	Interval Utilities	Midpoints	Weights
A1	[(l, −.0),(m, −.0)]	[2.00, 3.00]	2.50	0.28
A2	[(vl, −.0),(l, .17)]	[1.00, 2.17]	1.58	0.17
A3	[(h, .5),(vh, .5)]	[4.50, 5.50]	5.00	0.55

C4—Communication Skills	Linguistic Intervals	Interval Utilities	Midpoints	Weights
A1	[(m, .17),(h, −.33)]	[3.17, 3.67]	3.42	0.36
A2	[(vl, -.33),(l, −.17)]	[0.67, 1.83]	1.25	0.13
A3	[(h, .33),(vh, .5)]	[4.33, 5.50]	4.92	0.51

TABLE 5.29

The Priority Weights and the Ranking of the Alternatives

Alternatives	Priority Weight	Rank
A1	0.285	2
A2	0.219	3
A3	0.496	1

5.10 Conclusions

AHP is one of the most widely used MCDM methods due to the fact that it handles both tangibles and intangibles and also uses simple mathematical calculations. Uncertainty that can be categorized as vagueness and ambiguity is unavoidable in real-life applications and decision-making processes. A conventional or crisp AHP method uses crisp or objective mathematics to represent the subjective or personal preferences of an individual or a group in MCDM. The AHP procedure comprises both vagueness and ambiguity in assigning pairwise comparisons and evaluating alternatives. In order to overcome this drawback, fuzzy sets theory is the most convenient and most common approach. Fuzzy extensions of AHP are usually in the form of fuzzifying the pairwise comparisons by using fuzzy numbers. FAHP is an extension of crisp AHP in which fuzzy sets are incorporated with pairwise comparisons to model the uncertainty in human judgment and preference.

In addition to the ordinary fuzzy sets extensions, there have been some other generalizations and extensions of fuzzy sets for better modeling developed recently, such type 2 fuzzy sets, IFSs, and HFSs. In ordinary fuzzy sets, the membership degree of the element in a universe takes a single value between 0 and 1. Since these single values cannot represent adequately the lack of knowledge, ordinary fuzzy sets are not capable of co-opting the lack of knowledge with membership degrees. Therefore, IFSs, which are an extension of Zadeh's fuzzy sets, can be utilized for better simulating human decision-making processes and activities, which require human expertise and knowledge that are inevitably imprecise or not totally reliable. IFSs enable description of the opinions of decision makers from three sides, membership degree, nonmembership degree, and indeterminacy degree or hesitancy degree, which ensures more comprehensive preference information. Another important extension of fuzzy sets is HFSs which allows the membership degrees to have a set of possible values between 0 and 1. The main idea behind the usage of HFSs in decision-making is that when more than one decision maker assigns the membership degree of an element to a set, the difficulty of establishing a common membership degree rises.

Since there is a growing literature and interest in IFSs and HFSs, in the scope of this chapter, we presented important concepts and detailed literature review about the applications of these two methodologies within the context of the AHP method. Different approaches and techniques developed in the literature were presented. In addition to this, two approaches, an AHP method based on IVIFSs, and an HFLTS-based AHP method < were given in detail with numerical examples.

For further research, performing sensitivity analysis for the presented approaches and comparison with other approaches can be considered.

References

Abdullah, L., Jaafar, S., and Taib, I. (2009). A new analytic hierarchy process in multi-attribute group decision making. *International Journal of Soft Computing*, 4(5), 208–214.

Abdullah, L. and Najib, L. (2016a). A new preference scale MCDM method based on interval-valued intuitionistic fuzzy sets and the analytic hierarchy process. *Soft Computing*, 20(2), 511–523.

Abdullah, L. and Najib, L. (2016b). Sustainable energy planning decision using the intuitionistic fuzzy analytic hierarchy process: Choosing energy technology in Malaysia. *International Journal of Sustainable Energy*, 35(4), 360–377.

Atanassov, K. T. (1986). Intuitionistic fuzzy sets. *Fuzzy Sets and Systems*, 20(1), 87–96.

Atanassov, K. T. (1999). *Intuitionistic Fuzzy Sets: Theory and Applications*. New York, NY: Physica-Verlag.

Atanassov, K. T. (2012). *On Intuitionistic Fuzzy Sets Theory, Volume 283 of Studies in Fuzziness and Soft Computing*, Berlin/Heidelberg: Springer-Verlag.

Bali, O., Dagdeviren, M., and Gumus, S. (2015). An integrated dynamic intuitionistic fuzzy MADM approach for personnel promotion problem. *Kybernetes*, 44(10), 1422–1436.

Beskese, A., Kahraman, C., and Irani, Z. (2004). Quantification of flexibility in advanced manufacturing systems using fuzzy concept. *International Journal of Production Economics*, 89(1), 45–56.

Boender, C. G. E., De Graan, J. G., and Lootsma, F. A. (1989). Multicriteria decision analysis with fuzzy pairwise comparisons. *Fuzzy Sets and Systems*, 29(2), 133–143.

Buckley, J. J. (1985). Fuzzy hierarchical analysis. *Fuzzy Sets and Systems*, 17(3), 233–247.

Chang, D. Y. (1996). Applications of the extent analysis method on fuzzy AHP. *European Journal of Operational Research*, 95(3), 649–655.

Cheng, C. H. (1997). Evaluating naval tactical missile systems by fuzzy AHP based on the grade value of membership function. *European Journal of Operational Research*, 96(2), 343–350.

Chou, T. Y., Hsu, C. L., and Chen, M. C. (2008). A fuzzy multicriteria decision model for international tourist hotels location selection. *International Journal of Hospitality Management*, 27(2), 293–301.

Deepika, M. and Kannan, A. K. (2016, March). Global supplier selection using intuitionistic fuzzy analytic hierarchy process. In *International Conference on Electrical, Electronics, and Optimization Techniques (ICEEOT)* (pp. 1–5).

Dutta, B. and Guha, D. (2015). Preference programming approach for solving intuitionistic fuzzy AHP. *International Journal of Computational Intelligence Systems*, 8(5), 977–991.

Fahmi, A., Derakhshan, A., and Kahraman, C. (2015, August). Human resources management using interval valued intuitionistic fuzzy analytic hierarchy process. In *2015 IEEE International Conference on Fuzzy Systems (FUZZ-IEEE)* (pp. 1–5).

Feng, X., Qian, X., and Wu, Q. (2012). A DS-AHP approach for multi-attribute decision making problem with intuitionistic fuzzy information. *Information Technology Journal*, 11(12), 1764.

Kahraman, C., Öztayşi, B., and Onar, S. Ç. (2016). Intuitionistic fuzzy multicriteria evaluation of outsource manufacturers. *IFAC-Papers OnLine*, 49(12), 1844–1849.

Kahraman, C., Ruan, D., and Doğan, I. (2003). Fuzzy group decision-making for facility location selection. *Information Sciences*, 157, 135–153.

Kaur, P. (2014). Selection of vendor based on intuitionistic fuzzy analytical hierarchy process. *Advances in Operations Research*, 1–10. doi: http://dx.doi.org/10.1155/2014/987690.

Kaya, T. and Kahraman, C. (2011). Fuzzy multiple criteria forestry decision making based on an integrated VIKOR and AHP approach. *Expert Systems with Applications*, 38(6), 7326–7333.

Keshavarzfard, R., and Makui, A. (2015). An IF-DEMATEL-AHP based on triangular intuitionistic fuzzy numbers (TIFNs). *Decision Science Letters*, 4(2), 237–246.

Klir G. J. and Yuan, B. (1995). *Fuzzy Sets and Fuzzy Logic: Theory and Applications*. Upper Saddle River, NJ: Prentice Hall International.

Kulak, O., Cebi, S., and Kahraman, C. (2010). Applications of axiomatic design principles: A literature review. *Expert Systems with Applications*, 37(9), 6705–6717.

Kumar, P. S. and Hussain, R. J. (2014). A systematic approach for solving mixed intuitionistic fuzzy transportation problems. *International Journal of Pure and Applied Mathematics*, 92(2), 181–190.

Leung, L. C. and Cao, D. (2000). On consistency and ranking of alternatives in fuzzy AHP. *European Journal of Operational Research*, 124(1), 102–113.

Li, C., Sun, Y., and Du, Y. (2008, July). Selection of 3PL service suppliers using a fuzzy analytic network process. In *Control and Decision Conference, 2008. CCDC 2008*. China (pp. 2174–2179).

Li, D. F. (2011). Extension principles for interval-valued intuitionistic fuzzy sets and algebraic operations. *Fuzzy Optimization and Decision Making*, 10(1), 45–58.

Liao, H., Xu, Z., and Zeng, X. J. (2015). Hesitant fuzzy linguistic VIKOR method and its application in qualitative multiple criteria decision making. *IEEE Transactions on Fuzzy Systems*, 23(5), 1343–1355.

Mahapatra, G. S. and Roy, T. K. (2013). Intuitionistic fuzzy number and its arithmetic operation with application on system failure. *Journal of Uncertain Systems*, 7(2), 92–107.

Mikhailov, L. (2003). Deriving priorities from fuzzy pairwise comparison judgements. *Fuzzy Sets and Systems*, 134(3), 365–385.

Mousavi, S. M., Gitinavard, H., and Siadat, A. (2014, December). A new hesitant fuzzy analytical hierarchy process method for decision-making problems under uncertainty. In *2014 IEEE International Conference on Industrial Engineering and Engineering Management (IEEM)* (pp. 622–626).

Onar, S. Ç., Büyüközkan, G., Öztayşi, B., and Kahraman, C. (2016). A new hesitant fuzzy QFD approach: An application to computer workstation selection. *Applied Soft Computing*, 46, 1–16.

Onar, S. C., Oztaysi, B., Otay, İ., and Kahraman, C. (2015). Multi-expert wind energy technology selection using interval-valued intuitionistic fuzzy sets. *Energy*, 90, 274–285.

Öztaysi, B., Onar, S. Ç., Boltürk, E., and Kahraman, C. (2015, August). Hesitant fuzzy analytic hierarchy process. In *2015 IEEE International Conference on Fuzzy Systems (FUZZ-IEEE)* (pp. 1–7).

Ren, P., Xu, Z., and Liao, H. (2016). Intuitionistic multiplicative analytic hierarchy process in group decision making. *Computers and Industrial Engineering*, 101, 513–524.

Rodríguez, R. M., Martinez, L., and Herrera, F. (2012). Hesitant fuzzy linguistic term sets for decision making. *IEEE Transactions on Fuzzy Systems*, 20(1), 109–119.

Rodríguez, R. M., Martinez, L., and Herrera, F. (2013). A group decision making model dealing with comparative linguistic expressions based on hesitant fuzzy linguistic term sets. *Information Sciences*, 241, 28–42.

Saaty, T. L. (1980). *The Analytic Hierarchy Process*. New York, NY: McGraw Hill.

Saaty, T. L. and Vargas, L. G. (2001). How to make a decision. In *Models, Methods, Concepts & Applications of the Analytic Hierarchy Process* (pp. 1–21). New York, NY: Springer.

Sadiq, R. and Tesfamariam, S. (2009). Environmental decision-making under uncertainty using intuitionistic fuzzy analytic hierarchy process (IF-AHP). *Stochastic Environmental Research and Risk Assessment*, 23(1), 75–91.

Tavana, M., Zareinejad, M., Di Caprio, D., and Kaviani, M. A. (2016). An integrated intuitionistic fuzzy AHP and SWOT method for outsourcing reverse logistics. *Applied Soft Computing*, 40, 544–557.

Tolga, A. C., Tuysuz, F., and Kahraman, C. (2013). A fuzzy multicriteria decision analysis approach for retail location selection. *International Journal of Information Technology and Decision Making*, 12(4), 729–755.

Torra, V. (2010). Hesitant fuzzy sets. *International Journal of Intelligent Systems*, 25(6), 529–539.

Torra, V. and Narukawa, Y. (2009, August). On hesitant fuzzy sets and decision. In IEEE International Conference on Fuzzy Systems (FUZZ-IEEE) (pp. 1378–1382).

Tüysüz, F. and Çelikbilek, Y. (2016, May). Analysis of the factors for the evaluation of renewable energy resources using hesitant fuzzy linguistic term sets. In *The 8th International Ege Energy Syposium, Afyon Kocatepe University, Solar and Wind Research & Application Center, Turkey* (pp. 934–938).

Tüysüz, F. and Şimşek, B. (2016). Analysis of the factors affecting performance evaluation of branches in a cargo company by using hesitant fuzzy AHP. In *Uncertainty Modelling in Knowledge Engineering and Decision Making: Proceedings of the 12th International FLINS Conference, World Scientific Proceedings Series on Computer Engineering and Information Science* (pp. 866–871).

Van Laarhoven, P. J. M. and Pedrycz, W. (1983). A fuzzy extension of Saaty's priority theory. *Fuzzy Sets and Systems*, 11(1), 199–227.

Wang, T. C. and Chen, Y. H. (2008). Applying fuzzy linguistic preference relations to the improvement of consistency of fuzzy AHP. *Information Sciences*, 178(19), 3755–3765.

Wang, Y. M., Luo, Y., and Hua, Z. (2008). On the extent analysis method for fuzzy AHP and its applications. *European Journal of Operational Research*, 186(2), 735–747.

Wang, H., Qian, G., and Feng, X. (2011). An intuitionistic fuzzy AHP based on synthesis of eigenvectors and its application. *Information Technology Journal*, 10(10), 1850–1866.

Wang, J. Q. and Zhang, H. Y. (2013). Multicriteria decision-making approach based on Atanassov's intuitionistic fuzzy sets with incomplete certain information on weights. *IEEE Transactions on Fuzzy Systems*, 21(3), 510–515.

Wu, J., Huang, H. B., and Cao, Q. W. (2013). Research on AHP with interval-valued intuitionistic fuzzy sets and its application in multicriteria decision making problems. *Applied Mathematical Modelling*, 37(24), 9898–9906.

Xia, M. and Xu, Z. (2011). Hesitant fuzzy information aggregation in decision making. *International Journal of Approximate Reasoning*, 52(3), 395–407.

Xu, Z. and Liao, H. (2014). Intuitionistic fuzzy analytic hierarchy process. *IEEE Transactions on Fuzzy Systems*, 22(4), 749–761.

Xu, Z. and Xia, M. (2011). Distance and similarity measures for hesitant fuzzy sets. *Information Sciences*, 181(11), 2128–2138.

Yavuz, M., Oztaysi, B., Onar, S. C., and Kahraman, C. (2015). Multicriteria evaluation of alternative-fuel vehicles via a hierarchical hesitant fuzzy linguistic model. *Expert Systems with Applications*, 42(5), 2835–2848.

Yu, J. R. and Cheng, S. J. (2007). An integrated approach for deriving priorities in analytic network process. *European Journal of Operational Research*, 180(3), 1427–1432.

Zadeh, L. A. (1965). Fuzzy sets. *Information and Control*, 8(3), 338–353.

Zhang, C., Li, W., and Wang, L. (2011, July). AHP under the intuitionistic fuzzy environment. In 2011 8th International Conference on Fuzzy Systems and Knowledge Discovery (FSKD) (Vol. 1, pp. 583–587).

Zhou, W. and Xu, Z. (2016). Asymmetric hesitant fuzzy sigmoid preference relations in the analytic hierarchy process. *Information Sciences*, 358, 191–207.

Zhu, B. and Xu, Z. (2014). Analytic hierarchy process-hesitant group decision making. *European Journal of Operational Research*, 239(3), 794–801.

Zhu, B., Xu, Z., Zhang, R., and Hong, M. (2016). Hesitant analytic hierarchy process. *European Journal of Operational Research*, 250(2), 602–614.

6

An Integrated TOPSIS and FAHP

Antonio Rodríguez

CONTENTS

6.1 Introduction

The variety of proposals for multicriteria decision-making (MCDM) methods has increased in recent years. To solve various situations, trying to find the most appropriate techniques (Velásquez and Hester, 2013), new solutions are developed often as a combination of existing methods.

In many proposals, the fuzzy analytic hierarchy process (FAHP) appears in combination with other multicriteria decision methods (Kubler et al., 2016). These combined methods take advantage of the capability of FAHP to deal with vagueness and uncertainty assigning weights to parameters and/or evaluating alternatives. Depending on the approach, FAHP is used to select an option to discard the less significant parameters (or the less valuable options) or to rank parameters or options (Rodríguez et al., 2016).

Of all the methods accompanying FAHP in generating combined multicriteria decision methods, the technique for order of preference by similarity to ideal solution (TOPSIS) (Hwang and Yoon, 1981; Yoon, 1980) is, by far, the most frequently proposed method.

The idea behind TOPSIS is to select the option that best achieves a balance between two conditions: to be as close as possible to the positive-ideal solution and to be as far as possible from the negative-ideal solution. The positive-ideal solution represents the virtual best option that would have been composed by selecting the best performance for every parameter among the actual proposals; the negative-ideal solution represents the virtual worst option that would have been composed by selecting the worst performance for every parameter among the actual proposals.

TOPSIS can also be used to rank options in accordance to a unique variable that measures both their proximity to the virtual best option and their farness to the virtual worst option.

TOPSIS is often criticized for its inability to handle vague and uncertain problems (Yu et al., 2011); that is why some authors propose the use of fuzzy TOPSIS (Zyoud et al., 2016).

In this chapter, a mapping study of FAHP-TOPSIS proposals is performed to classify the approaches in accordance to their decision focus, objectives of application, and industry sectors. We describe the general structure of the classic FAHP-TOPSIS and we present a method that integrates FAHP, TOPSIS, and variable weights analysis (VWA), as an example of implementation. VWA permits the adaption of weights based on the values of parameters that describe options; this is especially useful when the procedures for the calculation of weights have to be set in advance. To demonstrate how this combined method works, we implemented a method for ranking offers in a public tender for the supply of customized equipment, presenting a numerical example.

The rest of this chapter continues as follows: Section 6.2 presents a literature review of the proposals for the combination of TOPSIS and FAHP; Section 6.3 presents a practical case of the combination of TOPSIS and FAHP by its application to a method for ranking offers in a public tender for the supply of customized equipment; Section 6.4 shows a numerical example of the practical case presented in the previous section; and Section 6.5 presents conclusions and proposes future lines of investigation.

6.2 Literature Review

To perform the literature review, we applied mapping study techniques (Kitchenham, 2010; Petersen et al., 2008), with search in titles (Dieste et al., 2009; Engström and Runeson, 2011) Scopus, ACM Digital Library, and IEEE Xplore, over a period of 8 years (2009–2016). After filtering articles that only presented simple comparisons or literature reviews, we found 278 documents proposing a combination of FAHP and, at least, another method or technique; 86 of them (30, 94%) propose combinations with TOPSIS.

Figure 6.1 shows the evolution of the quantity of documents proposing combinations of FAHP with other methods, for the most frequently cited of these methods: data envelopment analysis (DEA) (Kumar et al., 2015), Delphi (Lee and Seo, 2016), VIKOR ("Višekriterijska Optimizacija I Kompromisno

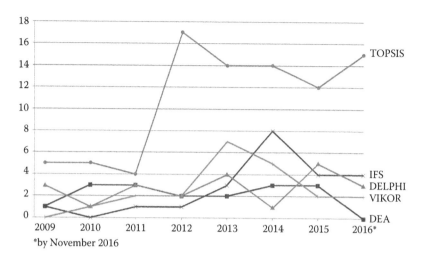

FIGURE 6.1
Number of articles reporting combinations of FAHP with other methods in their titles.

Rješenje", multicriteria optimization and compromise solution) (Ashtiani and Azgomi, 2016), intuitionistic fuzzy set (IFS) analysis (Rodríguez et al., 2017), and TOPSIS (Metaxas et al., 2016).

For FAHP-TOPSIS combinations, the search was widened with documents published before 2009, finding three more documents (Imanipour and Osati, 2007; Wang and Liu, 2007; Wang et al., 2008). To include articles on FAHP-TOPSIS combinations in which the methods are not mentioned in the title, we performed a "forward snowballing search" (Bakker, 2010; Wohlin, 2014), parting from the most frequently cited articles (Gumus, 2009; Ertuğrul and Karakaşoğlu, 2009; Sun, 2010) to find 55 additional papers for a total of 144 documents on FAHP-TOPSIS combination.

Figure 6.2 shows the map of proposals on FAHP-TOPSIS combinations.

For the evaluation and comparison of different options, TOPSIS selects, among the actual alternatives, the best performance for every criterion to construct a positive-ideal solution and the worst performance for every criterion to construct a negative-ideal solution. Then, for every option, the separation from the positive-ideal solution and the separation from the negative-ideal solution are calculated. Options are ranked according to their closeness to the positive-ideal solution and the farness from the negative-ideal solution. For the implementation of these comparisons, a weight is assigned to every criterion.

In the classic implementation of the FAHP-TOPSIS combination, FAHP is generally used to implement a criteria hierarchy to analyze the decision problem and assign weights to every criterion (Figure 6.3).

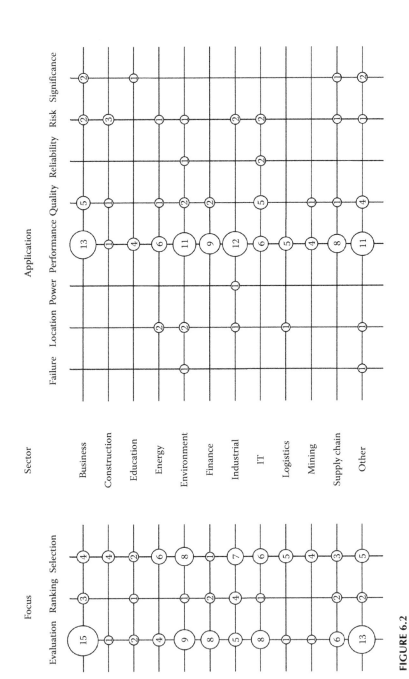

FIGURE 6.2

Map of articles on FAHP technique for order of preference by similarity to ideal solution (FAHP-TOPSIS) combinations.

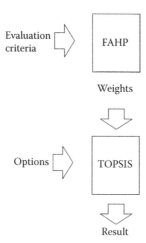

FIGURE 6.3
Classic implementation of the FAHP-TOPSIS combination.

6.3 Implementation of an FAHP-VWA-TOPSIS Method

To show how the FAHP-TOPSIS combination works, we implemented a method for ranking offers in a public tender for the supply of customized equipment; for this purpose, we adapted results from previous research (Rodríguez et al., 2013; Rodríguez, 2015) and integrated a procedure for the compensation of weights: VWA.

VWA (Li et al., 2004; Zeng et al., 2016) is a method that implements a procedure for the adaption of weights based on a state variable weight vector (Li and Li, 2004; Li and Hao, 2009).

Once the necessity of the supply has been determined, the acquisition procedure usually contemplates the steps of a bidding process: draft of the technical specifications; generation of the request for proposal (RFP) including technical, logistic, economic, and legal requisites; publication of the RFP; reception of offers by the deadline detailed in the RFP; discard of those proposals not fulfilling one or more of the compulsory requisites stated in the RFP; analysis and evaluation of the remaining proposals and selection of the winning proposal; and finally, writing of the supply contract in accordance to the RFP.

A set of different types of features has to be considered as input variables of the evaluation process: binary variables (like the presence or absence of desirable characteristics), numerical variables (like physical parameters specifications), or linguistic variables (qualitative evaluations).

TOPSIS transforms scales and units into common measurable units to allow comparisons across the criteria. TOPSIS takes, from the options under analysis, the best and the worst value from each characteristic to compose a theoretical positive-ideal solution and a theoretical negative-ideal solution. The best real solution will be in a balanced position as far as possible from the negative-ideal solution and as close as possible to the positive-ideal solution.

Weights for every criterion were calculated using FAHP, assigning values according to the importance of criteria for the evaluation.

Nevertheless, even with a good knowledge of the required technology, it is difficult to foresee the variability of some specifications among offers (e.g., some important characteristics may be equally specified by all of the potential suppliers). This may result in an unbalanced assignation of weights to the evaluation criteria, conferring high weights to nonsignificant criteria with the consequent loss of resolution in the evaluation, and, in some cases, the selection of an option that is not necessarily the best. The direct solution to this problem consists of the modification of weights during the evaluation process but, in some cases, this is not acceptable if, according to the legislation, the scoring rules and weights must be set in advance. That is why VWA was introduced as a way to automatically emphasize significant criteria (Figure 6.4).

The steps of the proposed method are as follows:

Step 1: The characteristics to evaluate the options are selected and organized in a hierarchy. If a characteristic shows the same value for all the proposals under evaluation, then that characteristic (no matter how important it is) should be discarded as an evaluation criterion.

Step 2: A weight is assigned to every criterion. Weights are obtained by FAHP from pairwise comparison of the importance of the evaluation criteria.

A fuzzy triangular number is assigned to every comparison:

$$\tilde{a}_{ij} = \left(l_{ij}, m_{ij}, u_{ij}\right), \frac{1}{9} \leq m_{ij} \leq 9 \tag{6.1}$$

For $m_{ij} \geq 1$:

$$l_{ij} = \begin{cases} m_{ij} - \dfrac{d}{2}, & m_{ij} - \dfrac{d}{2} \geq 1 \\[4mm] 1, & m_{ij} - \dfrac{d}{2} < 1 \end{cases}, \ 0 \leq d \leq 8 \tag{6.2}$$

$$u_{ij} = \begin{cases} m_{ij} + \dfrac{d}{2}, & m_{ij} + \dfrac{d}{2} \leq 9 \\[4mm] 9, & m_{ij} + \dfrac{d}{2} > 9 \end{cases}, \ 0 \leq d \leq 8 \tag{6.3}$$

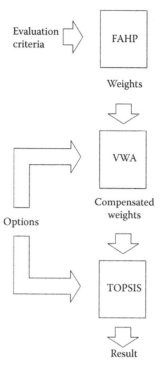

FIGURE 6.4
Implementation of the proposed FAHP–variable weights analysis–technique for order of preference by similarity to ideal solution (FAHP-VWA-TOPSIS) method.

$$\tilde{a}_{ji} = \frac{1}{\tilde{a}_{ij}} = \left(\frac{1}{u_{ij}}, \frac{1}{m_{ij}}, \frac{1}{l_{ij}} \right) \tag{6.4}$$

$$\text{For } i = j, \tilde{a}_{ii} = (1,1,1) \tag{6.5}$$

In contrast to the typical FAHP implementation with triangular numbers, m_{ij} is not a fixed value, related to a fixed width triangular number, associated to a linguistic variable. In the proposed method, m_{ij} is a flexible value that measures how important criterion i is in relation to criterion j (criterion i is considered to be m_{ij} times more important than criterion j).

When criterion i is considered to be less important than criterion j, m_{ij} will be set to a value lesser than 1: $0 < m_{ij} < 1$.

When both criteria are considered equally important, m_{ij} will be set to a value equal to 1: $m_{ij} = 1$.

The fuzzy triangular number associated to m_{ij} has a flexible width d, a dispersion value that measures the uncertainty or lack of confidence in the value assigned to m_{ij}. The higher the level of confidence in the value assigned to m_{ij}, the lower the value of d.

To facilitate the consistency preservation, the dispersion value must be constant within the same group and level in the hierarchy, so that there will be only one value of d for every comparison matrix.

To ensure the consistency of the assigned values, a graphical procedure is proposed. An example of the application of this procedure is shown in Figure 6.5, in which the following conditions must be fulfilled:

$$m_{ij} = m_{i(j-1)} \times m_{(j-1)j} , i \geq 1; j \geq i+2 \qquad (6.6)$$

$$m_{ji} = \frac{1}{m_{ij}} \qquad (6.7)$$

All the comparisons for a level in the hierarchy are assembled in a comparison matrix:

$$\tilde{A} = \begin{bmatrix} \tilde{a}_{11} & \tilde{a}_{12} & \cdots & \tilde{a}_{1N} \\ \tilde{a}_{21} & \tilde{a}_{12} & \cdots & \tilde{a}_{2N} \\ \vdots & \vdots & \cdots & \vdots \\ \tilde{a}_{N1} & \tilde{a}_{N2} & \cdots & \tilde{a}_{NN} \end{bmatrix} \qquad (6.8)$$

The strict application of the proposed graphical procedure makes unnecessary any further consistency verification. Nevertheless, small inconsistencies may be permitted, in which case it would be necessary to implement the verification procedure usually applied in classic AHP analysis (Chan and Kumar, 2007; Kwong and Bai, 2003; Nurcahyo et al., 2003).

Fuzzy weights $\tilde{w}$ for every criterion are obtained from comparison fuzzy values $\tilde{a}$:

$$\tilde{a}_{ij} = \frac{\tilde{w}_i}{\tilde{w}_j} \qquad (6.9)$$

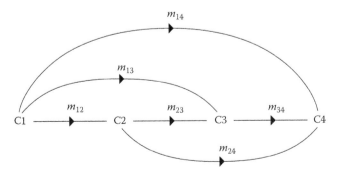

FIGURE 6.5
Graphical assignation of values to pairwise comparisons.

$$\tilde{w}_i = \frac{\sum_{j=1}^{n} l_{ij}}{\sum_{j=1}^{n}\sum_{i=1}^{n} u_{ij}}, \frac{\sum_{j=1}^{n} m_{ij}}{\sum_{j=1}^{n}\sum_{i=1}^{n} m_{ij}}, \frac{\sum_{j=1}^{n} u_{ij}}{\sum_{j=1}^{n}\sum_{i=1}^{n} l_{ij}} \qquad (6.10)$$

From the comparison matrix, a crisp weight is calculated:

$$w'_i = \frac{1}{3}\left(\frac{\sum_{j=1}^{n} l_{ij}}{\sum_{j=1}^{n}\sum_{i=1}^{n} u_{ij}} + \frac{\sum_{j=1}^{n} m_{ij}}{\sum_{j=1}^{n}\sum_{i=1}^{n} m_{ij}} + \frac{\sum_{j=1}^{n} u_{ij}}{\sum_{j=1}^{n}\sum_{i=1}^{n} l_{ij}} \right) \qquad (6.11)$$

It is then normalized:

$$w_i = \frac{w'_i}{\sum_{n=1}^{N} w'_n \sum_{n=1}^{N} w'_n} \qquad (6.12)$$

w_i values are no longer triangular fuzzy numbers, but numerical values represented by the column vector:

$$w = (w_1, w_2, ..., w_n)^T \qquad (6.13)$$

When, in the hierarchy, a criterion is made up of several subcriteria, the total weight of each final subcriterion w will be obtained from the partial weights W corresponding to every level in its line in the hierarchy:

$$W_{klm} = w_k \times w_{kl} \times w_{klm} \qquad (6.14)$$

Step 3: A value is assigned to each criterion for each option. An array of values for each option and criterion is obtained (Table 6.1).

For quantitative parameters, the corresponding value is assigned; if for each criterion, all the option values are given in the same units of measurement, no transformation is needed and the specified value of option i for quantitative criterion j will be directly assigned to x_{ij}.

For binary variables, logic "1" is assigned if the desirable feature is available, and logic "0" if it is unavailable. If option i fulfills binary criterion j, then $x_{ij} = 1$. Otherwise, $x_{ij} = 0$ (option i does not fulfill binary criterion j).

For qualitative variables, the evaluator assigns a fuzzy triangular number to option i for criterion j:

$$\tilde{x}_{ij} = (a_{ij}, b_{ij}, c_{ij}), 1 \le b_{ij} \le 5 \qquad (6.15)$$

TABLE 6.1

Array of Technique for Order of Preference by Similarity to Ideal Solution (TOPSIS) Values

Options			Criteria		
A_1	x_{11}	x_{12}	x_{13}	$\cdots$	x_{1M}
A_2	x_{21}	x_{22}	x_{23}	$\cdots$	x_{2M}
A_3	x_{31}	x_{32}	x_{33}	$\cdots$	x_{3M}
$\vdots$	$\vdots$	$\vdots$	$\vdots$	$\vdots$	$\vdots$
A_N	x_{N1}	x_{N2}	x_{M3}	$\cdots$	x_{NM}

$$a_{ij} = \begin{cases} b_{ij} - \dfrac{e_{ij}}{2}, & b_{ij} - \dfrac{e_{ij}}{2} \geq 1 \\ 1, & b_{ij} - \dfrac{e_{ij}}{2} < 1 \end{cases}, \ 0 \leq e_{ij} \leq 4 \qquad (6.16)$$

$$c_{ij} = \begin{cases} b_{ij} + \dfrac{e_{ij}}{2}, & b_{ij} + \dfrac{e_{ij}}{2} \leq 5 \\ 5, & b_{ij} + \dfrac{e_{ij}}{2} > 5 \end{cases}, \ 0 \leq e_{ij} \leq 4 \qquad (6.17)$$

where b_{ij} evaluates option i for criterion j. Values of b correspond to the linguistic values expressed in Table 6.2. Intermediate b values are allowed; e is a measure of the uncertainty or the lack of confidence in the value assigned to b; the value of e may vary from option to option for the same criterion.

Finally, for every qualitative variable, a quantitative crisp value x is assigned from the media of the assigned fuzzy triangular number:

$$x = \frac{a+b+c}{3} \qquad (6.18)$$

TABLE 6.2

Linguistic Variables

b	Linguistic Values
5	Very high
4	High
3	Medium
2	Low
1	Very low

Step 4: A new compensated total weight is calculated for every final subcriterion j:

$$W_j' = \frac{S(x_j)W_j}{\sum_{j=1}^{M} S(x_j)W_j} \tag{6.19}$$

$S(x)$ defines the state variable weight vector and w_j is the total weight calculated for subcriterion j in Step 2. We implement an exponent type incentive state variable weight vector:

$$S(x_j) = e^{\alpha \frac{\sigma_j}{|\overline{x_j}|}}; \alpha \geq 0 \tag{6.20}$$

α is the varying level of weights and σ is the dispersion calculated by the well-known standard deviation formula:

$$\sigma_j = \left[\frac{1}{N} \sum_{i=1}^{N} (x_{ij} - \overline{x_j})^2 \right]^{0.5} \tag{6.21}$$

N is the total number of valid options and $|\overline{x_j}|$ is the absolute value of the mean:

$$|\overline{x_j}| = \left| \frac{\sum_{i=1}^{N} x_{ij}}{N} \right| \tag{6.22}$$

When $\alpha = 0$, weights are not compensated ($W_n' = W_l$) and the procedure implements the classic approach.

Step 5: A normalized matrix is obtained. Normalized values are obtained by the application of the following formula to every column:

$$r_{ij} = \frac{x_{ij}}{\left(\sum_{i=1}^{M} (x_{ij})^2 \right)^{\frac{1}{2}}} \tag{6.23}$$

Each column of the normalized matrix is multiplied by a weight W_j', which corresponds to the value W' obtained for every final subcriterion in Step 4:

$$v_{ij} = W_j' r_{ij} \tag{6.24}$$

Thus, a normalized weighted matrix is generated:

$$[V] = \begin{bmatrix} v_{11} & v_{12} & v_{13} & \cdots & v_{1M} \\ v_{21} & v_{22} & v_{23} & \cdots & v_{2M} \\ v_{31} & v_{32} & v_{33} & \cdots & v_{3M} \\ \vdots & \vdots & \vdots & \vdots & \vdots \\ v_{N1} & v_{N2} & v_{N3} & \cdots & v_{NM} \end{bmatrix} \tag{6.25}$$

Step 6: The positive-ideal solution (A^+) and negative-ideal solution (A^-) are determined according to benefit criteria (J^+) and costs criteria (J^-), using the following formulas:

$$A^+ = \left\{ \left(\max_i v_{ij} / j \in J^+ \right), \left(\min_i v_{ij} / j \in J^- \right) \right\} \tag{6.26}$$

$$A^+ = \left\{ \left(\min_i v_{ij} / j \in J^+ \right), \left(\max_i v_{ij} / j \in J^- \right) \right\} \tag{6.27}$$

Step 7: For each option, its separation from the positive-ideal and negative-ideal values, using the following formulas, is estimated:

$$S_i^+ = \left[\sum_{j=1}^{N} (v_j^+ - v_{ij})^2 \right]^{\frac{1}{2}} \tag{6.28}$$

$$S_i^- = \left[\sum_{j=1}^{N} (v_j^- - v_{ij})^2 \right]^{\frac{1}{2}} \tag{6.29}$$

Step 8: For each option, the relative closeness to a theoretical best solution is calculated by the following formula:

$$R_i = \frac{S_i^-}{S_i^- + S_i^+} \tag{6.30}$$

Step 9: For every option, a proportional value is assigned in relation to the actual best solution:

$$V_i = \frac{R_i}{\max(R_i)} \tag{6.31}$$

Finally, we suggest the verification of the results to detect possible problems of rank reversal. Rank reversal is a known problem of evaluation methods consisting of an inversion in the position in the classification when an option is suppressed or a new one is added. We propose the use of a graphical scale, similar to the one presented by Huszák and Imre (2010).

6.4 Numerical Example

To demonstrate the application of the proposed method, we implemented it for the evaluation of offers for the supply of a display.

Step 1: After eliminating those characteristics equally valued for all the options, we obtained the following hierarchy:

C.1.   Image

C.1.1. Resolution

C.1.2. Horizontal angle

C.1.3. Vertical angle

C.2.   Price

C.3.   Environment

C.3.1. Maximum operating temperature

C.3.2. Minimum operating temperature

C.3.3. Minimum storage temperature

C.3.4. Maximum storage temperature

C.4.   Other

C.4.1. Connection detection

C.4.2. Switch off from host

C.4.3. Easy installation

Step 2: Weights are calculated for every criterion and subcriterion.

Table 6.3 shows comparisons and weights for criteria in the higher level of the hierarchy. For all the comparisons, we selected $d = 2$.

Table 6.4 shows comparisons and weights for C_1 subcriteria. For all the comparisons, we selected $d = 0$, which is equivalent to the implementation of a crisp AHP.

Table 6.5 shows comparisons and weights for C_3 subcriteria. For all the comparisons, we selected $d = 1$.

We considered the C_4 subcriteria to have the same weight so all of them were weighted equally:

$$w_{41} = w_{42} = w_{43} = 0.3333 \tag{6.32}$$

TABLE 6.3

Pairwise Comparison Matrix and Weights for the Higher Level Criteria

	C_1	C_2	C_3	C_4	Weights
C_1	(1, 1, 1)	(1, 2, 3)	(1, 2, 3)	(3, 4, 5)	$w_1 = 0.4300$
C_2	(0.33, 0.5, 1)	(1, 1, 1)	(1, 1, 2)	(1, 2, 3)	$w_2 = 0.2386$
C_3	(0.33, 0.5, 1)	(0.5, 1, 1)	(1, 1, 1)	(1, 2, 3)	$w_3 = 0.2134$
C_4	(0.2, 0.25, 0.33)	(0.33, 0.5, 1)	(0.33, 0.5, 1)	(1, 1, 1)	$w_4 = 0.1180$

TABLE 6.4

Pairwise Comparison Matrix and Weights for C_1 Subcriteria ($d = 0$; Crisp AHP)

	C_{11}	C_{12}	C_{13}	Weights
C_{11}	(1, 1, 1)	(1.5, 1.5, 1.5)	(3, 3, 3)	$w_{11} = 0.5000$
C_{12}	(0.67, 0.67, 0.67)	(1, 1, 1)	(2, 2, 2)	$w_{12} = 0.3333$
C_{13}	(0.33, 0.33, 0.33)	(0.5, 0.5, 0.5)	(1, 1, 1)	$w_{13} = 0.1667$

TABLE 6.5

Pairwise Comparison Matrix and Weights for C_3 Subcriteria

	C_{31}	C_{32}	C_{33}	C_{34}	Weights
C_{31}	(1, 1, 1)	(1.5, 2, 2.5)	(3.5, 4, 4.5)	(3.5, 4, 4.5)	$w_{31} = 0.4963$
C_{32}	(0.4, 0.5, 0.67)	(1, 1, 1)	(1.5, 2, 2.5)	(1.5, 2, 2.5)	$w_{32} = 0.2509$
C_{33}	(0.22, 0.25, 0.29)	(0.4, 0.5, 0.67)	(1, 1, 1)	(1, 1, 1.5)	$w_{33} = 0.1328$
C_{34}	(0.22, 0.25, 0.29)	(0.4, 0.5, 0.67)	(0.67, 1, 1)	(1, 1, 1)	$w_{34} = 0.1200$

Table 6.6 shows the preliminary global weight for every final criterion.

Step 3: We built the options-criteria array.

The fuzzy triangular values assigned to the only linguistic charac-teristic C_{43} are shown in Table 6.7 (for all the cases we selected $e = 2$).

Step 4: We calculated the compensated weights W_j'.

We used $\alpha = 1.5$. Table 6.9 shows the calculation of the new compen-sated weights:

TABLE 6.6

Preliminary Global Weight for Every Final Criterion

	$W_{11} = w_1 \times w_{11}$	0.2150
w_1	$W_{12} = w_1 \times w_{12}$	0.1433
	$W_{13} = w_1 \times w_{13}$	0.0717
w_2	$W_2 = w_2$	0.2386
	$W_{31} = w_3 \times w_{31}$	0.1059
	$W_{32} = w_3 \times w_{32}$	0.0535
w_3	$W_{33} = w_3 \times w_{33}$	0.0283
	$W_{34} = w_3 \times w_{34}$	0.0256
	$W_{41} = w_4 \times w_{41}$	0.0393
w_4	$W_{42} = w_4 \times w_{42}$	0.0393
	$W_{43} = w_4 \times w_{43}$	0.0393

TABLE 6.7

Fuzzy Triangular Numbers Assigned to Qualitative Variable C_{43} and Crisp Value x Obtained

Offer	C_{43}	x
O_1	$(4,5,5)$	4.67
O_2	$(3,4,5)$	4.00
O_3	$(2,3,4)$	3.00
O_4	$(3,4,5)$	4.00

TABLE 6.8

Values Assigned to Every Characteristic

Offer	C_{11}	C_{12}	C_{13}	C_2	C_{31}	C_{32}	C_{33}	C_{34}	C_{41}	C_{42}	C_{43}
O_1	786,432	160	140	70.80	45	5	−10	50	1	1	4.67
O_2	786,432	160	160	73.95	50	0	−20	60	0	0	4.00
O_3	480,000	160	140	72.40	45	5	−10	50	1	1	3.00
O_4	1,310,720	170	160	71.05	45	5	−10	60	1	1	4.00

Steps 5 and 6: We calculated the normalized weighted matrix [V], the positive-ideal solution A^+ and the negative-ideal solution A^-, considering that in some cases lower was better ($\downarrow$) (price for example) and, in other cases, the higher was better ($\uparrow$) (resolution for example) (Table 6.10).

TABLE 6.9

Calculus of Compensated Weights for Criteria for $\alpha = 1.5$

	C_{11}	C_{12}	C_{13}	C_2	C_{31}	C_{32}	C_{33}	C_{34}	C_{41}	C_{42}	C_{43}
$\|\overline{x}_j\|$	840,896.00	162.50	150.00	72.05	46.25	3.75	12.50	55.00	0.75	0.75	3.92
σ_j	298,711.05	4.33	10.00	1.25	2.17	2.17	4.33	5.00	0.43	0.43	0.60
$S(x_j)$	1.70	1.04	1.11	1.03	1.07	2.38	1.68	1.15	2.38	2.38	1.26
W_j'	0.263	0.107	0.057	0.176	0.081	0.091	0.034	0.021	0.067	0.067	0.035

TABLE 6.10

Normalized Weighted Matrix $[V]$, Positive-Ideal Solution A^+, and Negative-Ideal Solution A^-

	C_{11}	C_{12}	C_{13}	C_2	C_{31}	C_{32}	C_{33}	C_{34}	C_{41}	C_{42}	C_{43}
	↑	↑	↑	↓	↑	↓	↓	↑	↑	↑	↑
O_1	0.1158	0.0527	0.0265	0.0863	0.0396	0.0527	−0.0129	0.0095	0.0387	0.0387	0.0209
O_2	0.1158	0.0527	0.0302	0.0902	0.0440	0.0000	−0.0258	0.0114	0.0000	0.0000	0.0179
O_3	0.0707	0.0527	0.0265	0.0883	0.0396	0.0527	−0.0129	0.0095	0.0387	0.0387	0.0134
O_4	0.1930	0.0560	0.0302	0.0866	0.0396	0.0527	−0.0129	0.0114	0.0387	0.0387	0.0179
A^+	0.1930	0.0560	0.0302	0.0863	0.0440	0.0000	−0.0258	0.0114	0.0387	0.0387	0.0209
A^-	0.0707	0.0527	0.0265	0.0902	0.0396	0.0527	−0.0129	0.0095	0.0000	0.0000	0.0134

TABLE 6.11

Separations, Relative Closeness, and Proportional Value for Every Option

	S^+	S^-	R	V
O_1	0.0946	0.0714	0.4302	0.6049
O_2	0.0948	0.0710	0.4281	0.6020
O_3	0.1342	0.0548	0.2898	0.4075
O_4	0.0545	0.1343	0.7112	1.0000

Steps 7–9: We calculated, for every option, the separation (S^+) to the positive-ideal solution A^+, the separation (S^-) to the negative-ideal solution A^-, the relative closeness (R), and the proportional value (V) (Table 6.11).

6.4.1 Analysis of Results

Figure 6.6 shows an example of the effect of the variation in the width of the fuzzy triangular numbers on calculation of the weights. The central value

m_{ij} of every subcriterion in group C_1 (Table 6.4) remains constant. The graph shows the partial weights w that were obtained versus the values of d.

Figures 6.7 and 6.8 show the analysis of rank reversal. In every case, an option, or a pair of options, was suppressed. No rank reversal was detected.

Nevertheless, for higher values of α (meaning an increase in weight compensation), a case of rank reversal was observed; Figure 6.9 shows the case for $\alpha = 2.5$.

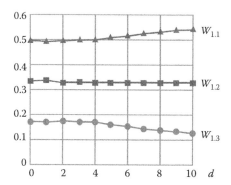

FIGURE 6.6
Effect of fuzzy values widths assigned to comparisons on calculation of weights.

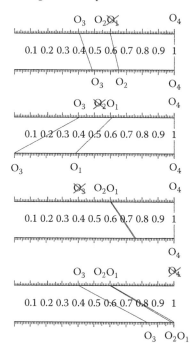

FIGURE 6.7
Verification of rank reversal suppressing one option.

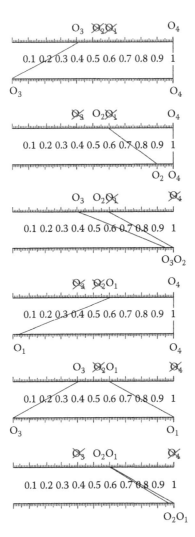

FIGURE 6.8
Verification of rank reversal suppressing two options.

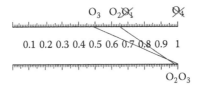

FIGURE 6.9
Case of rank reversal for $\alpha = 2.5$.

6.5 Conclusions

Many authors have proposed the use of FAHP in combination with other methods. Of all of them, TOPSIS is the most frequently proposed method to accompany FAHP. The FAHP-TOPSIS combination inherits the advantages of TOPSIS: it allows the use of evaluation parameters with varied scales and units, including negative values.

The integration of VWA for the adjustment of weights incorporates their compensation as part of the automatic evaluation process. The assigned weights of the evaluation criteria are automatically adjusted in relation to the dispersion of values for every criterion in the proposals, avoiding the loss of resolution in the evaluation when its criteria have close values.

The graphical procedure for pairwise comparison allows an easier implementation of FAHP that verifies the consistency, suppresses the necessity of mathematical checking, and facilitates the pairwise comparison iteration. The use of numerical values (instead of linguistic variables) to assess how many times one factor is more important than the rest of the factors provides flexibility and facilitates the consistency in FAHP. The introduction of adjustable widths in fuzzy triangular numbers permits taking into consideration the different levels of uncertainty in the calculation of weights by FAHP.

This research can open several future lines of work. A more in-depth investigation of the behavior of the exponential formula used to implement the VWA in the compensation of weights, and of the selection of the varying level of weights factor, should be performed. Further research can also propose different compensation formulas. Additional investigation should also be undertaken to analyze the consequences of a detected drawback: in some cases, the use of VWA may increase the risk of rank reversal. We also propose considering the use of fuzzy TOPSIS, instead of classic TOPSIS, to improve the management of uncertainty in the evaluation of some particular criteria.

References

Ashtiani, M. and Azgomi, M. A. (2016). Trust modeling based on a combination of fuzzy analytic hierarchy process and fuzzy VIKOR. *Soft Computing*, 20(1), 399–421.

Bakker, R. M. (2010). Taking stock of temporary organizational forms: A systematic review and research agenda. *International Journal of Management Reviews*, 12(4), 466–486.

Chan, F. T. S. and Kumar, N. (2007). Global supplier development considering risk factors using fuzzy extended AHP-based approach. *Omega*, 35(4), 417–431.

Dieste, O., Grimán, A., and Juristo, N. (2009). Developing search strategies for detecting relevant experiments. *Empirical Software Engineering*, 14(5), 513–539.

Engström, E. and Runeson, P. (2011). Software product line testing—A systematic mapping study. *Information and Software Technology*, 53(1), 2–13.

Ertuğrul, I. and Karakaşoğlu, N. (2009). Performance evaluation of Turkish cement firms with fuzzy analytic hierarchy process and TOPSIS methods. *Expert Systems with Applications*, 36(1), 702–715.

Gumus, A. T. (2009). Evaluation of hazardous waste transportation firms by using a two step fuzzy-AHP and TOPSIS methodology. *Expert Systems with Applications*, 36(2 Part 2), 4067–4074.

Huszák, Á. and Imre, S. (2010). Eliminating rank reversal phenomenon in GRA-based network selection method. In IEEE ICC 2010 Proceedings. IEEE International Conference on Communications, Cape Town, South Africa, 23–27 May 2010.

Hwang, C. L. and Yoon, K. (1981). Multiple Attribute Decision Making: Methods and Applications. New York, NY: Springer-Verlag.

Imanipour, N. and Osati, M. (2007). Ranking internal combustion engines using fuzzy AHP and TOPSIS. *37th International Conference on Computers and Industrial Engineering 2007*, Alexandria, Egypt, 2, 1231–1238.

Kitchenham, B. (2010). What's up with software metrics?—A preliminary mapping study. Journal of Systems *and Software*, 83(1), 37–51.

Kubler, S., Robert, J., Derigent, W., Voisin, A., and Le Traon, Y. (2016). A state-of the-art survey & testbed of fuzzy AHP (FAHP) applications. *Expert Systems with Applications*, 65, 398–422.

Kumar, A., Shankar, R., and Debnath, R. M. (2015). Analyzing customer preference and measuring relative efficiency in telecom sector: A hybrid fuzzy AHP/DEA study. *Telematics and Informatics*, 32(3), 447–462.

Kwong, C. K. and Bai, H. (2003). Determining the importance weights for the customer requirements in QFD using a fuzzy AHP with an extent analysis approach. *IIE Transactions*, 35(7), 619–626.

Lee, S. and Seo, K.-K. (2016). A hybrid multi-criteria decision-making model for a cloud service selection problem using BSC, fuzzy delphi method and fuzzy AHP. *Wireless Personal Communications*, 86(1), 57–75.

Li, D. Q. and Hao, F. L. (2009). Weights transferring effect of state variable weight vector. *Systems Engineering—Theory and Practice*, 29(6), 127–131.

Li, D. Q. and Li, H. X. (2004). Analysis of variable weights effect and selection of appropriate state variable weights vector in decision making. *Control and Decision*, 11, 1241–1245.

Li, H. X., Li, L. X., Wang, J. Y., Mo, Z. W., and Li, Y. D. (2004). Fuzzy decision making based on variable weights. *Mathematical and Computer Modelling*, 39(2), 163–179.

Metaxas, I. N., Koulouriotis, D. E., and Spartalis, S. H. (2016). A multicriteria model on calculating the Sustainable Business Excellence Index of a firm with fuzzy AHP and TOPSIS. *Benchmarking*, 23(6), 1522–1557.

Nurcahyo, G. W., Shamsuddin, S. M., Alias, R. A., and Noor, M. (2003). Selection of defuzzification method to obtain crisp value for representing uncertain data in a modified sweep algorithm. *Journal of Computer Science and Technology*, 3(2), 22–28.

Petersen, K., Feldt, R., Mujtaba, S., and Mattsson, M. 2008. Systematic mapping studies in software engineering, 12th International Conference on Evaluation and Assessment in Software Engineering (EASE). University of Bari, Italy, 26–27 June 2008.

Rodríguez, A. (2015). A fuzzy method for the selection of customized equipment suppliers in the public sector. *Artificial Intelligence Research*, 4(1), 36–44.

Rodríguez, A., Ortega, F., and Concepción, R. (2013). A method for the selection of customized equipment suppliers. *Expert Systems with Applications*, 40(4), 1170–1176.

Rodríguez, A., Ortega, F., and Concepción, R. (2016). A method for the evaluation of risk in IT projects. *Expert Systems with Applications*, 45, 273–285.

Rodríguez, A., Ortega, F., and Concepción, R. (2017). An intuitionistic method for the selection of a risk management approach to information technology projects. *Information Sciences*, 375, 202–218.

Sun, C.-C. (2010). A performance evaluation model by integrating fuzzy AHP and fuzzy TOPSIS methods. Expert Systems with Applications, 37(12), 7745–7754.

Velásquez, M. and Hester, P. T. (2013). An analysis of multi-criteria decision making methods. International Journal of Operations Research, 10(2), 56–66.

Wang, J., Fan, K., Su, Y., Liang, S., and Wang, W. (2008). Air combat effectiveness assessment of military aircraft using a fuzzy AHP and TOPSIS methodology. 2008 Asia Simulation Conference—7th International Conference on System Simulation and Scientific Computing, ICSC 2008, Beijing, China, 4675442, 655–662.

Wang, S. and Liu, P. (2007). The evaluation study on location selection of logistics center based on fuzzy AHP and TOPSIS. 2007 International Conference on Wireless Communications, Networking and Mobile Computing, WiCOM 2007, Shanghai, China, 4340710, 3774–3777.

Wohlin, C. (2014). Guidelines for snowballing in systematic literature studies and a replication in software engineering. In Proceedings of the 18th International Conference on Evaluation and Assessment in Software Engineering, London, England, 38.

Yoon, K. (1980). Systems selection by multiple attribute decision making, PhD dissertation. Kansas State University, Manhattan, Kansas.

Yu, X., Guo, S., Guo, J., and Huang, X. (2011). Rank B2C e-commerce websites in e-alliance based on AHP and fuzzy TOPSIS. Expert Systems with Applications, 38(4), 3550–3557.

Zeng, W., Li, D., and Wang, P. (2016). Variable weight decision making and balance function analysis based on factor space. International Journal of Information Technology & Decision Making, 15, 999.

Zyoud, S. H., Kaufmann, L. G., Shaheen, H., Samhan, S., and Fuchs-Hanusch, D. (2016). A framework for water loss management in developing countries under fuzzy environment: Integration of fuzzy AHP with fuzzy TOPSIS. *Expert Systems with Applications*, 61, 86–105.

7

An Integrated Fuzzy Delphi-AHP-TOPSIS with an Application to Logistics Service Quality Evaluation Using a Multistakeholder Multiperspective Approach

Anjali Awasthi and Hassan Mukhtar

CONTENTS

7.1 Introduction

According to the Council of Logistics Management, "Logistics can be defined as that part of the supply chain process that plans, implements, and controls the flow and storage of goods, services, and related information from the point of origin to the point of consumption in order to meet customers requirements." The origin of logistics service quality can be traced back to Perreault and Russ (1974) who maintained that logistics activities create time, place, and form utility, thereby enhancing product value (Bienstock et al., 2008). Two main viewpoints (Bienstock et al., 1997; Mentzer et al., 2001) have been commonly adopted for measuring quality. The first viewpoint, based on service providers, considers quality as adherence to specifications; this concept sees service as a physical object that is observable and with attributes that can be evaluated. The second viewpoint, based on customers, measures quality as the difference between the customers' perceptions and expectations for the service quality parameters. This thought is in line with Mentzer et al. (1989) who asserted that logistics service quality consisted of "customer service quality" and "physical distribution service quality." With growing awareness on environmental concerns in recent years, service quality can no longer be managed from the point of view of direct stakeholders; indirect stakeholders such as municipal administrators, traffic managers, residents, etc., should also be involved.

The logistics service quality problem has been investigated by several researchers, yet very few works address the point of view of all stakeholders in service quality evaluation. Some consider only the point of view of service providers, while some others consider the customer point of view. Our goal in this chapter is to consider the point of view of multiple stakeholders (e.g., shippers, carriers, 3PLs (party logistics), customers, municipal operators, traffic managers, city residents) in logistics service quality evaluation.

The rest of this chapter is organized as follows. Sections 7.2 and 7.3 present the literature review and the basics of the fuzzy Delphi (FDELPHI), fuzzy analytic hierarchy process (FAHP), and fuzzy technique for order of preference by similarity to ideal solution (TOPSIS) methods. Section 7.4 contains the solution approach. A numerical application is provided in Section 7.5. Finally, conclusions are drawn and future works are discussed in Section 7.6.

7.2 Literature Review

In the literature, a variety of methods have been used for logistics service quality evaluation. The most common ones are SERVQUAL, quality function deployment (QFD), fuzzy logic, Kano's model, AHP, Kansei engineering, the loss aversion method, the gray correlation method, goal programming, game theory, vague sets theory, and gap analysis. Aima et al. (2015) and Roslan et al. (2015) apply SERVQUAL for service quality evaluation in the logistics sector in Iskandar Malaysia. Baki et al. (2009) integrate SERVQUAL and Kano's model into QFD for quality evaluation of logistics services. Karpuzcu (2006) measures service quality in distribution logistics using SERVQUAL and AHP. Lee and Lin (2007) use QFD for generating customer voice-based quality improvement strategy for logistics service providers in Taiwan. Liao and Kao (2014) evaluate logistics service using fuzzy theory, QFD, and goal programming. Bottani and Rizzi (2006) proposed a fuzzy QFD approach for strategic management of logistics services. So et al. (2006) evaluate the service quality of third–party logistics service providers using AHP. Chan et al. (2006) apply an AHP approach for benchmarking the logistics performance of the postal industry. Chen et al. (2015) apply a Kansei engineering–based logistics service design approach to developing international express services. Esmaeili et al. (2015) apply fuzzy logic to assess service quality attributes in the logistics industry. Wu et al. (2014) perform a comprehensive evaluation of the logistics service quality based on vague sets theory. Hsu et al. (2010a) apply loss aversion to investigate service quality in logistics. Huiskonen and Pirttila (1998) investigate logistics customer service strategy planning by applying Kano's quality element classification. Jian and Zhenpeng (2008) perform logistics service quality analysis based on the gray correlation method. Limbourg et al. (2016) propose a SERVQUAL-based approach for evaluating logistics service quality in Da Nang City. Lin et al. (2016) propose an structural equation modeling (SEM)-based model for exploring service quality in the e-commerce context. Awasthi et al. (2016b) propose a hybrid approach based on gap analysis, QFD and AHP for evaluating logistics service quality.

It can be seen from above that QFD, SERVQUAL, and gap analysis are the most commonly used techniques reported in literature for evaluating logistics service quality. These techniques, however, require direct responses from customers and hence may not be suitable in situations with no or limited access to customer data. In our study, we are using FDELPHI, FAHP, and fuzzy TOPSIS for modeling logistics service quality. The FDELPHI technique is used to identify the requirements (criteria) for logistics service quality evaluation. The FAHP is used to attribute weights to these requirements, while fuzzy TOPSIS is used to rank the alternatives for logistics service quality improvement against the requirements.

The commonly reported service quality attributes include delivery of products on time, quality of material received, damage condition of goods, packaging of goods, completeness of orders, correctness in fulfilling customer orders, flexibility of service, delivery costs, technological soundness, recyclability of products, courteous employees, and responsiveness to customer queries. Thai (2013) emphasizes the inclusion of environmentally friendly attributes in logistics service quality evaluation such as use of recyclable material and energy-efficient vehicles. A detailed list of criteria for evaluating logistics services quality can be found in Caplice and Sheffi (1994), Millen et al. (1999), Sohal et al. (1999), Mentzer et al. (1999), Franceschini and Rafele (2000), Rafele (2004), Aguezzoul (2007), Rahman (2008), Saura et al. (2008), Werbińska-Wojciechowska (2011), and Meidutė-Kavaliauskienė et al. (2014).

From the literature review, we found that most of the studies on logistics service quality evaluation are based on physical distribution quality and customer satisfaction point of view. There are very few studies that integrate the perspective of all the stakeholders (Awasthi and Proth, 2006; Awasthi and Chauhan, 2012; Bhuiyan et al., 2015; Awasthi et al., 2016a). In this chapter, we propose a three-step approach for logistics service quality evaluation considering perspectives of different stakeholders. These stakeholders are logistics service providers, shippers, customers, municipal administrators, traffic managers, and city residents.

7.3 Basics of FDELPHI, FAHP, and Fuzzy TOPSIS

7.3.1 FDELPHI

The Delphi method was proposed by Dalkey and Helmer in 1963. It is a technique to obtain the most reliable consensus among a group of experts (Mereditha et al., 1989). The consensus is obtained by consultations that can go up to four rounds where each round receives information from the previous round on the mean, median, deviation, etc., of the results obtained. The experts can, therefore, revise their information toward the group average (Bueno and Salmeron, 2008). The advantage of this technique is that participants can base their response on feedback instead of direct confrontation. The limitation is that it suffers from low convergence of experts' opinions, high execution cost, and possible filtering of expert opinions (Kuo and Chen, 2008). The FDELPHI method integrates the Delphi method with fuzzy theory (Murry et al., 1985). In this technique, only one round of investigation with experts is required. The experts provide responses using membership degree functions such as triangular fuzzy numbers, trapezoidal fuzzy numbers, Gaussian fuzzy numbers, etc. (Hsu et al., 2010b). Fuzzy numbers work best for consolidating fragmented expert opinions (Tsai et al., 2010). The experts, opinions are combined using min, max, median, and mixed operators (Lin et al., 2007), or using the geometric mean in methods such as FAHP (Kuo and Chen, 2008; Ma et al., 2011; Pai, 2007).

7.3.2 FAHP

The FAHP approach performs AHP (Saaty, 1988) in a fuzzy environment to address uncertain, imprecise judgments of experts through the use of linguistic variables or fuzzy numbers. FAHP has been widely used in supplier selection problems due to its ability to integrate both qualitative and quantitative criteria and handling uncertainty (Wei et al., 2005). In FAHP, first the problem is decomposed into a hierarchical structure comprising of goal, criteria, subcriteria, and alternatives. Then, the elements are compared pairwise with respect to the importance to the goal, importance to the criterion, and importance to the subcriterion. Five triangular fuzzy numbers (TFNs), $\tilde{1}, \tilde{3}, \tilde{5}, \tilde{7}, \tilde{9}$, are used in our study where $\tilde{1}$ denotes equal importance and $\tilde{9}$ denotes extreme relative importance. Once all the pairwise comparisons are made at the individual level, group priority vectors are generated by aggregating the individual judgments. If the fuzzy ratings of k decision makers are described by triangular fuzzy number

$\ddot{R}_k = (a_k, b_k, c_k), k = 1, 2, .., K$, then the aggregated fuzzy rating is given by $R = (a, b, c), k = 1, 2, .., K$ where

$$a = \min_k \{a_k\}, b = \frac{1}{K} \sum_{k=1}^{K} b_k, c = \max_k \{c_k\} \qquad (7.1)$$

The crisp value (w) for a fuzzy TFN (a,b,c) is obtained using

$$w = \frac{a + (4 * b) + c}{6} \qquad (7.2)$$

Having obtained the aggregate judgment matrix of all the pairwise comparisons, the consistency is determined using the consistency ratio (CR) (Saaty, 1988). If CR ≤ 0.1, the judgment matrix is acceptable; otherwise it is considered inconsistent. To obtain a consistent matrix, judgments should be reviewed and improved. Normalized column totals and row sums are used to generate normalized priority scores (or weights). The final priorities of alternatives are obtained by multiplying the group priority vectors of criteria, subcriteria, and alternatives.

7.3.3 Fuzzy TOPSIS

The fuzzy TOPSIS approach involves fuzzy assessments of criteria and alternatives in TOPSIS (Hwang and Yoon, 1981). The TOPSIS technique chooses an alternative that is closest to the positive-ideal solution and farthest from the negative-ideal solution. A positive-ideal solution is composed of the best performance values for each criterion, whereas the negative-ideal solution consists of the worst performance values.

Let $\tilde{D}$ and $\tilde{W}$ denote the aggregate fuzzy decision matrix for the alternatives and the criteria. The individual evaluation of alternative $A_i (i = 1, 2, .., m)$ on criteria $C_j (j = 1, 2, .., n)$ is given by x_{ij}. The criteria weight is denoted by w_j. The aggregated evaluations are denoted by $\tilde{x}_{ij}$ and $\tilde{w}_j$, respectively.

$$\tilde{D} = \begin{matrix} & \begin{matrix} C_1 & C_2 & & C_n \end{matrix} \\ \begin{matrix} A_1 \\ A_2 \\ . \\ A_m \end{matrix} & \begin{bmatrix} \tilde{x}_{11} & \tilde{x}_{12} & \cdots & \tilde{x}_{1n} \\ \tilde{x}_{21} & \tilde{x}_{22} & \cdots & \tilde{x}_{2n} \\ \cdots & \cdots & \cdots & \cdots \\ \tilde{x}_{m1} & \tilde{x}_{m2} & \cdots & \tilde{x}_{mn} \end{bmatrix} \end{matrix}, i = 1, 2, \ldots, m; j = 1, 2, .., n \qquad (7.3)$$

$$\tilde{W} = (\tilde{w}_1, \tilde{w}_2, .., \tilde{w}_n) \tag{7.4}$$

First, the raw data are normalized using linear scale transformation to bring the various criteria scales into a comparable scale. The normalized fuzzy decision matrix $\tilde{R}$ is given by

$$\tilde{R} = [\tilde{r}_{ij}]_{mxn}, i = 1, 2, .., m; j = 1, 2, .., n \tag{7.5}$$

where

$$\tilde{r}_{ij} = \left(\frac{a_{ij}}{c_j^*}, \frac{b_{ij}}{c_j^*}, \frac{c_{ij}}{c_j^*} \right) \quad \text{and} \quad c_j^* = \max_i c_{ij} \quad \text{(benefit criteria)} \tag{7.6}$$

$$\tilde{r}_{ij} = \left(\frac{a_j^-}{c_{ij}}, \frac{a_j^-}{b_{ij}}, \frac{a_j^-}{a_{ij}} \right) \quad \text{and} \quad a_j^- = \min_i a_{ij} \quad \text{(cost criteria)} \tag{7.7}$$

Then, the weighted normalized matrix $\tilde{V}$ for criteria is computed by multiplying the weights ($\tilde{w}_j$) of evaluation criteria with the normalized fuzzy decision matrix $\tilde{r}_{ij}$:

$$\tilde{V} = [\tilde{v}_{ij}]_{mxn}, i = 1, 2, .., m; j = 1, 2, .., n \quad \text{where } \tilde{v}_{ij} = \tilde{r}_{ij}(.)\tilde{w}_j \tag{7.8}$$

The next step involves computing the fuzzy positive-ideal solution (FPIS) and fuzzy negative-ideal solution (FNIS) for each criteria:

$$A^* = (\tilde{v}_1^*, \tilde{v}_2^*, .., \tilde{v}_n^*) \quad \text{where } \tilde{v}_j^* = \max_i \{v_{ij3}\}, i = 1, 2.., m; j = 1, 2, .., n \tag{7.9}$$

$$A^- = (\tilde{v}_1, \tilde{v}_2, .., \tilde{v}_n) \quad \text{where } \tilde{v}_j^- = \min_i \{v_{ij1}\}, i = 1, 2.., m; j = 1, 2, .., n \tag{7.10}$$

Then, the distance (d_i^*, d_i^-) of each weighted alternative $i = 1, 2.., m$ from the FPIS and the FNIS is computed as follows:

$$d_i^* = \sum_{j=1}^{n} d_v(\tilde{v}_{ij}, \tilde{v}_j^*), i = 1, 2, .., m \tag{7.11}$$

$$d_i^- = \sum_{j=1}^{n} d_v(\tilde{v}_{ij}, \tilde{v}_j^-), i = 1, 2, .., m \tag{7.12}$$

where $d_v(\tilde{a}, \tilde{b})$ is the distance measurement between two fuzzy numbers $\tilde{a}$ and $\tilde{b}$:

$$d_v(\tilde{a},\tilde{b}) = \sqrt{\frac{1}{3}\left[(a_1 - b_1)^2 + (a_2 - b_2)^2 + (a_3 - b_3)^2\right]} \qquad (7.13)$$

Finally, the closeness coefficient (CC_i) of each alternative is computed as follows:

$$CC_i = \frac{d_i^-}{d_i^- + d_i^*}, i = 1, 2, .., m \qquad (7.14)$$

The closeness coefficient CC_i represents the distances to the FPIS (A^*) and the FNIS (A^-) simultaneously. The best alternative (highest rank) is closest to the FPIS and farthest from the FNIS.

7.4 Solution Approach

The proposed approach for logistics service quality evaluation consists of three main steps. The first step involves the identification of evaluation criteria and generating linguistic ratings using the FDELPHI technique. In the second step, FAHP is used to assign weights to the evaluation criteria. The third step involves evaluating logistics service quality providers against the identified criteria using fuzzy TOPSIS. Figure 7.1 shows the relationship between the involved steps and the techniques used.

Table 7.1 presents the various criteria identified by applying the FDELPHI technique with the logistics stakeholders. Both direct and indirect stakeholders were involved. The direct stakeholders are shippers, logistics service providers, and customers, whereas the indirect stakeholders include municipal administrators, traffic managers, and city residents. The direct stakeholders' perspective addresses physical distribution quality and customer

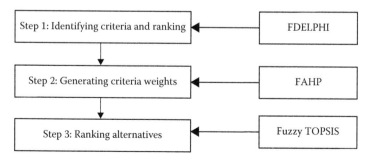

FIGURE 7.1
Proposed approach.

TABLE 7.1

Stakeholders, Perspectives, and Associated Criteria

Stakeholders	Stakeholder Category	Perspective	Requirements (Criteria)
Shippers, logistics service providers	Direct	Customer satisfaction (C1)	Undamaged packaging (C1.1) Timely goods delivery (C1.2) Accuracy of orders (C1.3)
Customers	Direct	Product and service (C2)	Product quality (C2.1) Service quality (C2.2) Reliability (C2.3) Flexibility (C2.4)
Municipal administrators, traffic managers, residents	Indirect	Environmental (C3)	Air pollution (C3.1) Emissions (C3.2) Noise (C3.3)
		Social (C4)	Occupation of public space (C4.1) Congestion (C4.2) Accidents (C4.3)

TABLE 7.2

Linguistic Assessments and Fuzzy Numbers

Criteria Importance	Alternative Score	Fuzzy Number	Crisp Value
Very low	Very poor	(1,1,3)	1.333333
Low	Poor	(1,3,5)	3
Medium	Medium	(3,5,7)	5
High	Good	(5,7,9)	7
Very high	Very good	(7,9,9)	8.666667

satisfaction criteria such as the product and service quality, whereas the indirect stakeholders are concerned with negative environmental and social impacts of goods distribution.

Linguistic rankings are used to perform pairwise comparisons among the identified criteria. The scales used for evaluation are presented in Table 7.2. These linguistic ratings were then converted into crisp numbers using Equation 7.2 and subject to FAHP for generating criteria weights.

Finally, the fuzzy TOPSIS method is used to generate final rankings for the logistics service providers (alternatives). A committee of decision makers (organizational representatives involved in goods distribution and logistics involved) is formed to assess the performance of logistics service providers against the weighted criteria. Overall performance scores are generated and the alternative with the highest score is finally chosen.

7.5 Numerical Application

Let us consider an organization ABC that is interested in evaluating the service quality of its three logistics operators A1, A2, and A3. A committee of three decision makers (D1, D2, and D3) is formed at ABC to rank qualitatively (Table 7.2) the criteria and the alternatives. These decision makers are technical members from the logistics department involved directly in shipping, distribution, and delivery of goods to customers. The FDELPHI technique is used to obtain the pairwise comparison scores for the various criteria identified in Table 7.1. The ratings obtained are presented in Tables 7.3 through 7.7.

The crisp values for linguistic scores are obtained using Table 7.2 and the FAHP technique is applied. Table 7.8 presents the normalized column sums,

TABLE 7.3

Pairwise Comparison Matrix for the Four Main Criteria (Perspective)

	C1	C2	C3	C4
C1	1	VH	H	H
C2	–	1	H	M
C3	–	–	1	H
C4	–	–	–	1

H, high; M, medium; VH, very high.

TABLE 7.4

Pairwise Comparison Matrix for the Customer Satisfaction Criteria

	C1.1	C1.2	C1.3
C1.1	1	H	VH
C1.2	–	1	H
C1.3	–	–	1

H, high; VH, very high.

TABLE 7.5

Pairwise Comparison Matrix for the Product and Process Quality Criteria

	C2.1	C2.2	C2.3	C2.4
C2.1	1	VH	H	VH
C2.2	–	1	H	VH
C2.3	–	–	1	H
C2.4	–	–	–	1

H, high; VH, very high.

TABLE 7.6

Pairwise Comparison Matrix for the Environmental Criteria

	C3.1	C3.2	C3.3
C3.1	1	H	VH
C3.2	–	1	VH
C3.3	–	–	1

H, high; VH, very high.

TABLE 7.7

Pairwise Comparison Matrix for the Social Criteria

	C4.1	C4.2	C4.3
C4.1	1	H	VH
C4.2	–	1	VH
C4.3	–	–	1

H, high; VH, very high.

TABLE 7.8

Main Criteria Weights

	C1	C2	C3	C4	C1	C2	C3	C4	Row Sum	Normalized % (Criteria Weight)
C1	1	8.667	7	7	0.713	0.865	0.462	0.35	2.391	0.597
C2	0.115	1	7	5	0.082	0.099	0.462	0.25	0.894	0.223
C3	0.142	0.142	1	7	0.101	0.014	0.066	0.35	0.532	0.133
C4	0.142	0.2	0.142	1	0.101	0.019	0.009	0.05	0.181	0.045
Column Total	1.401	10.009	15.142	20	1	1	1	1	4	–

row sums, and weight of the four main categories (criteria or perspective), namely customer satisfaction (C1), product and service (C2), environmental (C3), and social (C4).

In a similar way, local weights of the 13 subcriteria are generated. The global weights of the subcriteria are obtained by multiplying subcriteria weight with corresponding criteria weight. Table 7.9 presents the local and global weights for the 13 subcriteria. The fuzzy weight (a,b,c) for a given criteria is obtained by using the global weight value a and setting $a = b = c$.

Table 7.10 presents the qualitative ratings for the three alternatives provided by the decision-making committee.

The qualitative ratings are converted into fuzzy triangular numbers and then aggregate ratings are generated using Equation 7.1. For example, for criteria C1, the

TABLE 7.9

Subcriteria Weights

Requirements (Criteria)	ID	Local Weight	Global Weight	Fuzzy Weight
Product quality	C1.1	0.725	0.433	(0.433, 0.433, 0.433)
Service quality	C1.2	0.219	0.131	(0.131, 0.131, 0.131)
Service flexibility	C1.3	0.056	0.034	(0.034, 0.034, 0.034)
Timely goods delivery	C2.1	0.601	0.134	(0.134, 0.134, 0.134)
Undamaged goods	C2.2	0.247	0.055	(0.055, 0.055, 0.055)
Packaging of goods	C2.3	0.115	0.026	(0.026, 0.026, 0.026)
Accuracy of orders	C2.4	0.036	0.008	(0.008, 0.008, 0.008)
Air pollution	*C3.1*	*0.710*	*0.094*	*(0.094, 0.094, 0.094)*
Emissions	*C3.2*	*0.236*	*0.031*	*(0.031, 0.031, 0.031)*
Noise	*C3.3*	*0.053*	*0.007*	*(0.007, 0.007, 0.007)*
Occupation of public space	*C4.1*	*0.710*	*0.032*	*(0.032, 0.032, 0.032)*
Congestion	*C4.2*	*0.236*	*0.011*	*(0.011, 0.011, 0.011)*
Accidents	*C4.3*	*0.053*	*0.002*	*(0.002, 0.002, 0.002)*

Note: The values in italics are the cost category criteria while the other are benefit category criteria.

TABLE 7.10

Alternative Ratings

Criteria	A1			A2			A3		
	D1	D2	D3	D1	D2	D3	D1	D2	D3
C1.1	L	L	VH	VL	VH	VL	H	H	L
C1.2	VL	L	VL	VL	VL	M	H	L	VL
C1.3	H	VH	VL	VH	VH	M	VH	H	VL
C2.1	L	VL	VH	VL	VL	M	H	H	VL
C2.2	M	H	L	H	M	M	H	VL	VL
C2.3	VH	L	VH	M	M	M	L	VH	L
C2.4	VL	M	M	H	L	H	M	M	M
C3.1	M	VH	VH	M	M	VL	M	VL	VH
C3.2	VL	H	H	H	VL	VL	M	VH	L
C3.3	H	M	M	H	L	VH	H	M	L
C4.1	H	VL	VH	VH	H	L	L	VH	VL
C4.2	VL	M	VH	VL	VL	VL	H	VL	M
C4.3	L	H	M	VH	VH	M	L	VH	VH

L, low; H, high; M, medium; VL, very low; VH, very high.

qualitative ratings provided by the three decision makers are (L,L,VH) or ((1,3,5), (1,3,5), (7,9,9)). The aggregated fuzzy weight is given by $\tilde{w}_j = (w_{j1}, w_{j2}, w_{j3})$ where

$$w_{j1} = \min_k(1,1,7), w_{j2} = \frac{1}{3}(3+3+9), w_{j3} = \max_k(5,5,9)$$

$$\tilde{w}_j = (1,5,9)$$

For performing normalization, Equations 7.6 and 7.7 are used. For example, the normalized rating of alternative A1 for criteria C8 (air pollution), which is a cost category criteria, is given by

$$a_j^- = \min_i(3,1,1) = 1$$

$$\tilde{r}_{ij} = \left(\frac{1}{9}, \frac{1}{7.667}, \frac{1}{3}\right) = (0.111, 0.130, 0.333)$$

The normalized value of alternative A1 for criteria C1 (product quality), which is a benefit category criteria, is given by

$$c_j^* = \max_i(9,9,9) = 9$$

$$\tilde{r}_{ij} = \left(\frac{1}{9}, \frac{5}{9}, \frac{9}{9}\right) = (0.111, 0.556, 1)$$

Likewise, we compute the normalized values of the three alternatives for the remaining criteria to generate the normalized fuzzy decision matrix. Table 7.11 presents the fuzzy decision matrix and normalized scores for the three alternatives.

Table 7.12 presents the fuzzy weighted decision matrix for the three alternatives constructed using Equation 7.8. The $\tilde{r}_{ij}$ values and fuzzy criteria weight values ($\tilde{w}_j$) from Table 7.9 are used to compute the fuzzy weighted decision matrix for the alternatives. For example, for alternative A1, the fuzzy weight for criteria C1 (product quality) is given by

$$\tilde{v}_{ij} = (0.433, 0.433, 0.433)(.)(0.111, 0.556, 1) = (0.048, 0.241, 0.433)$$

Likewise, we compute the fuzzy weights for the remaining criteria for the three alternatives.

Based on these values, the FPIS (A^*) and the FNIS ($A-$) for each criteria are calculated using Equations 7.9 and 7.10. For example, for criteria C1 (product quality), $A^- = (0.048, 0.048, 0.048)$ and $A^* = (0.433, 0.433, 0.433)$. Then, we compute the distance $d_v(.)$ of each alternative from the fuzzy positive-ideal matrix (A^*) and fuzzy negative-ideal matrix ($A-$) using Equation 7.13. For example, for alternative A1 and criteria C1, the distances $d_v(A_1, A^*)$ and $d_v(A_1, A^-)$ are computed as follows:

$$d_v(A_1, A^*) = \sqrt{\frac{1}{3}\left[(0.048-0.048)^2 + (0.241-0.048)^2 + (0.433-0.048)^2\right]} = 0.249$$

$$d_v(A_1, A^-) = \sqrt{\frac{1}{3}\left[(0.048-0.433)^2 + (0.241-0.433)^2 + (0.433-0.433)^2\right]} = 0.249$$

TABLE 7.11

Fuzzy Decision Matrix and Normalized Scores

Criteria	A1	A2	A3	Min	Max	A1	A2	A3
C1.1	(1,5,9)	(1,3.667,9)	(1,5.667,9)	1	9	(0.111, 0.556, 1)	(0.111,0.407, 1)	(0.111,0.629,1)
C1.2	(1,1.667,5)	(1,2.333,7)	(1,3.667,9)	1	9	(0.111, 0.185, 0.556)	(0.111,0.259,0.778)	(0.111,0.407,1)
C1.3	(1,5.667,9)	(3,7.667,9)	(1,5.667,9)	1	9	(0.111, 0.629, 1)	(0.333,0.851,1)	(0.111,0.629,1)
C2.1	(1,4.333,9)	(1,2.333,7)	(1,5,9)	1	9	(0.111, 0.481, 1)	(0.111,0.259,0.778)	(0.111,0.556,1)
C2.2	(1,5,9)	(3,5.667,9)	(1,3,9)	1	9	(0.111, 0.556, 1)	(0.333,0.629,1)	(0.111,0.333,1)
C2.3	(1,7,9)	(3,5,7)	(1,5,9)	1	9	(0.111, 0.778, 1)	(0.333,0.556,0.778)	(0.111,0.556,1)
C2.4	(1,3.667,7)	(1,5.667,9)	(3,5,7)	1	9	0.111, 0.523, 0.778)	(0.111,0.629,1)	(0.333,0.556,0.778)
C3.1	(3,7.667,9)	(1,3.667,7)	(1,5,9)	1	9	(0.111, 0.130, 0.333)	(0.142,0.272,1)	(0.111,0.2,1)
C3.2	(1,5,9)	(1,3,9)	(1,5.667,9)	1	9	(0.111, 0.2,1)	(0.111,0.333,1)	(0.111,0.176,1)
C3.3	(3.5,6.667,9)	(1,6.333,9)	(1,5,9)	1	9	(0.111, 0.176, 0.333)	(0.111,0.157,1)	(0.111,0.2,1)
C4.1	(1,5.667,9)	(1,6.333,9)	(1,4.333,9)	1	9	(0.111, 0.176, 1)	(0.111,0.157,1)	(0.111,0.230,1)
C4.2	(1,5,9)	(1,1,3)	(1,4.333,9)	1	9	(0.111, 0.2, 1)	(0.333,1,1)	(0.111,0.230,1)
C4.3	(1,5,9)	(3,7.667,9)	(1,7,9)	1	9	(0.111, 0.2, 1)	(0.111,0.130,0.333)	(0.111,0.142,1)

Note: The values in italics are the cost category criteria while the other are benefit category criteria.

TABLE 7.12

Fuzzy Weighted Matrix

Criteria	A1	A2	A3
C1.1	(0.048,0.241,0.433)	(0.048,0.177,0.433)	(0.048,0.273,0.433)
C1.2	(0.015,0.024,0.073)	(0.015,0.034,0.102)	(0.015,0.053,0.131)
C1.3	(0.004,0.021,0.034)	(0.011,0.029,0.034)	(0.004,0.021,0.034)
C2.1	(0.015,0.065,0.134)	(0.015,0.035,0.105)	(0.015,0.075,0.134)
C2.2	(0.006,0.031,0.055)	(0.018,0.035,0.055)	(0.006,0.018,0.055)
C2.3	(0.003,0.020,0.026)	(0.009,0.014,0.020)	(0.003,0.014,0.026)
C2.4	(0.001,0.004,0.006)	(0.001,0.005,0.008)	(0.003,0.004,0.006)
C3.1	*(0.010,0.012,0.031)*	*(0.013,0.026,0.094)*	*(0.010,0.019,0.094)*
C3.2	*(0.003,0.006,0.031)*	*(0.003,0.010,0.031)*	*(0.003,0.006,0.031)*
C3.3	*(0.001,0.001,0.002)*	*(0.001,0.001,0.007)*	*(0.001,0.001,0.007)*
C4.1	*(0.004,0.006,0.032)*	*(0.004,0.005,0.032)*	*(0.004,0.007,0.032)*
C4.2	*(0.001,0.002,0.011)*	*(0.004,0.011,0.011)*	*(0.001,0.002,0.011)*
C4.3	*(0.000,0.000,0.002)*	*(0.000,0.000,0.001)*	*(0.000,0.000,0.002)*

Note: The values in italics are the cost category criteria while the other are benefit category criteria.

Likewise, we compute the distances for the remaining criteria for the three alternatives. Then, we compute the distances d_i^* and d_i^- using Equations 7.11 through 7.12 for the three alternatives. For example, for alternative A1 and criteria C1, the distances d_i^* and d_i^- are given by

$$d_i^* = 0.249 + 0.034 + 0.020 + \ldots + 0.006 + 0.001 = 0.482$$

$$d_i^- = 0.249 + 0.097 + 0.019 + \ldots + 0.007 + 0.002 = 0.631$$

Table 7.13 presents the results for the distances for the 13 criteria.

Then, we compute the closeness coefficient (CC_i) of the three alternatives using Equation 7.14. For example, for alternative A1, the closeness coefficient is given by

$$d_i^- \quad CC_i = d_i^- / (d_i^- + d_i^*) = 0.631 / (0.631 + 0.482) = 0.433 \quad d_i^+$$

Likewise, CC_i values for the other two alternatives are computed. The final results are shown in Table 7.14.

By comparing the CC_i values of the three alternatives, we find that A3 > A2 > A1. Therefore, alternative A3 is finally ranked as best and recommended for implementation.

To ensure the validity of results, it is important to involve the right number and type of decision makers in the evaluation process. Since the proposed

TABLE 7.13

Distances to Positive- and Negative-Ideal Solutions

Criteria	A^-	A^*	D^-			D^*		
			A1	A2	A3	A1	A2	A3
C1.1	0.048	0.433	0.249	0.234	0.257	0.249	0.267	0.241
C1.2	0.015	0.131	0.034	0.052	0.071	0.097	0.089	0.081
C1.3	0.004	0.034	0.020	0.023	0.020	0.019	0.013	0.019
C2.1	0.015	0.134	0.075	0.053	0.077	0.080	0.091	0.077
C2.2	0.006	0.055	0.032	0.034	0.029	0.032	0.024	0.035
C2.3	0.003	0.026	0.017	0.012	0.015	0.014	0.012	0.015
C2.4	0.001	0.008	0.004	0.005	0.004	0.005	0.004	0.004
C3.1	*0.010*	*0.094*	*0.012*	*0.049*	*0.049*	*0.077*	*0.061*	*0.065*
C3.2	*0.003*	*0.031*	*0.016*	*0.017*	*0.016*	*0.022*	*0.020*	*0.022*
C3.3	*0.001*	*0.007*	*0.001*	*0.004*	*0.004*	*0.006*	*0.005*	*0.005*
C4.1	*0.004*	*0.032*	*0.017*	*0.017*	*0.017*	*0.023*	*0.023*	*0.022*
C4.2	*0.001*	*0.011*	*0.006*	*0.008*	*0.006*	*0.007*	*0.004*	*0.007*
C4.3	*0.000*	*0.002*	*0.001*	*0.000*	*0.001*	*0.002*	*0.002*	*0.002*
		Total	0.482	0.507	0.565	0.631	0.618	0.595

Note: The values in italics are the cost category criteria while the other are benefit category criteria.

TABLE 7.14

CC_i Scores

	A1	A2	A3
D^-	0.482	0.507	0.565
D^*	0.631	0.618	0.595
CC_i	0.433	0.451	0.487
	A3 > A2 > A1		

approach is multistakeholder in nature, at least one representative of each stakeholder group that is directly or indirectly involved in generation or consumption of logistics services should be considered.

7.6 Conclusions and Direction for Future Works

In this chapter, we propose a three-step approach for logistics service quality evaluation considering perspectives of different stakeholders. These stakeholders are logistics service providers, shippers, customers, municipal administrators, traffic managers, and city residents. In the first step, we identify the requirements (criteria) for logistics service quality evaluation using the FDELPHI technique. The second step involves attributing weights to these requirements using FAHP. In the third and the last step, fuzzy TOPSIS is used to rank the alternatives for logistics service quality improvement. A numerical application is provided. Sensitivity analysis is conducted to determine the influence of modeling parameters on final results.

The strength of our work is the integration of multiple perspectives of different stakeholders in logistics service quality evaluation. The limitation is the lack of real data for testing the model.

The next step of our work involves developing models for evaluating strategies for improving the performance of poor quality logistics service providers such as collaboration, training, resource sharing, etc.

References

Aguezzoul A. (2007), The third party logistics selection: A review of literature, in *International Logistics and Supply Chain Congress*, Istanbul, Turkey, 1–7.

Aima R.N.A., Eta W., Hazana A.N. (2015), Service quality: A case study of logistics sector in Iskandar Malaysia using SERVQUAL model, *Procedia—Social and Behavioral Sciences*, 172, 457–462.

Awasthi A., Adetiloye T., Crainic T.G. (2016a), Collaboration partner selection for city logistics planning under municipal freight regulations, *Applied Mathematical Modelling*, 40(1), 510–525.

Awasthi A., Chauhan S.S. (2012), A hybrid approach integrating affinity diagram, AHP and fuzzy TOPSIS for sustainable city logistics planning, *Applied Mathematical Modelling*, 36(2), 573–584.

Awasthi A., Proth J.M. (2006), A systems-based approach for city logistics decision making, *Journal of Advances in Management Research*, 3(2), 7–17.

Awasthi A., Sayyadi R., Khabbazian A. (2016b), A combined approach integrating gap analysis, QFD and AHP for improving logistics service quality, *International Journal of Logistics Systems and Management*, Accepted.

Baki B, Basfirinci C.S., Cilingir Z., Murat A.R.I. (2009), An application of integrating SERVQUAL and Kano's model into QFD for logistics services: A case study from Turkey, *Asia Pacific Journal of Marketing and Logistics*, 21(1), 106–126.

Bhuiyan M.F.H., Awasthi A., Wang C. (2015), Investigating the impact of access–timing–sizing regulations on urban logistics, *International Journal of Logistics Systems and Management*, 20(2), 216–238.

Bienstock C. C., Mentzer J. T., Bird M. (1997), Measuring physical distribution service quality, *Journal of Academy of Marketing Science*, 25(1), 31–44.

Bottani E., Rizzi A. (2006), Strategic management of logistics service: A fuzzy QFD approach, *International Journal of Production Economics*, 103(2), 585–599.

Bueno S., Salmeron J.L. (2008), Fuzzy modeling enterprise resource planning tool selection, *Computer Standards and Interfaces*, 30(3), 137–147.

Caplice C., Sheffi Y. (1994), A review and evaluation of logistics metrics, *International Journal of Logistics Management*, 5(2), 11–28.

Chan F.T.S., Chan H K., Lau H.C.W., Ip R.W.L. (2006), An AHP approach in benchmarking logistics performance of the postal industry, *Benchmarking: An International Journal*, 13(6), 636–661.

Chen M.C., Chang K.C., Hsu C.L., Xiao J.H. (2015), Applying a Kansei engineering-based logistics service design approach to developing international express services, *International Journal of Physical Distribution & Logistics Management*, 45(6), 618–646.

Dalkey N.C., Helmer O. (1963), An experimental application method to the use of experts, *Management Science*, 9(3), 458–467.

Esmaeili A., Kahnali R.A., Rostamzadeh R., Zavadskas E.K., Ghoddami B. (2015), An application of fuzzy logic to assess service quality attributes in logistics industry, *Transport*, 30(2), 172–181.

Franceschini F., Rafele C. (2000), Quality evaluation in logistics services, *International Journal of Agile Management Systems*, 2(1), 49–53.

Hsu C.L., Chen M.C., Chang K.C., Chao C.M. (2010a), Applying loss aversion to investigate service quality in logistics: A moderating effect of service convenience, *International Journal of Operations and Production Management*, 30, 508–525.

Hsu Y., Lee C., Kreng V. (2010b), The application of fuzzy Delphi method and fuzzy AHP in lubricant regenerative technology selection, *Expert Systems with Applications*, 37, 419–425.

Huiskonen J., Pirttila T. (1998), Sharpening logistics customer service strategy planning by applying Kano's quality element classification, *International Journal of Production Economics*, 56–57, 253–260.

Hwang, C.L., Yoon, K. (1981), *Multiple Attributes Decision Making Methods and Applications*. Berlin Heidelberg: Springer.

Jian, X., Zhenpeng C. (2008), Logistics service quality analysis based on gray correlation method, *International Journal of Business and Management*, 3(1), 58–61.

Karpuzcu H. (2006), Measuring Service Quality in Distribution Logistics Using SERVQUAL and AHP: A Case Study in a Pharmaceutical Wholesaler in Turkey: A dissertation presented in part consideration for the degree of "MSc Operations Management." University of Nottingham.

Kuo Y.F., Chen P.C. (2008), Constructing performance appraisal indicators for mobility of the service industries using fuzzy Delphi method, *Expert Systems with Applications*, 35, 1930–1939.

Lee T.R., Lin J.-H (2007), Generating customer voice based quality improvement strategy for logistics service providers in Taiwan from the perspective of quality function deployment, *International Journal of Logistics Economics and Globalisation*, 1(1), 63–76.

Liao C.N., Kao H.P. (2014), An evaluation approach to logistics service using fuzzy theory, quality function development and goal programming, *Computers & Industrial Engineering*, 68, 54–64.

Limbourg, S., Giang, H.T.Q., Cools, M. (2016), Logistics service quality: The case of Da Nang City, *Procedia Engineering*, 142(1), 124–130.

Lin H.Y., Hsu P.Y., Sheen G.J. (2007), A fuzzy based decision making procedure for data warehouse system selection, *Expert Systems with Applications*, 32, 939–953

Lin, Y., Luo, J., Cai, S., Ma, S., Rong, K. (2016), Exploring the service quality in the e-commerce context: A triadic view, *Industrial Management and Data Systems*, 116(3), 388–415.

Ma Z., Shao C., Ma S., Ye Z. (2011), Constructing road safety performance indicators using fuzzy Delphi method and grey Delphi method, *Expert Systems with Applications*, 38, 1509–1514.

Meidutė-Kavaliauskienė I., Aranskis A., Litvinenko M. (2014), Consumer satisfaction with the quality of logistics services, *Procedia—Social and Behavioral Sciences*, 110, 330–340.

Mentzer J.T., Flint D.J., Hult T.M. (2001), Logistics service quality as a segment-customized process, *Journal of Marketing*, 65(4), 82–104.

Mentzer J.T., Flint D.J., Kent J.L. (1999), Developing a logistics service quality scale, *Journal of Business Logistics*, 20(1), 9–32.

Mentzer J.T., Gomes R., Krapfel, R.E. (1989), Physical distribution service: A fundamental marketing concept? *Journal of the Academy of Marketing Science*, 17(1), 53–62.

Mereditha J.R., Raturia A., Amoako-Gyampahb K., Kaplana B. (1989), Alternative research paradigms in operations, *Journal of Operations Management*, 8(4), 297–326.

Millen R., Sohal A., Moss S. (1999), Quality management in the logistics function: An empirical study, *International Journal of Quality & Reliability Management*, 16(2), 166–180.

Murry T.J., Pipino L.L., Gigch J.P. (1985), A pilot study of fuzzy set modification of Delphi, *Human Systems Management*, 5(1), 76–80.

Pai J. (2007), A fuzzy MCDM evaluation framework based on humanity oriented transport for transforming scheme of major arterial space in Taipei metropolitan, *Journal of Eastern Asia Society for Transportation Studies*, 7, 1731–1744.

Perrault, Jr., W.D., Frederick A.R. (1974), Physical distribution service: A neglected aspect of marketing management, *MSU Business Topics*, 22(2), 37–45.

Rafele C. (2004), Logistic service measurement: A reference framework, *Journal of Manufacturing Technology Management*, 15(3), 280–290.

Rahman S.-U. (2008), Quality management in logistics services: A comparison of practices between manufacturing companies and logistics firms in Australia, *Total Quality Management & Business Excellence*, 19(5), 535–550.

Roslan N.A.A., Wahab E., Abdullah N.H. (2015), Service quality: A case study of logistics sector in Iskandar Malaysia using SERVQUAL model, *Procedia—Social and Behavioral Sciences*, 172, 457–462.

Saaty T.L. (1988), *The Analytic Hierarchy Process: Planning, Priority Setting, Resource Allocation*. Pittsburgh, PA: RWS Publications.

Saura I.G., Frances D.S., Contri G.B., Blasco M.F. (2008), Logistics service quality: A new way to loyalty, *Industrial Management & Data Systems*, 108(5), 650–668.

So S.H., Kim J.J., Cheong K.J., Cho G. (2006), Evaluating the service quality of third party logistics service providers using the analytic hierarchy process, *Journal of Information Systems and Technology Management*, 3(3), 261–270.

Sohal A.S., Millen R., Maggard M., Moss S. (1999), Quality in logistics: A comparison of practices between Australian and North American/European firms, *International Journal of Physical Distribution & Logistics, Management*, 29(4), 267–280.

Thai, V.V. (2013), Logistics service quality: Conceptual model and empirical evidence, *International Journal of Logistics Research and Applications*, 16(2), 114–131.

Tsai H., Chang C., Lin H. (2010), Fuzzy hierarchy sensitive with Delphi method to evaluate hospital organization performance, *Expert Systems with Applications*, 37(8), 5533–5541.

Werbińska-Wojciechowska S. (2011), On logistics service quality evaluation – case study, *Logistics and Transport*, 13(2), 45–56.

Wei C.C., Chien C.F., Wang M.J. (2005). An AHP-based approach to ERP system selection. *International Journal of Production Economics*, 96, 47–62.

Wu A.H., Su J.Q., Wang F. (2014), A comprehensive evaluation of the logistics service quality based on vague sets theory, *International Journal of Shipping and Transport Logistics*, 6(1), 69–87.

8

The Inclusion of Logical Interaction between Criteria in FAHP

Ksenija Mandić, Vjekoslav Bobar, and Boris Delibašić

CONTENTS

8.1 Introduction

This chapter presents a methodology that allows modeling of logical interaction between criteria in *multicriteria decision-making* (MCDM). We propose a hybrid model for supporting the selection of suppliers in the telecommunications sector, combining the fuzzy analytic hierarchy process (FAHP) and *interpolative Boolean algebra* (IBA). The baseline model will be FAHP where each criterion is supposed to be independent, and the aggregation function is supposed to be linear, that is, a weighted sum of criteria is a good aggregation function for ranking. The competing approach is the application of IBA in order to establish the logical interactions than can be expressed by decision makers. Once the logical functions are set, they are translated into a *generalized Boolean polynomial* (GBP), after which the GBP is reduced to a numerical value through logical aggregation (LA). What makes IBA more flexible

181

and efficient is that all the structural transformations—that is, the logical dependencies among criteria—are established before the introduction of a numerical value, which is not the case with conventional *fuzzy multicriteria decision-making* (FMCDM) methods. In this way, one can overcome the complex matrix for a large number of criteria, subcriteria, and alternatives that are inherent in fuzzy logic, thereby simplifying the computational effort. In addition, the proposed approach treats contradiction differently than other FMCDM methods (i.e., the negated variable is not transformed immediately into a value) and respects the law of the excluded middle.

When solving real problems, in order to make a good decision, it is necessary to take into account a large number of complex parameters. Therefore, there has been a significant growth in the development of tools for modeling the decision-making process recorded in recent years. Among various methods, multicriteria models are in particular useful for solving complex and conflicting decision situations. Multicriteria models allow decision makers to find an optimal solution in situations where there are many various and often conflicting criteria (Roy, 1990). The main aim of MCDM is to develop a methodology that allows the aggregation of criteria, including the subjective preferences of decision makers (Doumpos and Zopounidis, 2002). This methodology is widely used in the evaluation, assessment, and ranking of decision alternatives in various fields of research.

One of the most commonly used MCDM methods is the *analytic hierarchy process* (AHP) (Saaty, 1980). The main advantage of this method is that it adequately handles the unquantifiable (qualitative) information that is present in almost every instance of decision-making. In the literature, however, this method is often criticized because it does not take into account the fact that human assessments are often vague and imprecise. For this reason, researchers proposed FMCDM and a fuzzy version of AHP. FAHP is a technique that is based on AHP and fuzzy logic (Zadeh, 1965), and it is commonly implemented by using triangular fuzzy numbers.

However, FMCDM methods such as FAHP have certain disadvantages. These methods do not take into account the fact that criteria may be mutually dependent and interacting. In real situations, the goal of decision-making is usually created in accordance with the logical requirements of the decision makers, whereas the criteria are often interrelated or conflicting. In some cases, decision makers might also want to trade off one criterion against another. Correlation and trade-offs can be presented using logical functions. Logical functions result in a new structure of components that differs from the weighted sum approach. Therefore, as classical FMCDM methods do not allow the setting of logical interactions among criteria—that is, they do not obey all Boolean laws—consistent fuzzy logic is introduced by Radojevic (2000a). The basis of the proposed approach is the interpolative realization of Boolean algebra (IBA) that transforms logical functions among criteria into a GBP, then aggregates logical conditions using a logic aggregation function (Radojevic, 2000b).

This chapter provides a description of the integration of Boolean algebra and Boolean operators into FAHP. This type of integration allows for logical dependencies and relationships among decision-making criteria. The basic concept of this chapter is the inclusion of logical interactions between criteria in FAHP (Chang, 1996) using IBA (Radojevic, 2000a). These methods will be applied to the case of supplier selection in the telecommunications sector. In this chapter, we show how the linguistic requirements of decision makers in the telecommunications sector can be modeled with GBPs. The selection of suppliers is a critical and challenging task for companies that participate in the telecommunications market. Selection of the best possible supplier is a strategic decision that ensures profitability and long-term survival of the company. The telecommunications sector develops rapidly on a daily basis and under such conditions companies must choose quality suppliers in order to stay competitive in the market. The company's goal is to carefully choose the right suppliers that will provide the purchaser with the right quality products at the right price, at the right quantity, and at the right time. In most cases, the strengths and weaknesses of suppliers vary over time, so decision makers must make complex decisions in the selection of suppliers.

This chapter is organized as follows: Section 8.2 provides an explanation of the main concepts such as IBA, GBP, and LA. Also, it presents a detailed literature review (combined AHP–IBA approach). Section 8.3 presents a brief description of the supplier selection process in the telecommunication sector. Section 8.4 analyzes the problem of selecting suppliers by using FAHP and IBA. Finally, the chapter concludes with Section 8.5 where the final considerations are presented.

8.2 Literature Review

8.2.1 Interpolative Boolean Algebra

Classical Boolean algebra (Boole, 1848) is based on statements that are true or false, yes or no, and black or white. There are situations, however, in which the classical binary realization of Boolean algebra is not adequate. Often it is impossible to express something in absolute terms, and we are forced to use vague generalizations. This issue led to the development of many-valued logic that includes a third logical value, in addition to true and false—that is, alongside the values 1 and 0, the value ½ is introduced (Lukasiewicz, 1970). However, the main disadvantage of many-valued logic is that because it does not comply with the laws of the excluded middle and contradiction, it does not work in a Boolean framework.

In the literature, fuzzy logic is recognized as the next form of generalization which uses the principle of many-valued logic (Zadeh, 1965). The main advantage of fuzzy logic is that it is very close to human perception and does not require exact data. However, the main disadvantage of many-valued logic is transferred to fuzzy logic: it, too, ignores the laws of the excluded middle and contradiction and thus even fuzzy logic does not work in a Boolean framework (Figure 8.1) (Dubois et al., 2005; Zhang, 2011; Milošević, 2012).

These problems have necessitated the development of a methodology that allows the placing of fuzzy logic into a Boolean framework. The newly developed approach allows for the consistent treatment of gradation and fuzzy variables based on IBA. IBA is proposed by Radojević (2000a,b) as a many-valued realization of Boolean algebra that preserves the laws on which Boolean algebra is based (Milošević et al., 2014). Radojević (2000a, 2002, 2005, 2013a,b) points out that this new framework is for dealing with generalized realizations of values. IBA is a real-valued and/or [0,1] value realization of Boolean algebra (Dragović et al., 2013). The difference between this and other many-valued logics is the preservation of all Boolean axioms and theorems (Radojević, 2006b).

IBA has a finite number of elements and it is a substantially different gradation method than the fuzzy approach. In general, IBA represents an atomic algebra (as it has a finite number of elements). Atoms, being the simplest elements of algebra, play a fundamental role (Radojević, 2006a). IBA, for the first time, clearly separates the structure and the value of the elements in Boolean algebra. IBA consists of two levels: (1) symbolic or qualitative—at this level the structure of elements is determined, and defines the final Boolean framework—and (2) semantic or valued—at this level the values are introduced, preserving all the laws set at the symbolic level; generally it is a matter of interpolation (Radojević, 2008a,b; Radojević et al., 2008c).

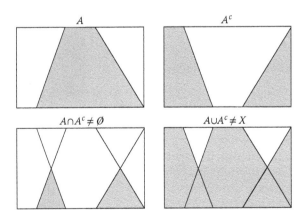

FIGURE 8.1
A fuzzy set and its complement according to conventional fuzzy logic. (From Milošević, P., *Softverska podrška za primenu inerpolativne Bulove algebre*, Master rad, Fakultet organizacionih nauka, 2012.)

The symbolic or qualitative level: Objects at this level are the laws of Boolean algebra that must be fulfilled. One of the basic concepts of the symbolic level is the structure of IBA elements and the principle of structural functionality. The structure of the analyzed elements determines which atom (of the final set of IBA elements) is included or not included in the analysis. The principle of structural functionality specifies that the structure of any element of IBA may be directly calculated based on the structure of its components (Radojević, 2002; Poledica et al., 2013). Within this level, all structural transformations are performed before the introduction of value. The structure is value independent and that is the key to preserving all Boolean laws in both the symbolical and valued levels (Radojević, 2006b, 2007).

One characteristic is that contradiction is treated in a different way, that is, a negated variable is not transformed immediately into a value as is the case in conventional fuzzy logic. Proper observation of the contradiction of elements presents a new framework of how the negation of a fuzzy set should be treated. Figure 8.2 is a sequence of graphs showing the fulfilment of the laws of the excluded middle (a $\lor \neg a = 1$) and contradiction according to IBA (Milošević et al., 2013). From Figure 8.2, it can be concluded that the union of any IBA element and its complement is equal to the structure of the IBA element "1." The structure of the intersection of any IBA element and its complement is equal to the structure of the IBA element "0." The laws of the excluded middle and contradiction are valid for the structure of IBA (on the symbolic level) but are also applied for value realization (at the valued level) (Radojević, 2006a).

The valued or semantic level: This is a tangible version of the symbolic level in terms of value. An element from the symbolic level preserves all its characteristics (described by Boolean axioms and laws) on the valued level

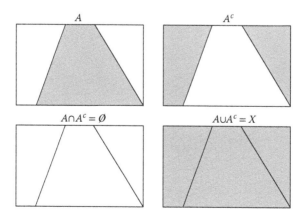

FIGURE 8.2
A fuzzy set and its complement according to interpolative Boolean algebra. (From Milošević, P., *Softverska podrška za primenu inerpolativne Bulove algebre*, Master rad, Fakultet organizacionih nauka, 2012.)

(Radojević, 2002). On the valued level, the values are introduced so as to preserve all the laws set symbolically. Generally speaking, it is a matter of interpolation (Radojević, 2008a,b). Elements take values from an interval and a suitable operator for the generalized product (GP) is chosen (Milošević et al., 2013).

IBA is identical to Boolean algebra with a finite number of elements, which is represented by Equation 8.1 (Radojević, 2006a):

$$\langle BA, \wedge, \vee, \neg \rangle \tag{8.1}$$

where BA is the set of finite elements, and the remaining symbols are the binary operators of conjunction $\wedge$ and disjunction $\vee$, and the unary negation operator $\neg$, for which all Boolean axioms and theorems are valid.

Boolean algebra is often represented graphically using a Hasse diagram (Figure 8.3). In fact, it visualizes Boolean algebra generated with $n = 2$ independent variables a and b, as is represented in Figure 8.3. Thus, generated algebra has $2^{2^n} = 2^{2^n} = 16$ elements (Radojević, 2005; Milošević, 2012).

The Boolean algebra generated by the two independent variables a and b has $2^n = 2^2 = 4$ atoms. The elements $a \wedge b$, $a \wedge \bar{b}$, $\bar{a} \wedge b$, and $\bar{a} \wedge \bar{b}$ are the atomic elements. From Figure 8.3, it can be observed that there are operations not inherent to conventional fuzzy logic such as equivalence and exclusive disjunction where $a \Leftrightarrow b = (a \wedge b) \vee (\bar{a} \wedge \bar{b})$ and $a \underline{\vee} b = (a \wedge \bar{b}) \vee (\bar{a} \wedge b)$ (Radojević, 2010).

Figure 8.4 shows how each of the operations between the two IBA elements is illustrated using classical sets. Their sum equals 1 regardless of the value realization of the variables a and b. This assertion is true for atomic elements in IBA; however, these elements do not only have to be either 0 or 1; each can take a value from the interval [0,1].

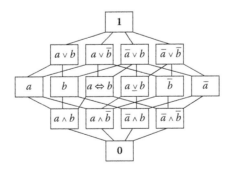

FIGURE 8.3
Hasse diagram showing the Boolean algebra for two elements a and b. (Adapted from Radojević, D., *Yugoslav Journal of Operations Research—YUJOR*, 15, 171–189, 2005.)

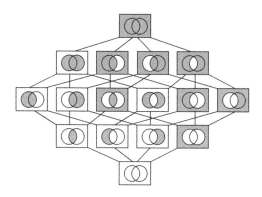

FIGURE 8.4
Hasse diagram showing the classical sets. (Adapted from Radojević, D., *International Journal of Computers, Communications & Control*, 3, 121–131, 2008a.)

8.2.2 Generalized Boolean Polynomials

IBA is technically based on a GBP (Radojević, 2006a, 2008a). This means that if any element of Boolean algebra can be represented as a canonical disjunction, it can also be represented by an appropriate GBP. A GBP is a polynomial whose variables are elements of Boolean algebra and thus it allows the corresponding element of Boolean algebra to be processed into a value on the real interval [0,1] using operators such as classical (+), classical (−), and GP ($\otimes$) (Milošević et al., 2013). The GP can be any function that maps $[0,1] \times [0,1] \to [0,1]$ and is a subclass of the conventional fuzzy t-norm satisfying all four axioms (commutativity, associativity, monotonicity, and boundary condition) and the additional nonnegativity condition, represented by Equations 8.2 through 8.6 (Radojević, 2008a, 2013c):

1. Commutativity:

$$A_i \otimes A_j = A_j \otimes A_i \tag{8.2}$$

2. Associativity:

$$A_i \otimes (A_j \otimes A_k) = (A_i \otimes A_j) \otimes A_k \tag{8.3}$$

3. Monotonicity:

$$A_t \leq A_j \Rightarrow A_i \otimes A_k \leq A_j \otimes A_k \tag{8.4}$$

4. Boundary:

$$A_t \otimes 1 = A_t \tag{8.5}$$

5. Nonnegativity:

$$\sum_{K \in P(\Omega/S)} (-1)^{|K|} \otimes A_i(x) \geq 0,\ S \in P(\Omega),\ A_i(x) \in [0,1],\ A_i \in \Omega,\ i = 1 \quad (8.6)$$

In order to calculate all of the elements of IBA, first the polynomials should be made for the atoms using a GP as shown in Equations 8.7 through 8.10 ($\neg a = 1 - a$ is used as a function of negation):

$$a \wedge b \rightarrow a \otimes b \quad (8.7)$$

$$a \wedge \neg b \rightarrow a \otimes (1 - b) \rightarrow a - a \otimes b \quad (8.8)$$

$$\neg a \wedge b \rightarrow (1 - a) \otimes b \rightarrow b - a \otimes b \quad (8.9)$$

$$\neg a \wedge \neg b \rightarrow (1 - a) \otimes (1 - b) \rightarrow 1 - a - b + a \otimes b \quad (8.10)$$

By entering these values into the Hasse diagram for consistent fuzzy logic (Figure 8.3), the polynomials corresponding to the elements of Boolean algebra are obtained (Figure 8.5).

The theory of interpolative sets represents the realization of the concept of fuzzy sets within a Boolean framework. The main characteristics are illustrated using the example of two interpolative sets a and b. A GBP or appropriate sets are represented by the expressions given in Figure 8.5 (Radojević, 2006a).

8.2.3 Logical and Pseudo-LA

In IBA, the method of unifying factors is referred to as LA. The main goal of LA is to merge the primary criteria $\Omega = \{a_1 \cdots, a_n\}$ into a single criterion, which represents a given set, by using a logical/pseudological function. For MCDM, the subject of this chapter, LA can be realized with two steps (Mirković et al., 2006):

1. The normalization of criteria values, represented by Equation 8.11:

$$\|\cdot\| : \Omega \rightarrow [0,1] \quad (8.11)$$

2. The aggregation of normalized criteria values, using an LA or pseudological function as an LA operator, defined by Equation 8.12 (Radojević, 2008a):

$$\text{Aggr}[0,1]^n \rightarrow [0,1] \quad (8.12)$$

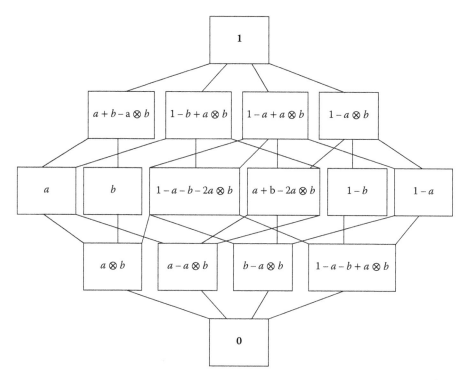

FIGURE 8.5
Generalized Boolean polynomial corresponding to the elements of the Hasse diagram (Figure 8.3) for two elements *a* and *b*. (Adapted from Radojević, D., *Yugoslav Journal of Operations Research—YUJOR*, 15, 171–189, 2005.)

A Boolean logical function enables the aggregation of factors, that is, it is an expression that transforms its input into a GBP. Partial demands are usually logical demands that can only be adequately described by logical functions (Radojević, 2008b; Radojević et al., 2008c). This approach treats the logical function (partial aggregation demand) as a GBP, which can process the value of the real interval [0,1]. A GBP performs the role of a logical combined element in LA (Radojević, 2010). LA is in general a weighted sum of partial demands. LA is mostly used in measuring performance and modeling uncertainty (Dobrić et al., 2010; Milošević et al., 2013; Poledica et al., 2012).

A pseudological function (pseudo-GBP) is a convex combination of GBP. It allows one to define several objectives and merge them into a single representative outcome using the appropriate aggregation tools (Radojević, 2013c, 2014).

8.2.4 Literature Review (Combined AHP–IBA Approach)

The subject of combining AHP and Boolean consistent fuzzy logic has been dealt with by several authors. Among them it is worth singling out the following.

Dragović et al. (2012) combine Boolean consistent fuzzy logic and AHP. They point out that classical AHP does not take into account the fact that decision-making criteria are interconnected. Therefore, they suggest the incorporation of logical functions into AHP, which will enable the aggregation of criteria into a single objective function. The proposed approach is applied to the case of web service selection. The criteria that are considered for the analysis of the problem are availability (A), reliability (R), response time (RT), and security (S). The logical condition that is set by the decision makers is as follows: "if a service has a high level of A then it may be more important to take into consideration the value of RT and if it is low then R is more important. On the other hand, the decision maker might prefer more S over a less secure one regardless of the value of the other criteria." The logical function that is derived in relation to the logical condition is as follows in Equation 8.13:

$$((A \wedge RT) \vee (\neg A \wedge R)) \wedge S \qquad (8.13)$$

Additionally, Dragović et al. (2013) came to similar conclusions in their paper, which illustrates the expansion of AHP by introducing Boolean consistent fuzzy logic. In this paper, the proposed approach is also applied in the case of web service selection. They considered two cases. In the first case, the following criteria are analyzed: RT, A, and R. The logical condition that is set in this case is "if a service has a high level of A then it may be more important to take into consideration the value of the RT and if it is low then R is more important." So, the logical function is as follows in Equation 8.14:

$$(A \wedge RT) \vee (\neg A \wedge R) \qquad (8.14)$$

In the second case, the criteria security (S) and encryption (E) are considered and the logical condition set is "a high level of either S or E indicates that the particular service is a secure one." From the above condition arises the following logical function, Equation 8.15:

$$S \vee E \qquad (8.15)$$

The authors conclude that the correlation and interdependence between criteria can be adequately represented using logical functions only, that is, LA that will result with a new structure of the components compared to the weighted sum approach.

Martinović and Delibašić (2014) dealt with the problem of selecting the best consultant for an Enterprise Resource Planning Software developed by the German company SAP SE (SAP ERP) project using a combined AHP–IBA approach. The criteria analyzed are cost (CO), work experience (WE), education level (EL), and communication ability (CA), whereas the subcriteria are technical cost (TC), consultancy cost (CC), company where consultant is employed (CE), projects completed (PC), references (RE), customer recommendation (CR), occupational seminars (OS), awareness of responsibility (AR), and ability to persuade (AP). Logical conditions set by the decision makers are "CC is equivalent with CR" and "RE excludes OS." Therefore, the logical functions are as follows in Equations 8.16 and 8.17, respectively:

$$CC \Leftrightarrow CR \qquad (8.16)$$

$$RE \veebar OS \qquad (8.17)$$

The authors conclude that a combined AHP–IBA approach introduces a new generated criterion that includes interdependencies between subcriteria. The new structure of the components gives better results in comparison with the weighted sum approach.

The problem of inclusion of logical interactions in decision-making processes is still a current topic, which is proved by recent research: Jeremić et al. (2015) proposed a multicriteria algorithm, which determines the path by using IBA, which satisfies quality-of-service requirements. The decision is made at each node, based on the ranking of available options and considering multiple constraints. Lilić et al. (2016) presented a software solution for analyzing the reliability of a system. The proposed software solution uses a GBP to determine an algebraic representation of the system structural function and thus evaluate its reliability. Dragović et al. (2015) used IBA for diagnosing diseases and its application for determining peritonitis likelihood. Poledica et al. (2015) explored the possibility of using logic-based similarity measures for modeling consensus. They proposed a soft consensus model for calculating the consensus and proximity degrees on two different levels. Dobrić et al. (2015) used IBA for the formalization of the human categorization process.

8.3 Supplier Selection in the Telecommunications Sector

The telecommunications industry records constant growth in both investment and income. This growth has undergone a significant reduction in recent years, however, and equipment manufacturers are therefore seeking new ways and new approaches to maintain and improve business. Globalization and the increased uncertainty of companies have prompted a growth in outsourcing services. Under such circumstances, closer attention is paid to supply and decision-making related to the supply process. These decisions entail large-scale investment in the telecommunications industry and influence the strategic standing of companies in the sector. The selection of an appropriate supplier in these conditions is one of the most important problems telecommunications companies face.

8.3.1 The Supplier Selection Process

The world is becoming an increasingly global marketplace and in this global environment companies must take into account almost all aspects of the business simultaneously when making decisions. One of the most important aspects of business is the selection of suppliers. As a fundamental component of supply chain management, supplier selection plays an essential role in the production and logistics management of many companies (Han and Liu, 2011). A poor choice of supplier, therefore, may cause a significant disturbance to a business, both in terms of finance and operations. On the other hand, selecting the right supplier cuts supply costs significantly, increases competitiveness, and greatly improves the fulfillment of end user requests.

Supplier selection compares a widespread network of suppliers by employing a common set of criteria and measurements in order to identify the supplier with the greatest potential. The interest of decision makers in the selection process of suppliers is constantly growing as a reliable supplier reduces costs and improves the quality of products and services.

The selection process involves outlining a model and method of analysis and performance measurement for a group of suppliers in order to increase competitiveness. Various methodologies of supplier selection are put forward in different studies, which are rather simplified compared to real-world experiences of decision-making (Chen et al., 2006). Thus, the number of studies dedicated to this problem increases as does the development of various methodologies for solving this type of problem. Studies have been examining the problem of selecting suppliers since the 1960s (Dickson, 1966; Weber et al., 1991).

The selection of suppliers is a process that involves a large number of both quantitative and qualitative criteria. Therefore, in order to select the best suppliers it is necessary to take into account both material (tangible) and immaterial (intangible) factors (Pi and Low, 2005). Essentially, it may be reduced to MCDM. Specifically, a number of quantitative (price, distance, time) and qualitative (quality, design, technological performance) criteria, often conflicting, are considered (Bhutta and Huk, 2002; Bhutta, 2003; Choy et al., 2003; Ramanathan, 2007; Ordoobadi, 2009). Relevant texts propose various criteria for supplier performance measurement. The first attempts to provide a comprehensive framework of criteria were made by Dickson (1966), whose framework was adopted by Weber et al. (1991). Ha and Krishnan (2008) updated this set of criteria, concluding that cost, quality, and delivery were the three most frequently employed criteria. Furthermore, they established that the nature of criteria is individualistic and cannot always be transferred into a measurable value.

8.4 Results

8.4.1 The Application of FAHP to Supplier Selection

For the purpose of this example, FAHP is applied to the selection of a supplier for a telecommunications company. A company that specializes in manufacturing the equipment needed to build, supervise, and maintain telecommunication systems wants to select the most suitable supplier of transmission frequency repeaters, which allow for area coverage without a Global System for Mobile Communications (GSM) signal or with a very weak signal.

In this chapter, for solving the supplier selection problem we used the FAHP methodology proposed by Chang (1996). Chang's FAHP method is implemented through the use of triangular fuzzy numbers. The application of FAHP to supplier selection can be broken down into three stages:

Stage I: Defining the objective and essential criteria and building a hierarchical tree. Figure 8.6 shows the principal objective, which is the "Selection of the Most Suitable Supplier." Four main criteria are determined: product performance, supplier capability, financial aspect, and support and services.

- **Product performance (C_1)**—a very important criterion, which includes all important product features that must be taken into account when purchasing, such as technical specifications, quality, quantity, and on-time delivery.
- **Supplier capability (C_2)**—market standing, references, and business relations with suppliers play an important role in the decision-making process for the most suitable supplier.
- **Financial aspect (C_3)**—a key factor in the relationship between companies and suppliers.
- **Support and Services (C_4)**—high-quality support service during the warranty and postwarranty periods, which includes the servicing and maintenance of equipment supplied, as well as training for the company's employees for proper handling and maintenance of the equipment, is an important criterion for the company when analyzing suppliers.

Stage II: Calculating priority weights for each criterion and alternative using FAHP. Comparison of criteria and alternatives is enabled by a linguistic of scale importance (Table 8.1). In Table 8.1 (Kilincci and Onal, 2011), linguistic variables are converted into triangular fuzzy numbers, that is, triangular fuzzy reciprocal numbers. Table 8.2 represents a fuzzy comparison matrix for four main criteria according to the primary objective.

From Table 8.2, we can see the most important criterion in the supplier selection process is product performance (C_1) as it has the highest priority

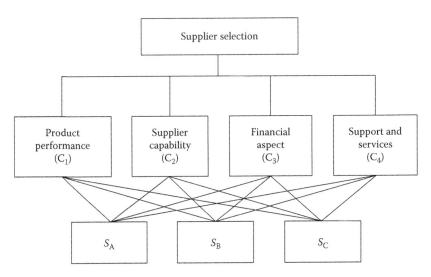

FIGURE 8.6
Hierarchical tree for the problem of supplier selection.

TABLE 8.1

Linguistic Scale of Importance

	Triangular Fuzzy Numbers	Triangular Fuzzy Reciprocal Numbers
Equal	(1,1,1)	(1,1,1)
Weak	(1/2,1,3/2)	(2/3,1,2)
Fairly strong	(3/2,2,5/2)	(2/5,1/2,2/3)
Very strong	(5/2,3,7/2)	(2/7,1/3,2/5)
Absolute	(7/2,4,9/2)	(2/9,1/4,2/7)

TABLE 8.2

Fuzzy Comparison Matrix for Four Main Criteria and Their Priority Vectors

Criteria	C_1	C_2	C_3	C_4	Priority Vectors (Wc)
C_1	(1,1,1)	(1/2,1,3/2)	(1,1,1)	(3/2,2,5/2)	0.2938
C_2	(2/3,1,2)	(1,1,1)	(2/3,1,2)	(1/2,1,2/3)	0.2474
C_3	(1,1,1)	(1/2,1,3/2)	(1,1,1)	(3/2,2,5/2)	0.2938
C_4	(2/5,1/2,2/3)	(2/3,1,2)	(2/5,1/2,2/3)	(1,1,1)	0.1650

TABLE 8.3

Overall Weights of the Alternatives with Respect to the Objective

Alternatives (Vectors)	C_1 (0.2938)	C_2 (0.2474)	C_3 (0.2938)	C_4 (0.1650)	Priority Weights
S_A	0.3333	0.1450	0.4572	0.5321	0.3558
S_B	0.3333	0.5321	0.0857	0.3229	0.3080
S_C	0.3333	0.3229	0.4572	0.1450	0.3360

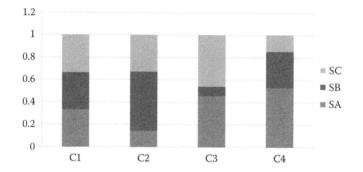

FIGURE 8.7
Overall weights of the alternatives for each criterion.

vector. This criterion is followed by supplier capability (C_2), financial aspect (C_3), and finally support and services (C_4). The identical FAHP method is applied to the comparison of alternatives (Table 8.3).

The overall alternative weights have been calculated with respect to the essential criteria.

It has been determined that the aggregate alternative weights are (0.36, 0.31, 0.33). According to the final result, the most suitable supplier is Supplier A (S_A) with the highest priority weight of 0.36, even if it is the worst in terms of the criterion supplier capability (C_2) (Figure 8.7).

8.4.2 The Inclusion of Logical Interaction between Criteria in Supplier Selection

This chapter shows how IBA can help decision makers incorporate their preferences into the decision-making process in a more sophisticated way than the weighted sum method. Many techniques of (conventional fuzzy) decision-making use a weighted approach, which exclusively allows linear relationships between the criteria. When solving problems using MCDM, however, such as the problem of supplier selection, criteria are often interdependent and logical interactions must be established between them. Logical interactions are based on the introduction of the Boolean algebra operators ∧, ∨, and ¬, by which decision makers can more accurately show dependence and comparisons between criteria. In this way, a large number of real problems can be expressed using Boolean algebra.

When selecting suppliers in real situations, decision makers often want to integrate their knowledge into a decision support model, which requires establishing logical relationships among criteria. Classical fuzzy methods of MCDM, such as FAHP, do not allow the setting of logical interactions among criteria, that is, they do not obey all Boolean laws. Boolean consistent fuzzy logic is introduced by Radojević (2000a). The basis of this approach is the IBA that transforms logical conditions among criteria into a GBP, then aggregates logical conditions using a logic aggregation function (Radojević, 2000b). This chapter shows how classical FAHP can be extended by applying IBA to merge the criteria into a single objective function. Also, it demonstrates how the linguistic requirements of decision makers in a telecommunications company can be modeled using Boolean polynomials.

8.4.3 The Inclusion of Logical Interaction in FAHP

IBA does not treat logical expressions in the same way as conventional fuzzy logic does. In effect, structure and value are separated into two different levels of logic. Contrary to conventional fuzzy logic, IBA is based on the principle of structural functionality—the structure of any element may be directly calculated based on the structure of its components (Radojević, 2006a). In accordance with this principle, negation is treated differently. This allows the preservation of the laws of the excluded middle and contradiction (Milošević et al., 2014).

Since certain criteria may influence or be influenced by other criteria and the importance of criteria may vary depending upon the measurement of other criteria, it is necessary to take all these interdependencies into account when choosing the best alternative (Dragović et al., 2013). In order to arrive at the best possible decision in real situations, decision makers often want to set mutual relationships among the criteria. Therefore, it is proposed that a logical function is used to define the importance of the criteria with respect to the goal, instead of comparison matrices. The logical function, which takes into account the correlation among the criteria, should be defined within a Boolean framework. In other words, the main goal of LA is to combine the initial criteria into a single global criterion using a logical function as an LA operator (Dragović et al., 2012). This is enabled using the following logical conditions:

Condition 1: "If product performance is at a high level, then the supplier is acceptable; if it is not at a high level, pay attention to the supplier capability, financial aspect, and support and services" (Equation 8.18):

$$k_1 \vee (\neg k_1 \wedge k_2 \wedge k_3 \wedge k_4) \tag{8.18}$$

Condition 2: "If supplier capability is satisfactory, the supplier should also have good product performance; if supplier capability is not satisfactory, closer attention should be paid to the financial aspect and support and services" (Equation 8.19):

$$(k_2 \wedge k_1) \vee (\neg k_2 \wedge k_3 \wedge k_4) \tag{8.19}$$

Condition 3: "If the financial aspect is high, close attention should be paid to product performance; if not high, close attention should be paid to supplier capability" (Equation 8.20):

$$(k_3 \wedge k_1) \vee (\neg k_3 \wedge k_2) \tag{8.20}$$

Each of these logical conditions is transformed into a GBP, by using the standard product as an appropriate operator of GP. In LA, a GBP performs the role of a logical combined element. As an introduction to the process of transformation of logical conditions into GBP, the following overview of the basic examples of Boolean operators and polynomials for two elements a and b is given (Equations 8.21 through 8.24):

$$a(x) = a(x); \, b(x) = b(x \tag{8.21}$$

$$a\neg(x) = 1 - a(x); \, b\neg(x) = 1 - b(x) \tag{8.22}$$

$$(a \wedge b)(x) = a(x) \otimes b(x) \tag{8.23}$$

$$(a \vee b)(x) = a(x) + b(x) - a(x)b(x) \tag{8.24}$$

These operators and polynomials will be used for the purpose of this study but with more variables (criteria).

The transformation for Condition 1 is given in Equation 8.25:

$$k_1 \vee (\neg k_1 \wedge k_2 \wedge k_3 \wedge k_4) = k_1 + (\neg k_1 \wedge k_2 \wedge k_3 \wedge k_4) - k_1 \otimes (\neg k_1 \wedge k_2 \wedge k_3 \wedge k_4)$$
$$= k_1 + ((1 - k_1) \otimes k_2 \otimes k_3 \otimes k_4) - k_1 \otimes ((1 - k_1) \otimes k_2 \otimes k_3) \tag{8.25}$$
$$= k_1 + k_2 \otimes k_3 \otimes k_4 - k_1 \otimes k_2 \otimes k_3 \otimes k_4$$

In the same way, the remaining logical Conditions 2 and 3 are transformed, as represented by Equations 8.26 and 8.27:

$$(k_2 \wedge k_1) \vee (\neg k_2 \wedge k_3 \wedge k_4) = k_2 \otimes k_1 + k_3 \otimes k_4 - k_2 \otimes k_3 \otimes k_4 \tag{8.26}$$

$$(k_3 \wedge k_1) \vee (\neg k_3 \wedge k_2) = k_2 - k_2 \otimes k_3 + k_3 \otimes k_1 \tag{8.27}$$

Only after the transformations have been conducted and the final structure established are the values introduced and computed (Dragović et al., 2013). This is the main difference between the conventional and Boolean consistent approaches. All tautologies and contradictions on the symbolic level are tautologies and contradictions, respectively, on the valued level as well (Milošević et al., 2014).

Table 8.4 specifies the priority weights of alternatives obtained by individual criteria using FAHP.

The criteria weights from Table 8.4 are introduced into the GBP Equations 8.25 through 8.27, based on which we obtain the values in Table 8.5 using LA. For example, Equation 8.28 demonstrates the incorporation of value into Equation 8.25 for supplier S_A:

$$k_1 \vee (\neg k_1 \wedge k_2 \wedge k_3 \wedge k_4) = k_1 + k_2 \otimes k_3 \otimes k_4 - k_1 \otimes k_2 \otimes k_3 \otimes k_4$$

$$= 0.3333 + 0.1450 \otimes 0.4572 \otimes 0.5321 - 0.3333 \otimes 0.1450 \otimes 0.4572 \otimes 0.5321$$

$$= 0.3568 \tag{8.28}$$

The resulting values for the suppliers for all three conditions are shown in Table 8.5.

Table 8.6 compares the values obtained by logical Condition 1 with the overall weights obtained using FAHP. It can be concluded that both methods give an equivalent ranking of alternatives.

From Table 8.7, it can be stated that logical Condition 2 gave a different order of alternatives in comparison with FAHP.

From Table 8.8, it can be concluded that logical Condition 3 gave a completely different ranking of alternatives than FAHP.

The final ranking of suppliers can be obtained by using another weighted function (pseudo-LA), where logical conditions are used rather than individual values (Equation 8.29). Decision makers determine weights for each logical condition:

$$0.5 \otimes (k_1 \vee (\neg k_1 \wedge k_2 \wedge k_3 \wedge k_4)) + 0.2 \otimes ((k_2 \wedge k_1) \vee (\neg k_2 \wedge k_3 \wedge k_4))$$

$$+ 0.3 \otimes ((k_3 \wedge k_1) \vee (\neg k_3 \wedge k_2)) = p \tag{8.29}$$

TABLE 8.4

The Weights of Suppliers for the Four Basic Criteria Using the Fuzzy Analytic Hierarchy Process (FAHP)

Alternatives	Product Performance (C_1)	Supplier Capability (C_2)	Financial Aspect (C_3)	Support and Services (C_4)
S_A	0.3333	0.1450	0.4572	0.5321
S_B	0.3333	0.5321	0.0857	0.3229
S_C	0.3333	0.3229	0.4572	0.1450

TABLE 8.5

Values for the Suppliers for Three Logical Conditions

Alternatives	Condition 1	Condition 2	Condition 3
S_A	0.3568	0.2563	0.2311
S_B	0.3431	0.1903	0.5150
S_C	0.3476	0.1525	0.3276

TABLE 8.6

Ranking of Alternatives Using Logical Condition 1 and FAHP

Alternatives	FAHP	Rank	Condition 1	Rank
S_A	0.3558	1	0.3568	1
S_B	0.3080	3	0.3431	3
S_C	0.3360	2	0.3476	2

TABLE 8.7

Ranking of Alternatives Using Logical Condition 2 and FAHP

Alternatives	FAHP	Rank	Condition 2	Rank
S_A	0.3558	1	0.2563	1
S_B	0.3080	3	0.1903	2
S_C	0.3360	2	0.1525	3

TABLE 8.8

Ranking of Alternatives Using Logical Condition 3 and FAHP

Alternatives	FAHP	Rank	Condition 3	Rank
S_A	0.3558	1	0.2311	3
S_B	0.3080	3	0.5150	1
S_C	0.3360	2	0.3276	2

The weights for the conditions are 0.5, 0.2, and 0.3, respectively, and p represents the final supplier weight.

By entering the values for the logical conditions obtained in Table 8.5 into the expression and applying pseudo-LA the resulting values are as shown in Table 8.9.

Table 8.9 shows that the ranking of suppliers using pseudological conditions is as follows: $S_B > S_C > S_A$, which is a completely different order than the results of FAHP: $S_A > S_C > S_B$. The main reason for the differences in the ranking of alternatives in the FAHP and IBA approach is in the way decision makers compared the criteria. The FAHP method allows decision makers only pairwise comparison of criteria, while by using the IBA approach decision makers can establish a logical interaction between criteria.

TABLE 8.9

Final Ranking of Suppliers Using FAHP and the Three Pseudological Conditions

Alternatives	FAHP	Rank	Pseudological Condition	Rank
S_A	0.3558	1	0.298	3
S_B	0.3080	3	0.364	1
S_C	0.3360	2	0.302	2

8.5 Conclusions

Many MCDM methods are used effectively to solve a large number of decision-making problems in different fields of research. However, some of the most famous classical MCDM methods, such as AHP, only allow the establishment of linear relationships between factors of decision-making. Only treating the relationship between criteria as a linear one, ignoring any mutual interaction or interdependence, has proven not to be adequate in practice.

Conventional fuzzy logic is a more advanced approach in comparison to classical (binary) logic because it allows logical operations on data that carry more information and can furthermore process values from the entire interval [0,1] and not just binary values of 1 or 0. However, the main disadvantage of this logic is that it does not have a Boolean framework. This indicates that the fuzzy approach considers the values but not the variable structure. Thus, the contradiction law is not respected.

Such conditions motivated the development of Boolean consistent fuzzy logic, which is based on IBA. A characteristic feature of IBA is that it focuses on the structure of elements and not just on the value, that is, it is based on the principle of structural functionality. It consists of two levels, symbolic and valued. At the symbolic level, the expression is transformed into a GBP so that all Boolean axioms and theorems are met. At the valued level, a numerical value is assigned to variables and through LA the value of the expression is calculated.

The goal of this chapter is to propose a model for supporting decision makers in choosing suppliers in the telecommunications sector. In order to solve the aforementioned problems, FAHP is combined with Boolean consistent fuzzy logic. FAHP was used to calculate a priority weight for the main criteria and alternatives and the following ranking of alternatives was obtained: $S_A > S_C > S_B$. In the next step, IBA was applied to the analyzed problem. This approach allows decision makers to set logical functions. IBA enables the transformation of logical functions into a GBP, while the use of LA reduces these polynomials to values. For this purpose, the FAHP priority weights of criteria are used and FAHP is extended by applying IBA to generate new criteria, which include some interdependencies. Ultimately, by using IBA the ranking $S_B > S_C > S_A$

was achieved. It can be concluded that IBA reached a completely different ranking of alternatives in comparison to FAHP. What makes IBA logic a more suitable way to solve these types of problems compared to conventional fuzzy methods is that the structural transformations are performed before the introduction of values. This approach to problem solving allows decision makers to model a greater range of problems, while at the same time complying with the Boolean laws of the excluded middle and contradiction.

In order to select the best alternative during supplier selection, criteria are considered that are often interdependent and conflicting. As these dependencies among criteria are often presented verbally by decision makers, logical expressions are a natural way to represent these logical conditions. IBA gives more reliable results by viewing the aggregation of criteria as a nonlinear structure where elements are interrelated. Most MCDM methods, in structuring complex decision-making models, view criteria as independent elements. In many complex real-world decision problems, however, there are relationships and interdependencies between criteria. Moreover, the value of criteria for an appropriate action can be influenced by a number of factors that are external to the decision system and cannot be controlled by the decision maker.

This chapter highlights the importance of modeling the interactions between criteria in decision-making because otherwise decision-making models may suggest bad decisions according to false assumptions of linearity and independence. In addition, with the use of logical conditions, decision makers are able to present their requirements more clearly. In this way, they can make a more comprehensive and better informed decision than is the case with conventional fuzzy methods that do not work in a Boolean framework and treat negation differently.

Further research will be directed toward the inclusion of logical interactions into the MCDM methods (such as technique for order of preference by similarity to ideal solution TOPSIS [The Technique for Order of Preference by Similarity to Ideal Solution], ELECTRE [ELimination and Choice Expressing REality], DELFI) and comparison of the obtained results with the proposed FAHP-IBA model.

References

Bhutta, M. K. S. (2003). Supplier selection problem: Methodology literature review. *Journal of International Technology and Information Management*, 12(2), 53–72.

Bhutta, K. B. and Huk, F. (2002). Supplier selection problem: A comparison of the total cost of ownership and analytic hierarchy process approaches. *Supply Chain Management: An International Journal*, 7(3), 126–135.

Boole, R. G. (1848). The calculus of logic. *Cambridge and Dublin Mathematical Journal*, 3, 183–198.

Chang, D. (1996). Applications of the extent analysis method on fuzzy AHP. *European Journal of Operational Research*, 95, 649–655.

Chen, C. T., Lin, C. T., and Huang, S. F. (2006). A fuzzy approach for supplier evaluation and selection in supply chain management. *International Journal of Production Economics*, 102(2), 289–301.

Choy, K.L., Lee, W.B., and Lo, V. (2003). Design of an intelligent supplier relationship management system: A hybrid case based neural network approach. *Expert Systems with Applications*, 24, 225–237.

Dickson, G. W. (1966). An analysis of vendor selection systems and decisions. *Journal of Purchasing and Supply Management*, 2(1), 5–17.

Dobrić, V., Kovaćević, D., Petrović, B., Radojević, D., and Milošević, P. (2015). Formalization of human categorization process using interpolative Boolen algebra. *Mathematical Problems in Engineering*. Cairo, Egypt: Hindawi Publishing Corporation, Vol. (2015), 8.

Dobrić, V., Poledica, A., and Petrović, B. (2010). Supply chain performance measurement using logical aggregation, In D. Ruan, T. Li (Eds.), Computational Intelligence. Foundation and Application: Proceedings of the 9th International FLINS Conference, World Scientific Publishing Company, Singapore, 616–621.

Doumpos, M. and Zopounidis, C. (2002). *Multicriteria Decision Aid Classification Methods*. Dordrecht: Kluwer Academic Publishers.

Dragović, I., Turajlić, N., Pilćević, D., Petrović, B., and Radojević, D. (2015). A Boolean consistent fuzzy inference system for diagnosing diseases and its application for determining peritonitis likelihood. *Computational and Mathematical Methods in Medicine*. Cairo, Egypt: Hindawi Publishing Corporation, Vol. (2015), 10.

Dragović, I., Turajlić, N., and Radojević, D. (2012). Extending AHP with Boolean consistent fuzzy logic and its application in web service selection. Proceedings of the 12th International FLINS Conference, World Scientific Publishing Company, Istanbul, pp. 576–591.

Dragović, I., Turajlić, N., Radojević, D., and Petrović, B. (2013). Combining Boolean consistent fuzzy logic and AHP illustrated on the web service selection problem. *International Journal of Computational Intelligence Systems*, 7(1), 84–93.

Dubois, D., Gottwald, S., Hajek, P., Kacprzyk, J., and Prade, H. (2005). Terminological difficulties in fuzzy set theory—The case of "intuitionistic fuzzy sets." *Fuzzy Set Systems*, 156, 485–491.

Ha, H. S. and Krishnan, R. (2008). A hybrid approach to supplier selection for the maintenance of a competitive supply chain. *Expert Systems with Applications*, 34(2), 1303–1311.

Han, S. and Liu, X. (2011). A perceptual computer based method for supplier selection problem. In *Conference on Fuzzy Systems*, Taipei, Taiwan, 27-30 June 2011, Vol. (2011), 1201–1207.

Jeremić, M., Rakićević, A., and Dragović, I. (2015). Interpolative Boolean algebra based multicriteria routing algorithm. *Yugoslav Journal of Operations Research*, 25(3), 397–412.

Kilincci, O. and Onal, S. A. (2011). Fuzzy AHP approach for supplier selection in a washing machine company. *Expert Systems with Applications*, 38(8), 9656–9664.

Lilić, N., Petrović, B., and Milošević, P. (2016). Software solution for reliability analysis based on interpolative Boolean algebra. Soft computing applications. *Advances in Intelligent Systems and Computing*, 356, 185–198.

Lukasiewicz, J. (1970). The shortest axiom of the implicational calculus of propositions. In: L. Borowski (Ed.), *Selected Works*. Amsterdam and PWN, Warsaw: North-Holland Publishing Company, pp. 295–305.

Martinović, N. and Delibašić, B. (2014). Selection of the best consultant for SAP ERP project using combined AHP-IBA approach. *Yugoslav Journal of Operation Research*, 24, 383–398.

Milošević, P. (2012). Softverska podrška za primenu inerpolativne Bulove algebre. Master rad, Fakultet organizacionih nauka.

Milošević, P., Nesić, I., Poledica, A., Radojević, D., and Petrović, B. (2013). Models for ranking students: Selecting applicants for a master of science studies. *Soft Computing Applications—Advances in Intelligent Systems and Computing*, 195, 93–103.

Milošević, P., Petrović, B., Radojević, D., and Kovačević, D. (2014). A software tool for uncertainty modeling using interpolative Boolean algebra. *Knowledge-Based Systems*, 62, 1–10.

Mirković, M., Hodolić, J., and Radojević, D. (2006). Aggregation for quality management. *Yugoslav Journal for Operational Research—YUJOR*, 16(2), 177–188.

Ordoobadi, S. (2009). Application of Taguchi loss functions for supplier selection. *Supply Chain Management: An International Journal*, 14(1), 22–30.

Pi, W. N. and Low, C. (2005). Supplier evaluation and selection using Taguchi loss functions. *International Journal of Advanced Manufacturing Technology*, 26(1–2), 155–160.

Poledica, A., Milošević, P., Dragović, I., Petrović, B., and Radojević, D. (2015). Modeling consensus using logic-based similarity measures. *Soft Computing*, 19(11), 3209–3219.

Poledica, A., Milošević, P., Dragović, I., Radojević, D., and Petrović, B. (2013). A consensus model in group decision making based on interpolative Boolean algebra. *8th Conference of the European Society for Fuzzy Logic and Technology (EUSFLAT-13)*, 11–13 September, Milan, Italy.

Poledica, A., Rakićević, A., and Radojević, D. (2012). Multi-expert decision making using logical aggregation, In C. Kahraman, E.E. Kerre, F.T. Bozbura (Eds.), *Uncertainty Modeling in Knowledge Engineering and Decision Making: Proceedings of the 10th International FLINS Conference*, World Scientific Publishing Company, Singapore, pp. 561–566.

Radojević, D. (2000a). Logical measure of continual logical function. *8th International Conference IPMU—Information Processing and Management of Uncertainty in Knowledge-Based Systems*, Madrid, 574–578.

Radojević, D. (2000b). New [0, 1]-valued logic: A natural generalization of Boolean logic. *Yugoslav Journal of Operational Research—YUJOR*, 10(2), 185–216.

Radojević, D. (2002). Logical measure-structure of logical formula. In: Bouchon-Meunier, B., Gutierrez-Rios, J., Magdalena, L:, Yager, R.R. (eds.). In: *Technologies for Constructing Intelligent Systems 2. Studies in Fuzziness and Soft Computing*. Springer, Heidelberg, 90, 417–430. Springer, 90, 417–430.

Radojević, D. (2005). Interpolative relations and interpolative preference structures. *Yugoslav Journal of Operations Research—YUJOR*, 15(2), 171–189.

Radojević, D. (2006a). Interpolative realization of Boolean algebra. *Proceedings of the NEUREL 2006, The 8th Neural Network Applications in Electrical Engineering*, 25–27 September 2006, Belgrade, Serbia, pp. 201–206.

Radojević, D. (2006b). Boolean frame is adequate for treatment of gradation or fuzziness equally as for two-valued or classical case. 4th Serbian-Hungarian Joint Symposium on Intelligent Systems, SISY 2006, pp. 43–57.

Radojević, D. (2007). Interpolative realization of Boolean algebra as consistent frame for gradation and/or fuzziness. In: Nikravesh, M., Kacprzyk, J, Zadeh, L.A. (eds.), *Studies in Fuzziness and Soft Computing: Forging New Frontiers: Fuzzy Pioneers II*, Heidelberg: Springer, pp. 295–318.

Radojević, D. (2008a). Fuzzy set theory in Boolean frame. *International Journal of Computers, Communications & Control*, 3, 121–131.

Radojević, D. (2008b). Logical aggregation based on interpolative Boolean algebra. *Mathware & Soft Computing*, 15, 125–141.

Radojević, D. (2010). Logical aggregation—Why and how. Proceedings of the 10th International FLINS Conference, World Scientific Publishing Company, Chengdu, pp. 511–517.

Radojević, D. (2013a). Real-valued realizations of Boolean algebras are a natural frame for consistent fuzzy logic. In R. Sesing, E. Trillas, C. Moraga, S. Termini (Eds.), *On Fuzziness: A Homage to Lotfi A. Zadeh, Studies in Fuzziness and Soft Computing*. Berlin: Springer Verlag, 299, 559–565.

Radojević, D. (2013b). Real-valued implication as generalized Boolean polynomial. *New Concepts and Applications in Soft Computing. Studies in Computational Intelligence*, 471, 57–69.

Radojević, D. (2013c). Real-valued implication function based on real-valued realization of Boolean algebra. 1st Serbian-Hungarian Joint Symposium on Intelligent systems, SISY 2013, pp. 45–50.

Radojević, D. (2014). Structural functionality as a fundamental property of Boolean algebra and base for its real-valued realizations. Paper Presented at the 15th International Conference on Information Processing and Management of Uncertainty in Knowledge-Based Systems IPMU 2014, 15–19 July 2014, Montpellier, France, pp. 28–36.

Radojević, D., Perović, A., Ognjanović, Z., and Rašković, M. (2008c). Interpolative Boolean logic, In D. Dochev, M. Pistore, P. Traverso (Eds.), The International Conference on Artificial Intelligence: Methodology, Systems, Applications, AIMSA 2008, LNCS (LNAI), Springer, Heidelberg, 5253, 209–219.

Ramanathan, R. (2007). Supplier selection problem: Integrating DEA with the approaches of total cost of ownership and AHP. *Supply Chain Management: An International Journal*, 12(4), 258–261.

Roy, B. (1990). Decision-aid and decision-making. *European Journal of Operational Research*, 45, 324–331.

Saaty, T. L. (1980). *The Analytic Hierarchy Process: Planning, Priority Setting, Resource Alocation*. New York, NY: McGraw-Hill.

Weber, C. A., Current, J. R., and Benton, W. C. (1991). Vendor selection criteria and methods. *European Journal of Operational Research*, 50, 2–18.

Zadeh, L. A. (1965). Fuzzy sets. *Information and Control*, 8, 338–353.

Zhang, X. (2011). Duality and pseudoduality of dual disjunctive normal forms. *Knowledge Based Systems*, 24(7), 1033–1036.

9

Interval Type-2 FAHP: A Multicriteria Wind Turbine Selection

Cengiz Kahraman, Başar Öztayşi, and Sezi Çevik Onar

CONTENTS

9.1 Introduction

The rising awareness of environmental issues, especially global warming and emission of greenhouse gases, security of supplying fossil fuels, and uncertainties in the price of the fossil fuels has increased the attention on renewable energy. Many authors have emphasized the importance of transforming the usage of fossil fuels to environmentally friendly energy sources that are new, clean, affordable, and include inexhaustible energy sources obtained from nature such as sun, wind, and hot water springs (Acikgoz, 2011; Dursun and Gokcol, 2014). Wind energy accounts for almost 36% of total generated renewable energy.

The usage of wind energy is increasing dramatically. Between 2007 and 2011, around 70% of the total global wind energy capacity was generated (Sahu et al., 2013). By the end of 2014, there were a total of 24 countries in Europe (16), the Asia-Pacific region (4), including China, India, Japan, and Australia, North America (3) and Latin America (1) with more than 1000 MW installed capacity. In addition, China (114,604 MW), the United States (65,879 MW), Germany (39,165 MW), Spain (22,987 MW), India (22,465

MW), and the United Kingdom (12,440 MW) have more than 10,000 MW of installed capacity as of the beginning of 2015 (web 1). More specifically, in Europe, Germany has the largest installed capacity is followed by Spain, the United Kingdom, and France, respectively (EWEA Report, 2014).

The World Wind Energy Association (WWEA) has predicted that global wind power capacity will be 600 GW in 2015 and 1500 GW in 2020 (Sahu et al., 2013). An International Energy Agency (IEA) report (2014) indicates that the wind industry will add 750 MW (annually) and have a cumulative installed capacity of 3817 MW in 2020. According to the Global Wind Energy Council (GWEC), a total number of 520,000 people will be employed by the wind industry in the European Union by 2020 and 45% of global electricity is expected to be supplied from wind energy by 2030 (web 2).

Turkey has one of the best growth rates considering operational, ongoing, and planning projects with regard to wind energy in Europe. The wind energy potential of the country being almost 48 GW attracts many investors. Figure 9.1 shows the cumulative installations for wind power plants in Turkey (Turkish Wind Energy Statistics Report, 2015). In 2014, the total installed wind energy capacity in Turkey was 3.762 GW. In the last 5 years, this installed capacity has increased by 183% with a 28% average growth rate.

In the literature, there are many studies that focus on decision-making problems related to wind energy such as wind farm sites, project plans, and wind turbine design selection (Kahraman et al., 2016a). Wu and

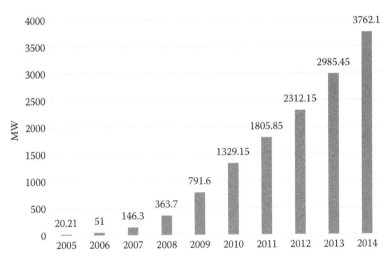

FIGURE 9.1
Cumulative installations for wind power plants in Turkey (Turkish wind energy statistics report).

Geng (2014) use an analytic hierarchy process (AHP) for the site selection problem for solar–wind hybrid power station. Sanchez-Lozano et al. (2014) utilize lexicographic order and ELimination Et Choix Traduisant la Realité (ELECTRE)-TRI methods for the wind farm site selection problem. Wu et al. (2014) use an intuitionistic fuzzy Choquet (IFC) operator and generalized intuitionistic fuzzy ordered geometric averaging (GIFOGA) operator for a wind farm project plan selection problem. Similarly, Lee et al. (2009) use an AHP-based multicriteria decision-making (MCDM) approach in order to select the appropriate wind farm project. Kayal and Chanda (2015) develop an optimization model for both location and site selection for solar and wind energy. Perkin et al. (2015) utilize evolutionary algorithms and blade element momentum theory with a realistic cost model in order to optimize theoretical wind turbine designs. Montoya et al. (2014) propose multiobjective evolutionary algorithm for selecting wind turbines.

Wind turbine selection is a MCDM problem that includes both tangible and intangible criteria, which necessitates usage of linguistic evaluations. The wind turbine evaluation process contains ambiguity, vagueness, and subjectivity in the human judgments. In this chapter, the usage of type-2 fuzzy sets in the AHP method will be explained in detail. Then an interval type-2 (IT2) fuzzy AHP (FAHP) method is used in technology selection for wind energy.

The rest of the chapter is organized as follows. Section 9.2 presents the basics of IT2 fuzzy sets. Section 9.3 gives a literature review on type-2 fuzzy sets in multicriteria decision-making. Section 9.4 introduces wind energy technologies and selection criteria. Section 9.5 includes the IT2 FAHP method and its application. Finally, Section 9.6 presents the conclusions.

9.2 IT2 Fuzzy Sets

In type-1 (ordinary) fuzzy sets, each element has a degree of membership, which is given by a membership function valued in the interval [0, 1] (Zadeh, 1965). The concept of a type-2 fuzzy set was introduced by Zadeh (1975) as an extension of the concept of an ordinary fuzzy set called a type-1 fuzzy set. Such sets are fuzzy sets whose membership grades themselves are type-1 fuzzy sets; they are very useful in circumstances where it is difficult to determine an exact membership function for a fuzzy set; hence, they are useful for incorporating linguistic uncertainties, for example, the words that are used in linguistic knowledge can mean different things to different people (Karnik and Mendel, 2001). While the membership functions of ordinary fuzzy sets are two dimensional, the

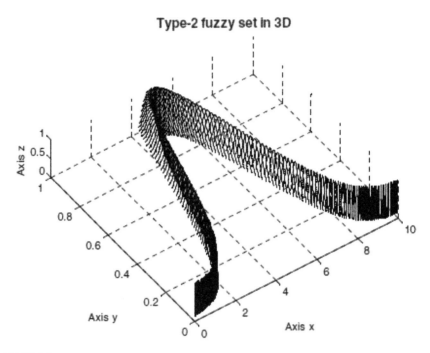

FIGURE 9.2
Interval type-2 fuzzy set.

membership functions of type-2 fuzzy sets are three dimensional. It is the new third dimension that provides additional degrees of freedom that make it possible to directly model uncertainties. Figure 9.2 illustrates an IT2 fuzzy set (Castillo and Melin, 2008).

A type-2 fuzzy set $\tilde{\tilde{A}}$ in the universe of discourse X can be represented by a type-2 membership function $\mu_{\tilde{\tilde{A}}}$, shown as follows (Zadeh, 1975):

$$\tilde{\tilde{A}} = \left\{ \left((x,u), \mu_{\tilde{\tilde{A}}}(x,u) \right) \mid \forall x \in X, \forall u \in J_x \subseteq [0,1], 0 \le \mu_{\tilde{\tilde{A}}}(x,u) \le 1 \right\} \qquad (9.1)$$

where J_x denotes an interval [0,1]. The type-2 fuzzy set $\tilde{\tilde{A}}$ also can be represented as follows (Mendel et al., 2006):

$$\tilde{\tilde{A}} = \int_{x \in X} \int_{u \in J_x} \mu_{\tilde{\tilde{A}}}(x,u)/(x,u) \qquad (9.2)$$

where $J_x \subseteq [0,1]$ and $\iint$ denotes union over all admissible x and u.

Let $\tilde{\tilde{A}}$ be a type-2 fuzzy set in the universe of discourse X represented by the type-2 membership function $\mu_{\tilde{\tilde{A}}}$. If all $\mu_{\tilde{\tilde{A}}}(x,u)=1$, then $\tilde{\tilde{A}}$ is called an IT2 fuzzy set (Buckley, 1985). An IT2 fuzzy set $\tilde{\tilde{A}}$ can be regarded as a special case of a type-2 fuzzy set, represented as follows (Mendel et al., 2006):

$$\tilde{\tilde{A}} = \int_{x \in X} \int_{u \in J_x} 1/(x,u) \tag{9.3}$$

where $J_x \subseteq [0,1]$.

Arithmetic operations with trapezoidal IT2 fuzzy sets are given in the following.

Definition 9.2.1: The upper and lower membership functions of an IT2 fuzzy set are type-1 membership functions.

A trapezoidal IT2 fuzzy set is illustrated as $\tilde{\tilde{A}}_i = \left(\tilde{A}_i^U ; \tilde{A}_i^L \right) = \left(\left(a_{i1}^U, a_{i2}^U, a_{i3}^U, \right. \right.$ $a_{i4}^U ; H_1\left(\tilde{A}_i^U\right), H_2\left(\tilde{A}_i^U\right) \right), \left(a_{i1}^L, a_{i2}^L, a_{i3}^L, a_{i4}^L ; H_1\left(\tilde{A}_i^L\right), H_2\left(\tilde{A}_i^L\right) \right) \right)$ where $\tilde{A}_i^U$ and $\tilde{A}_i^L$ are type-1 fuzzy sets, $a_{i1}^U, a_{i2}^U, a_{i3}^U, a_{i4}^U, a_{i1}^L, a_{i2}^L, a_{i3}^L$, and a_{i4}^L are the reference points of the IT2 fuzzy set $\tilde{\tilde{A}}_i$, $H_j\left(\tilde{A}_i^U\right)$ denotes the membership value of the element $a_{j(j+1)}^U$ in the upper trapezoidal membership function $\left(\tilde{A}_i^U\right)$, $1 \le j \le 2$, $H_j\left(\tilde{A}_i^L\right)$ denotes the membership value of the element $a_{j(j+1)}^L$ in the lower trapezoidal membership function $\tilde{A}_i^L$, $1 \le j \le 2$, $H_1\left(\tilde{A}_i^U\right) \in [0,1]$, $H_2\left(\tilde{A}_i^U\right) \in [0,1]$, $H_1\left(\tilde{A}_i^L\right) \in [0,1]$, $H_2\left(\tilde{A}_i^L\right) \in [0,1]$, and $1 \le i \le n$ (Chen and Lee, 2010a).

Definition 9.2.2: The addition operation between the trapezoidal IT2 fuzzy sets $\tilde{\tilde{A}}_1 = \left(\left(a_{11}^U, a_{12}^U, a_{13}^U, a_{14}^U ; H_1\left(\tilde{A}_1^U\right), H_2\left(\tilde{A}_1^U\right) \right), \left(a_{11}^L, a_{12}^L, a_{13}^L, a_{14}^L, H_1\left(\tilde{A}_1^L\right), H_2\left(\tilde{A}_1^L\right) \right) \right)$ and $\tilde{\tilde{A}}_2 = \left(\left(a_{21}^U, a_{22}^U, a_{23}^U, a_{24}^U ; H_1\left(\tilde{A}_2^U\right), H_2\left(\tilde{A}_2^U\right) \right), \left(a_{21}^L, a_{22}^L, a_{23}^L, a_{24}^L ; H_1\left(\tilde{A}_2^L\right), H_2\left(\tilde{A}_2^L\right) \right) \right)$ is defined as follows (Chen and Lee, 2010a):

$$\tilde{\tilde{A}}_1 \oplus \tilde{\tilde{A}}_2 = \left(\left(a_{11}^U + a_{21}^U, a_{12}^U + a_{22}^U, a_{13}^U + a_{23}^U, a_{14}^U + a_{24}^U ; \right. \right.$$

$$\min\left(H_1\left(\tilde{A}_1^U\right); H_1\left(\tilde{A}_2^U\right) \right), \min\left(H_2\left(\tilde{A}_1^U\right); H_2\left(\tilde{A}_2^U\right) \right) \right),$$

$$\left(a_{11}^L + a_{21}^L, a_{12}^L + a_{22}^L, a_{13}^L + a_{23}^L, a_{14}^L + a_{24}^L ; \right. \tag{9.4}$$

$$\left. \left. \min\left(H_1\left(\tilde{A}_1^L\right); H_1\left(\tilde{A}_2^L\right) \right), \min\left(H_2\left(\tilde{A}_1^L\right); H_2\left(\tilde{A}_2^L\right) \right) \right) \right)$$

Definition 9.2.3: The subtraction operation between the trapezoidal IT2 fuzzy sets $\tilde{\tilde{A}}_1 = \left(\left(a_{11}^U, a_{12}^U, a_{13}^U, a_{14}^U; H_1\left(\tilde{A}_1^U\right), H_2\left(\tilde{A}_1^U\right)\right), \left(a_{11}^L, a_{12}^L, a_{13}^L, a_{14}^L, H_1\left(\tilde{A}_1^L\right), H_2\left(\tilde{A}_1^L\right)\right)\right)$ and $\tilde{\tilde{A}}_2 = \left(\left(a_{21}^U, a_{22}^U, a_{23}^U, a_{24}^U; H_1\left(\tilde{A}_2^U\right), H_2\left(\tilde{A}_2^U\right)\right), \left(a_{21}^L, a_{22}^L, a_{23}^L, a_{24}^L; H_1\left(\tilde{A}_2^L\right), H_2\left(\tilde{A}_2^L\right)\right)\right)$ is defined as follows (Chen and Lee, 2010a):

$$
\begin{aligned}
\tilde{\tilde{A}}_1 \ominus \tilde{\tilde{A}}_2 = & \Big(\left(a_{11}^U - a_{24}^U, a_{12}^U - a_{23}^U, a_{13}^U - a_{22}^U, a_{14}^U - a_{21}^U;\right. \\
& \min\left(H_1\left(\tilde{A}_1^U\right); H_1\left(\tilde{A}_2^U\right)\right), \min\left(H_2\left(\tilde{A}_1^U\right); H_2\left(\tilde{A}_2^U\right)\right)\Big), \\
& \left(a_{11}^L - a_{24}^L, a_{12}^L - a_{23}^L, a_{13}^L - a_{22}^L, a_{14}^L - a_{21}^L;\right. \\
& \min\left(H_1\left(\tilde{A}_1^L\right); H_1\left(\tilde{A}_2^L\right)\right), \min\left(H_2\left(\tilde{A}_1^L\right); H_2\left(\tilde{A}_2^L\right)\right)\Big)
\end{aligned}
\tag{9.5}
$$

Definition 9.2.4: The multiplication operation between the trapezoidal IT2 fuzzy sets $\tilde{\tilde{A}}_1 = \left(\left(a_{11}^U, a_{12}^U, a_{13}^U, a_{14}^U; H_1\left(\tilde{A}_1^U\right), H_2\left(\tilde{A}_1^U\right)\right), \left(a_{11}^L, a_{12}^L, a_{13}^L, a_{14}^L, H_1\left(\tilde{A}_1^L\right), H_2\left(\tilde{A}_1^L\right)\right)\right)$ and $\tilde{\tilde{A}}_2 = \left(\left(a_{21}^U, a_{22}^U, a_{23}^U, a_{24}^U; H_1\left(\tilde{A}_2^U\right), H_2\left(\tilde{A}_2^U\right)\right), \left(a_{21}^L, a_{22}^L, a_{23}^L, a_{24}^L; H_1\left(\tilde{A}_2^L\right), H_2\left(\tilde{A}_2^L\right)\right)\right)$ is defined as follows (Chen and Lee, 2010a):

$$
\begin{aligned}
\tilde{\tilde{A}}_1 \otimes \tilde{\tilde{A}}_2 \cong & \Big(\left(a_{11}^U \times a_{21}^U, a_{12}^U \times a_{22}^U, a_{13}^U \times a_{23}^U, a_{14}^U \times a_{24}^U;\right. \\
& \min\left(H_1\left(\tilde{A}_1^U\right); H_1\left(\tilde{A}_2^U\right)\right), \min\left(H_2\left(\tilde{A}_1^U\right); H_2\left(\tilde{A}_2^U\right)\right)\Big), \\
& \left(a_{11}^L \times a_{21}^L, a_{12}^L \times a_{22}^L, a_{13}^L \times a_{23}^L, a_{14}^L \times a_{24}^L;\right. \\
& \min\left(H_1\left(\tilde{A}_1^L\right); H_1\left(\tilde{A}_2^L\right)\right), \min\left(H_2\left(\tilde{A}_1^L\right); H_2\left(\tilde{A}_2^L\right)\right)\Big)
\end{aligned}
\tag{9.6}
$$

Definition 9.2.5: The arithmetic operations between the trapezoidal IT2 fuzzy set $\tilde{\tilde{A}}_1 = \left(\left(a_{11}^U, a_{12}^U, a_{13}^U, a_{14}^U; H_1\left(\tilde{A}_1^U\right), H_2\left(\tilde{A}_1^U\right)\right), \left(a_{11}^L, a_{12}^L, a_{13}^L, a_{14}^L, H_1\left(\tilde{A}_1^L\right), H_2\left(\tilde{A}_1^L\right)\right)\right)$ and the crisp value k are defined as follows (Chen and Lee, 2010a):

$$
\begin{aligned}
k\tilde{\tilde{A}}_1 = & \Big(\left(k \times a_{11}^U, k \times a_{12}^U, k \times a_{13}^U, k \times a_{14}^U\right); H_1\left(\tilde{A}_1^U\right), H_2\left(\tilde{A}_1^U\right), \\
& \left(k \times a_{11}^L, k \times a_{12}^L, k \times a_{13}^L, k \times a_{14}^L; H_1\left(\tilde{A}_1^L\right), H_2\left(\tilde{A}_1^L\right)\right)\Big)
\end{aligned}
\tag{9.7}
$$

$$\frac{\tilde{\tilde{A}}_1}{k} = \left(\left(\frac{1}{k} \times a_{11}^U, \frac{1}{k} \times a_{12}^U, \frac{1}{k} \times a_{13}^U, \frac{1}{k} \times a_{14}^U \right); H_1\left(\tilde{A}_1^U \right), H_2\left(\tilde{A}_1^U \right), \right.$$

$$\left. \left(\frac{1}{k} \times a_{11}^L, \frac{1}{k} \times a_{12}^L, \frac{1}{k} \times a_{13}^L, \frac{1}{k} \times a_{14}^L; H_1\left(\tilde{A}_1^L \right), H_2\left(\tilde{A}_1^L \right) \right) \right)$$

(9.8)

where $k > 0$.

9.3 Literature Review: Type-2 Fuzzy Sets in Multicriteria Decision-Making

In the literature, fuzzy sets have been extensively used for MCDM (Kahraman et al., 2015, 2016b,c; Oztaysi et al., 2015). Type-2 fuzzy sets are one of the fuzzy set extensions frequently used in the literature.

Kahraman et al. (2014) develop an IT2 FAHP method together with a new ranking method for type-2 fuzzy sets. They apply the proposed method to a supplier selection problem.

Kiliç and Kaya (2015) propose a new evaluation model for investment projects presented to regional development agencies operating in Turkey. In this model, type-2 fuzzy sets and crisp sets are simultaneously used. The proposed model is composed of type-2 FAHP and type-2 fuzzy technique for order of preference by similarity to ideal solution (TOPSIS) methods. Abdullah and Najib (2014) propose a new IT2 FAHP method based on linguistic variables.

Chen et al. (2013) develop an extended QUALIFLEX method for handling multiple criteria decision-making problems in the context of IT2 fuzzy sets. QUALIFLEX, a generalization of Jacquet–Lagreze's permutation method, is a useful outranking method in decision analysis because of its flexibility with respect to cardinal and ordinal information. Using the linguistic rating system converted into IT2 trapezoidal fuzzy numbers, the extended QUALIFLEX method investigates all possible permutations of the alternatives with respect to the level of concordance of the complete preference order. Based on a signed distance-based approach, they propose the concordance/discordance index, the weighted concordance/discordance index, and the comprehensive concordance/discordance index for ranking the alternatives. This paper extends QUALIFLEX using type-2 fuzzy sets for the first time.

Chen (2014) develops an interactive method for handling multiple criteria group decision-making problems, in which information about criteria weights is incompletely (imprecisely or partially) known and the criteria values are expressed as IT2 trapezoidal fuzzy numbers. With respect to the relative importance of multiple decision makers (DMs) and group consensus of fuzzy opinions, a hybrid averaging approach combining weighted averages and ordered weighted averages is employed to construct the collective decision matrix. An integrated programming model is then established based on the concept of signed distance-based closeness coefficients to determine the importance weights of criteria and the priority ranking of alternatives. Chen (2013) develops a new linear assignment method to produce an optimal preference ranking of the alternatives in accordance with a set of criterion-wise rankings and a set of criterion importance within the context of IT2 trapezoidal fuzzy numbers. The proposed IT2 fuzzy linear assignment method utilizes signed distances and does not require a complicated computation procedure.

Wang et al. (2012) investigate the group decision-making problems in which all the information provided by the DMs is expressed as IT2 fuzzy decision matrices, and the information about attribute weights is partially known. They first use the IT2 fuzzy weighted arithmetic averaging operator to aggregate all individual IT2 fuzzy decision matrices provided by the DMs into the collective IT2 fuzzy decision matrix, then they utilize the ranking-value measure and construct the ranking-value matrix of the collective IT2 fuzzy decision matrix.

Chen and Lee (2010b) present a new method for handling fuzzy multiple criteria hierarchical group decision-making problems based on arithmetic operations and fuzzy preference relations of IT2 fuzzy sets. Because the time complexity of the proposed method is $O(nk)$, where n is the number of criteria and k is the number of decision makers, it is more efficient than Wu and Mendel's method, whose time complexity is $O(mnk)$, where m is the number of α-cuts.

Balezentis and Zeng (2013) extend the MULTIMOORA method with generalized interval-valued type-2 trapezoidal fuzzy sets. Utilization of aggregation operators also enables to facilitate group MCDM.

Çelik et al. (2014) address the problems of public transportation customers in Istanbul and their satisfaction levels are evaluated by using a customer satisfaction survey and statistical analyses. A novel IT2 fuzzy multiattribute decision-making (MADM) method is proposed based on TOPSIS and Grey-related analysis (GRA) to evaluate and improve customer satisfaction in Istanbul public transportation.

9.4 Wind Turbine Investments

Wind energy has great potential but it cannot be fully utilized due to expensive initial investment costs and risks associated with these investments. In wind turbine investment decisions, many factors such as wind turbine characteristics, environmental conditions, and the coherence between these two factors should be taken into account. By considering the complex and conflicting factors, selecting the most efficient wind turbine investment can increase the wind energy usage.

The environmental and technological factors are used to calculate available wind energy potential in a given area. Wind speed and air density are the main environmental factors whereas swept area, rotor diameter, cut-in speed, cut-out speed, power curve, and hub height are the technological characteristics that define the wind energy potential in a given area with the given wind turbine (Cevik Onar et al., 2015; Cevik Onar and Kilavuz, 2015; Perkin et al., 2015; Chowdhury et al., 2013; Sarja and Halonen, 2013; Kaya and Kahraman, 2010).

Depending on the characteristics of the average wind speeds the wind turbines are classified. For the lower average speed values class IVA and IVB wind turbines are used and for the higher average speed values class IA and IB are used. The initial decision is to define the appropriate classes of the wind turbines for the selected area. Also different wind turbine providers supply turbines with different characteristics. Figure 9.3 shows the turbine manufacturers according to their installed capacities in Turkey.

Along with the wind energy potential, other cost and benefit factors should be considered for financially optimal wind turbine selection. Usually, the bounded rationality of the DMs limits the financially optimal decisions (Perkin et al., 2015). Reliability, technical characteristics, performance, cost factors, availability, maintenance, cooperation, and domesticity are the main factors that should be considered for appropriate wind turbine selection (Cevik Onar et al., 2015).

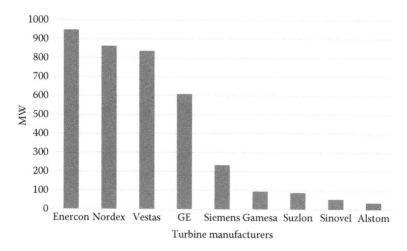

FIGURE 9.3
Turbine manufacturers according to their installed capacities in Turkey. (From Turkish Wind Energy Statistics Report. [2015] http://www.tureb.com.tr/en/twea-announcements/434-turkish-wind-energy-statistics-report-january-2015 [Accessed date July 26, 2015].)

The reliability factor shows the overall reputation of the manufacturer. A manufacturer's closeness to the investment area and the warranty provided by the manufacturer is considered under reliability. Usually, investors tend to select the manufacturers with which they have previous business connections. Reliability also represents this path dependency of the wind energy investor (Cevik Onar et al., 2015).

Technical characteristics are characteristics such as swept area, rotor diameter, cut-in speed, cut-out speed, and hub height. These characteristics affect maximum wind energy production. The performance represents the wind energy produced by the turbine at a given height and can be measured with the power curve (Cevik Onar et al., 2015). Cost factors are the items that constitute the overall cost of the investment. The main cost items are initial investment, installation, infrastructure such as earthmoving, tower foundations, and high voltage substations, operation costs, and maintenance costs (Cevik Onar et al., 2015).

Usually, it takes time to provide the desired wind turbine models with the desired characteristics. The availability factor represents the wind turbine manufacturer's availability to supply the desired wind turbine models with the desired characteristics and spare parts on time (Cevik Onar et al., 2015). The maintenance factor represents the quality and the punctuality of the maintenance service given by the manufacturer. The cooperation factor shows the cooperation capabilities of the manufacturer with the other parties involved in wind energy production such as electricity grid companies, contractors, and authorities (Cevik Onar et al., 2015). Usually, the domesticity of a wind turbine production creates additional cost or benefit advantages. In many countries, domesticity increases the value of feed–in tariffs (the guaranteed price provided by the government for wind energy investments) and provides additional compensation such as lower licensing fees (Cevik Onar et al., 2015, Cevik Onar and Kilavuz, 2015).

9.5 Multicriteria Wind Turbine Technology Selection Using Type-2 FAHP

In this section, we first give the steps of our methodology and then apply it to a real investment problem.

9.5.1 Methodology

In this section, Kahraman et al. (2014) approach is introduced. The procedure of this type-2 FAHP method is explained as follows:

Step 1: Construct fuzzy pairwise comparison matrices $(\tilde{A})$ of the criteria as in Equation 9.9:

$$
\tilde{A} = \begin{bmatrix} 1 & \tilde{a}_{12} & \cdots & \tilde{a}_{1n} \\ \tilde{a}_{21} & 1 & \cdots & \tilde{a}_{2n} \\ \vdots & \vdots & \ddots & \vdots \\ \tilde{a}_{n1} & \tilde{a}_{n2} & \cdots & 1 \end{bmatrix} = \begin{bmatrix} 1 & \tilde{a}_{12} & \cdots & \tilde{a}_{1n} \\ 1/\tilde{a}_{12} & 1 & \cdots & \tilde{a}_{2n} \\ \vdots & \vdots & \ddots & \vdots \\ 1/\tilde{a}_{1n} & 1/\tilde{a}_{2n} & \cdots & 1 \end{bmatrix} \tag{9.9}
$$

where

$$
1/\tilde{a} = \left(\left(\frac{1}{a_{14}^U}, \frac{1}{a_{13}^U}, \frac{1}{a_{12}^U}, \frac{1}{a_{11}^U}; H_1(a_{12}^U), H_2(a_{13}^U) \right), \left(\frac{1}{a_{24}^L}, \frac{1}{a_{23}^L}, \frac{1}{a_{22}^L}, \frac{1}{a_{21}^L}; H_1(a_{22}^L), H_2(a_{23}^L) \right) \right)
$$

The linguistic variables and their triangular and trapezoidal IT2 fuzzy scales that can be used in IT2 FAHP are given in Table 9.1.

Step 2: Check the consistency of each fuzzy pairwise comparison matrix. Assume $\tilde{A} = \left[\tilde{a}_{ij} \right]$ is a fuzzy positive reciprocal matrix and $A = \left[a_{ij} \right]$ is its defuzzified positive reciprocal matrix. If the result of the comparisons of $A = \left[a_{ij} \right]$ is consistent, then this can imply that the result of the comparisons of $\tilde{A} = \left[\tilde{a}_{ij} \right]$ is also consistent. In order to check the consistency of the fuzzy pairwise comparison matrices, the proposed DTriT or DTraT approach is used.

Step 3: Calculate the geometric mean of each row and then compute the fuzzy weights by normalization.

TABLE 9.1

The Linguistic Variables and Their Triangular and Trapezoidal Interval Type-2 Fuzzy Scales

Linguistic Variables	Triangular Interval Type-2 Fuzzy Scales	Trapezoidal Interval Type-2 Fuzzy Scales
Absolutely Strong (AS)	(7.5,9,10.5;1) (8.5,9,9.5;0.9)	(7,8,9,9;1,1) (7.2,8.2,8.8,9;0.8,0.8)
Very strong (VS)	(5.5,7,8.5;1) (6.5,7,7.5;0.9)	(5,6,8,9;1,1) (5.2,6.2,7.8,8.8;0.8,0.8)
Fairly strong (FS)	(3.5,5,6.5;1) (4.5,5,5.5;0.9)	(3,4,6,7;1,1) (3.2,4.2,5.8,6.8;0.8,0.8)
Slightly strong (SS)	(1.5,3,4.5;1) (2.5,3,3.5;0.9)	(1,2,4,5;1,1) (1.2,2.2,3.8,4.8;0.8,0.8)
Exactly equal (E)	(1,1,1;1)(1,1,1;1)	(1,1,1,1;1,1)(1,1,1,1;1,1)
If factor *i* has one of the above linguistic variables assigned to it when compared with factor *j*, then *j* has the reciprocal value when compared with *i*.	Reciprocals of above	Reciprocals of above

The geometric mean of each row $\tilde{r}_i$ is calculated as follows:

$$\tilde{r}_i = \left[\tilde{a}_{i1} \otimes \ldots \otimes \tilde{a}_{in}\right]^{1/n} \tag{9.10}$$

where

$$\sqrt[n]{\tilde{a}_{ij}} = \left(\left(\sqrt[n]{a_{ij1}^U}, \sqrt[n]{a_{ij2}^U}, \sqrt[n]{a_{ij3}^U}, \sqrt[n]{a_{ij4}^U}; H_1^u(a_{ij}), H_2^u(a_{ij})\right),\right.$$
$$\left.\left(\sqrt[n]{a_{ij1}^L}, \sqrt[n]{a_{ij2}^L}, \sqrt[n]{a_{ij3}^L}, \sqrt[n]{a_{ij4}^L}; H_1^L(a_{ij}), H_2^L(aa_{ij})\right)\right)$$

The fuzzy weight $\tilde{w}_i$ of the ith criterion is calculated as follows:

$$\tilde{w}_i = \tilde{r}_i \otimes \left[\tilde{r}_1 \oplus \ldots \oplus \tilde{r}_i \oplus \ldots \oplus \tilde{r}_n\right]^{-1} \tag{9.11}$$

where

$$\frac{\tilde{a}_{ij}}{\tilde{b}_{ij}} = \left(\frac{a_1^u}{b_4^u}, \frac{a_2^u}{b_3^u}, \frac{a_3^u}{b_2^u}, \frac{a_4^u}{b_1^u}, \min\left(H_1^u(a), H_1^u(b)\right), \min\left(H_2^u(a), H_2^u(b)\right)\right)$$

Step 4: Aggregate the fuzzy weights and fuzzy performance scores as follows:

$$\tilde{U}_i = \sum_{j=1}^{n} \tilde{w}_j \tilde{r}_{ij}, \forall i. \tag{9.12}$$

where $\tilde{U}_i$ is the fuzzy utility of alternative i, $\tilde{w}_j$ is the weight of the criterion j, and $\tilde{r}_{ij}$ is the performance score of alternative i with respect to criterion j.

Step 5: Apply the classical AHP method's procedure to determine the best alternative.

In order to find the crisp weights, the defuzzification formula given in Equation 9.13 is applied. For an IT2 trapezoidal fuzzy number $\tilde{A}_1 = \left(\left(a_{11}^U, a_{12}^U, a_{13}^U, a_{14}^U; H_1\left(\tilde{A}_1^U\right), H_2\left(\tilde{A}_1^U\right)\right), \left(a_{11}^L, a_{12}^L, a_{13}^L, a_{14}^L, H_1\left(\tilde{A}_1^L\right), H_2\left(\tilde{A}_1^L\right)\right)\right)$, the defuzzification formula is as follows:

$$DTtrT = \cfrac{\cfrac{\left(a_{14}^U - a_{11}^U\right) + \left(H_1\left(\tilde{A}_1^U\right) \cdot a_{12}^U - a_{11}^U\right) + \left(H_2\left(\tilde{A}_1^U\right) \cdot a_{13}^U - a_{11}^U\right)}{4} + }{2}$$

$$a_{11}^U + \cfrac{\left[\cfrac{\left(a_{14}^L - a_{11,}^L\right) + \left(H_1\left(\tilde{A}_1^L\right) \cdot a_{12}^L - a_{11,}^L\right) + \left(H_2\left(\tilde{A}_1^L\right) \cdot a_{13}^L - a_{11,}^L\right)}{4} + a_{11,}^L\right]}{} \tag{9.13}$$

9.5.2 An Illustrative Application

A wind energy technology selection problem will be handled in this section. Since wind energy technology selection is an MCDM problem under uncertainty, which involves many linguistic evaluations and incomplete and vague information, we prefer using the type-2 FAHP method. The wind energy investments in Turkey have accelerated in recent years. Because of its high average wind speeds and availability of the land, the western region of Turkey is very attractive for wind energy investors. Soma is a district in the city of Manisa, which is located in the western region of Turkey, where some investors have already invested in wind energy. A new investor in the area wants to make a wind energy investment in a selected area in Soma. The considered criteria and alternatives are displayed in Figure 9.4 (Cevik Onar et al., 2015).

The considered eight criteria are reliability, technical characteristics, performance, cost factors, availability, maintenance, cooperation, and domesticity. The features of the alternative WTE82 are as follows: its rated power (MW) is 2; its wind class is IIA; its cut–in wind speed is 2; its cut–out wind speed is 28; and its swept area (m2) is 5281. The same features for the alternative WTS101 are 2, IIA, 4, 25, and 6800, respectively. These features are 2, IIA, 3, 20, and 7850 for the alternative WTV100 and 2, IIA, 3, 25, and 5945 for the alternative WTG87, respectively.

First of all, the equally weighted experts make pairwise comparisons for the criteria by using the scale in Table 9.1. The assigned linguistic evaluations are given in Table 9.2.

Applying Equation 9.10, the aggregated comparison matrix has been obtained from Table 9.2 as demonstrated in Table 9.3. For example, the bold numbers $(5.59,6.60,8.32,9;1,1)(5.79,6.80,8.12,8.83;0.8,0.8)$ in Table 9.3 is calculated as follows.

The experts compare reliability and technical characteristics and evaluate that reliability is more important as {absolutely strong, very strong, very strong}. The type-2 fuzzy numbers associated with these linguistic terms are $\{(7,8,9,9;1,1)(7.2,8.2,8.8,8.9;0.8,0.8);\ (5,6,8,9;1,1);(5.2,6.2,7.8,8.8;0.8,0.8);\ (5,6,8,9;1,1),(5.2,6.2,7.8,8.8;0.8,0.8)\}$

$$\widetilde{\widetilde{Agg}}_{Rel\text{-}TC} = [(7,8,9,9;1,1)(7.2,8.2,8.8,8.9;0.8,0.8) \otimes (5,6,8,9;1,1);$$

$$(5.2,6.2,7.8,8.8;0.8,0.8) \otimes (5,6,8,9;1,1),(5.2,6.2,7.8,8.8;0.8,0.8)]^{\frac{1}{3}}$$

$$= (5.59,6.60,8.32,9;1,1)(5.79,6.80,8.12,8.83;0.8,0.8)$$

Next, the geometric values of aggregated values in each row are determined. The geometric mean for reliability is obtained as follows and the inverse of the sum of geometric means is calculated:

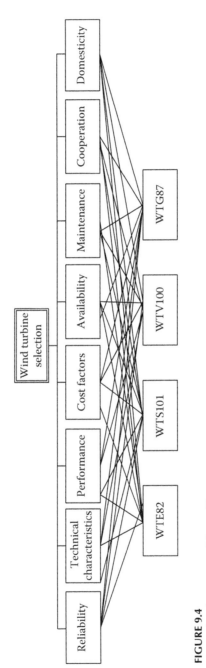

FIGURE 9.4
Hierarchical structure of the problem.

TABLE 9.2

Pairwise Comparisons of Criteria by Experts

Expert 1	Reliability	Technical Characteristics	Performance	Cost Factors	Availability	Maintenance	Cooperation	Domesticity
Reliability	EE	AH	H	MH	VH	EE	EE	E
Technical characteristics	–	EE	L	VL	E	E	MH	AL
Performance	–	–	EE	E	E	AH	EE	ML
Cost factors	–	–	–	EE	VH	AH	EE	ML
Availability	–	–	–	–	EE	H	AH	VL
Maintenance	–	–	–	–	–	EE	MH	AL
Cooperation	–	–	–	–	–	–	–	AL
Domesticity	–	–	–	–	–	–	–	EE
Expert 2	**Reliability**	**Technical Characteristics**	**Performance**	**Cost Factors**	**Availability**	**Maintenance**	**Cooperation**	**Domesticity**
Reliability	EE	VH	MH	MH	H	VH	AH	ML
Technical characteristics	–	EE	VL	VL	E	MH	H	VL
Performance	–	–	EE	E	ML	H	H	VL
Cost factors	–	–	–	EE	MH	ML	H	AL
Availability	–	–	–	–	EE	H	MH	AL
Maintenance	–	–	–	–	–	EE	EE	AL
Cooperation	–	–	–	–	–	–	–	AL
Domesticity	–	–	–	–	–	–	–	EE
Expert 3	**Reliability**	**Technical Characteristics**	**Performance**	**Cost Factors**	**Availability**	**Maintenance**	**Cooperation**	**Domesticity**
Reliability	EE	VH	H	E	H	VH	AH	E
Technical characteristics	–	EE	VL	VL	E	E	MH	AL
Performance	–	–	EE	ML	MH	VH	VH	VL
Cost factors	–	–	–	EE	MH	VH	AH	EE
Availability	–	–	–	–	EE	H	VH	L
Maintenance	–	–	–	–	–	EE	E	AL
Cooperation	–	–	–	–	–	–	EE	AL
Domesticity	–	–	–	–	–	–	–	EE

TABLE 9.3

Aggregated Comparison Matrix

	Reliability	Technical Characteristics	Performance	Cost Factors
Reliability	(1,1,1;1,1) (1,1,1;1,1)	(5.59,6.60,8.32,9;1,1) (5.79,6.80,8.12,8.83;0.8,0.8)	(2.08,3.17,5.24,6.25;1,1) (2.30,3.38,5.03,6.05;0.8,0.8)	(1.1.58,2.51,2.92;1,1) (1.12,1.69,2.43,2.84;0.8,0.8)
Technical characteristics	(0.11,0.12,0.15,0.17;1,1) (0.11,0.12,0.14,0.17;0.8,0.8)	(1,1,1;1,1) (1,1,1;1,1)	(0.12,0.13,0.19,0.23;1,1) (0.12,0.14,0.18,0.22;0.8,0.8)	(0.11,0.12,0.16,0.2;1,1) (0.11,0.12,0.16,0.19;0.8,0.8)
Performance	(0.15,0.19,0.31,0.48;1,1) (0.16,0.19,0.29,0.43;0.8,0.8)	(4.21,5.24,7.26,8.27;1,1) (4.42,5.44,7.06,8.07;0.8,0.8)	(1,1,1;1,1) (1,1,1;1,1)	(0.58,0.62,0.79;1,1) (0.59,0.64,0.76,0.94;0.8,0.8)
Cost factors	(0.34,0.39,0.62;1,1) (0.35,0.41,0.59,0.88;0.8,0.8)	(5,6,8,9;1,1) (5.2,6.2,7.8,8.8;0.8,0.8)	(1,1.25,1.58,1.70;1,1) (1.06,1.30,1.56,1.68;0.8,0.8)	(1,1,1;1,1) (1,1,1;1,1)
Availability	(0.13,0.15,0.21,0.28;1,1) (0.13,0.15,0.20,0.26;0.8,0.8)	(1,1,1;1,1) (1,1,1;1,1)	(0.58,0.79,1.25,1.70;1,1) (0.62,0.83,1.19,1.58;0.8,0.8)	(0.16,0.19,0.34,0.58;1,1) (0.17,0.20,0.32,0.51;0.8,0.8)
Maintenance	(0.23,0.25,0.30,0.34;1,1) (0.23,0.25,0.29,0.33;0.8,0.8)	(0.58,0.62,0.79;1,1) (0.59,0.64,0.76,0.94;0.8,0.8)	(0.12,0.13,0.17,0.21;1,1) (0.12,0.13,0.16,0.20;0.8,0.8)	(0.23,0.30,0.43,0.52;1,1) (0.24,0.31,0.42,0.50;0.8,0.8)
Cooperation	(0.23,0.23,0.25,0.27;1,1) (0.23,0.23,0.24,0.26;0.8,0.8)	(0.17,0.21,0.39,0.69;1,1) (0.18,0.22,0.36,0.60;0.8,0.8)	(0.25,0.27,0.34,0.40;1,1) (0.25,0.28,0.33,0.39;0.8,0.8)	(0.25,0.26,0.31,0.36;1,1) (0.25,0.26,0.30,0.35;0.8,0.8)
Domesticity	(1,1.25,1.58,1.70;1,1) (1.06,1.30,1.56,1.68;0.8,0.8)	(6.25,7.26,8.65,9;1,1) (6.45,7.47,8.45,8.86;0.8,0.8)	(2.92,4.16,6.34,7.39;1,1) (3.18,4.38,6.13,7.19;0.8,0.8)	(1,1.58,2.51,2.92;1,1) (1.12,1.69,2.43,2.84;0.8,0.8)

	Availability	Maintenance	Cooperation	Domesticity
Reliability	(3.55,4.57,6.60,7.61;1,1) (3.76,4.78,6.40,7.41;0.8,0.8)	(2.92,3.30,4.4,4.32;1,1) (3.00,3.37,3.93,4.26;0.8,0.8)	(3.65,4.4,4.32,4.32;1,1) (3.72,4.06,4.26,4.29;0.8,0.8)	(0.58,0.62,0.79;1,1) (0.59,0.64,0.76,0.94;0.8,0.8)
Technical characteristics	(1,1,1;1,1) (1,1,1;1,1)	(1,1.25,1.58,1.70;1,1) (1.06,1.30,1.56,1.68;0.8,0.8)	(1.44,2.51,4.57,5.59;1,1) (1.66,2.72,4.37,5.39;0.8,0.8)	(0.11,0.11,0.13,0.15;1,1) (0.11,0.11,0.13,0.15;0.8,0.8)
Performance	(0.58,0.79,1.25,1.70;1,1) (0.62,0.83,1.19,1.58;0.8,0.8)	(4.71,5.76,7.55,8.27;1,1) (4.92,5.97,7.35,8.10;0.8,0.8)	(2.46,2.88,3.63,3.97;1,1) (2.55,2.96,3.56,3.91;0.8,0.8)	(0.13,0.15,0.24,0.34;1,1) (0.13,0.16,0.22,0.31;0.8,0.8)
Cost factors	(1.70,2.88,5.03,6.08;1,1) (1.95,3.10,4.82,5.87;0.8,0.8)	(1.91,2.28,3.30,4.32;1,1) (1.98,2.37,3.14,4.02;0.8,0.8)	(2.75,3.17,3.77,3.97;1,1) (2.84,3.25,3.70,3.92;0.8,0.8)	(0.34,0.39,0.62;1,1) (0.35,0.41,0.59,0.88;0.8,0.8)

(Continued)

TABLE 9.3 (*Continued*)

Aggregated Comparison Matrix

	Reliability	Technical Characteristics	Performance	Cost Factors
Availability	(1,1,1,1;1,1) (1,1,1,1;1,1)	(3,4,6,7;1,1) (3.2,4.2,5.8,6.8;0.8,0.8)	(4.71,5.76,7.55,8.27;1,1) (4.92,5.97,7.35,8.10;0.8,0.8)	(0.12,0.13,0.17,0.21;1,1) (0.12,0.13,0.16,0.20;0.8,0.8)
Maintenance	(0.14,0.16,0.25,0.33;1,1) (0.14,0.17,0.23,0.31;0.8,0.8)	(1,1,1,1;1,1) (1,1,1,1;1,1)	(1,1.58,2.51,2.92;1,1) (1.12,1.69,2.43,2.84;0.8,0.8)	(0.11,0.11,0.12,0.14;1,1) (0.11,0.11,0.12,0.13;0.8,0.8)
Cooperation	(0.12,0.13,0.17,0.21;1,1) (0.12,0.13,0.16,0.20;0.8,0.8)	(0.34,0.39,0.62,1;1,1) (0.35,0.41,0.59,0.88;0.8,0.8)	(1,1,1,1;1,1) (1,1,1,1;1,1)	(0.11,0.11,0.12,0.14;1,1) (0.11,0.11,0.12,0.13;0.8,0.8)
Domesticity	(4.71,5.76,7.55,8.27;1,1) (4.92,5.97,7.35,8.10;0.8,0.8)	(7,8,9,9;1,1) (7.2,8.2,8.8,8.9;0.8,0.8)	(7,8,9,9;1,1) (7.2,8.2,8.8,8.9;0.8,0.8)	(1,1,1,1;1,1) (1,1,1,1;1,1)

$$\tilde{r}_{\text{reliability}} = \begin{bmatrix} (1,1,1,1;1,1)(1,1,1,1;1,1) \\ \otimes \ (5.59,6.60,8.32,9;1,1)(5.79,6.80,8.12,8.83;0.8,0.8) \\ \otimes \ (2.08,3.17,5.24,6.25;1,1)(2.30,3.38,5.03,6.05;0.8,0.8) \\ \otimes \ (1,1.58,2.51,2.92;1,1)(1.12,1.69,2.43,2.84;0.8,0.8) \\ \otimes \ (3.55,4.57,6.60,7.61;1,1)(3.76,4.78,6.40,7.41;0.8,0.8) \\ \otimes \ (2.92,3.30,4,4.32;1,1)(3.00,3.37,3.93,4.26;0.8,0.8) \\ \otimes \ (3.65,4,4.32,4.32;1,1)(3.72,4.06,4.26,4.29;0.8,0.8) \\ \otimes \ (0.58,0.62,0.79,1;1,1)(0.59,0.64,0.76,0.94;0.8,0.8) \end{bmatrix}^{\frac{1}{8}}$$

$$= (2.00,2.44,3.16,3.51;1,1)(2.09,2.52,3.08,3.43;0.8,0.8)$$

Table 9.4 presents all of the calculated geometric means for the criteria. The sum of geometric means in Table 9.4 is calculated as follows:

$$\sum \tilde{r} = (2.00,2.44,3.16,3.51;1,1)(2.09,2.52,3.08,3.43;0.8,0.8)$$

$$\oplus \ (0.35,0.40,0.51,0.58;1,1)(0.36,0.41,0.50,0.56;0.8,0.8)$$

$$\oplus (0.88,1.03,1.40,1.72;1,1)(0.91,1.06,1.35,1.63;0.8,0.8)$$

$$\oplus (1.23,1.49,2.05,2.51;1,1)(1.28,1.54,1.98,2.39;0.8,0.8)$$

$$\oplus \ (0.61,0.72,0.96,1.16;1,1)(0.64,0.74,0.93,1.11;0.8,0.8)$$

$$\oplus \ (0.29,0.34,0.44,0.51;1,1)(0.30,0.35,0.42,0.50;0.8,0.8)$$

$$\oplus (0.24,0.26,0.33,0.41;1,1)(0.24,0.26,0.32,0.39;0.8,0.8)$$

$$\oplus \ (2.83,3.49,4.37,4.66;1,1)(2.98,3.61,4.27,4.58;0.8,0.8)$$

$$= (8.46,10.19,13.24,15.09;1,1)(8.84,10.52,12.88,14.63;0.8,0.8)$$

Finally in order to determine the weight of each criterion, the geometric mean of each criterion is divided by sum of geometric means. For *reliability* criteria the weight $(0.13,0.18,0.30,0.41;1,1)(0.14,0.19,0.29,0.38;0.8,0.8)$ is calculated as follows:

$$\tilde{w}_{\text{reliability}} = \tilde{r}_{\text{reliability}} \otimes \left[\sum \tilde{r}\right]^{-1}$$

TABLE 9.4

Geometric Means and Sum of Geometric Means

Criteria	Type-2 Fuzzy Numbers
Reliability	(2.00,2.44,3.16,3.51;1,1)(2.09,2.52,3.08,3.43;0.8,0.8)
Technical characteristics	(0.35,0.40,0.51,0.58;1,1)(0.36,0.41,0.50,0.56;0.8,0.8)
Performance	(0.88,1.03,1.40,1.72;1,1)(0.91,1.06,1.35,1.63;0.8,0.8)
Cost factors	(1.23,1.49,2.05,2.51;1,1)(1.28,1.54,1.98,2.39;0.8,0.8)
Availability	(0.61,0.72,0.96,1.16;1,1)(0.64,0.74,0.93,1.11;0.8,0.8)
Maintenance	(0.29,0.34,0.44,0.51;1,1)(0.30,0.35,0.42,0.50;0.8,0.8)
Cooperation	(0.24,0.26,0.33,0.41;1,1)(0.24,0.26,0.32,0.39;0.8,0.8)
Domesticity	(2.83,3.49,4.37,4.66;1,1)(2.98,3.61,4.27,4.58;0.8,0.8)
Sum of geometric means	(8.46,10.19,13.24,15.09;1,1)(8.84,10.52,12.88,14.63;0.8,0.8)

$$\tilde{\tilde{w}}_{\text{reliability}} = (2.00, 2.44, 3.16, 3.51; 1, 1)(2.09, 2.52, 3.08, 3.43; 0.8, 0.8)$$

$$\otimes \left[(8.46, 10.19, 13.24, 15.09; 1, 1)(8.84, 10.52, 12.88, 14.63; 0.8, 0.8) \right]^{-1}$$

$$\tilde{\tilde{w}}_{\text{reliability}} = (2.00, 2.44, 3.16, 3.51; 1, 1)(2.09, 2.52, 3.08, 3.43; 0.8, 0.8)$$

$$\otimes (0.066, 0.075, 0.098, 0.11; 1, 1)(0.06, 0.077, 0.095, 0.11; 0.8, 0.8)$$

$$= (0.13, 0.18, 0.30, 0.41; 1, 1)(0.14, 0.19, 0.29, 0.38; 0.8, 0.8)$$

For each criterion, the same calculations are executed and the type-2 fuzzy weights are calculated as given in Table 9.5.

In a similar way, the experts also compared each alternative with respect to eight criteria. Due to the page limitations we only demonstrate the resulting priorities of these evaluations as shown in Table 9.6.

As the priorities of the alternatives and importance of each criterion are obtained, next the global priorities of the alternatives should be calculated. To this end, the priorities are multiplied by the weights of the considered criteria. For example, the global priority of alternative WTE82 with respect to the criterion *reliability* is (0.031,0.070,0.24,0.47;1,1)(0.03,0.080,0.21,0.40;0.8,0.8). This value is calculated as follows:

$$\tilde{\tilde{p}}_{\text{global}} = \tilde{\tilde{p}}_{\text{local}} \otimes \tilde{\tilde{w}}_{\text{reliability}}$$

$$\tilde{\tilde{p}}_{\text{global}} = (0.23, 0.38, 0.77, 1.14; 1, 1)(0.26, 0.41, 0.72, 1.05; 0.8, 0.8)$$

$$\otimes (0.13, 0.18, 0.30, 0.41; 1, 1)(0.14, 0.19, 0.29, 0.38, 0.8, 0.8)$$

$$= (0.031, 0.070, 0.24, 0.47; 1.1)(0.03, 0.080, 0.21, 0.40; 0.8, 0.8)$$

TABLE 9.5

Type-2 Fuzzy Weights of the Criteria with Respect to the Goal

Criteria	Type-2 Fuzzy Weights
Reliability	(0.13,0.18,0.30,0.41;1,1)(0.14,0.19,0.29,0.38;0.8,0.8)
Technical characteristics	(0.02,0.030,0.050,0.068;1,1)(0.024,0.032,0.04,0.064;0.8,0.8)
Performance	(0.058,0.078,0.13,0.20;1,1)(0.062,0.082,0.12,0.18;0.8,0.8)
Cost factors	(0.081,0.11,0.20,0.29;1,1)(0.087,0.11,0.18,0.27;0.8,0.8)
Availability	(0.04,0.054,0.094,0.13;1,1)(0.043,0.05,0.088,0.12;0.8,0.8)
Maintenance	(0.019,0.025,0.04,0.061;1,1)(0.021,0.027,0.040,0.056;0.8,0.8)
Cooperation	(0.016,0.019,0.032,0.048;1,1)(0.016,0.02,0.030,0.044;0.8,0.8)
Domesticity	(0.18,0.26,0.42,0.55;1,1)(0.20,0.28,0.40,0.51;0.8,0.8)

After applying the same procedure for all the local priorities, the global priority of each alternative is obtained as shown in Table 9.7.

In order to select the best alternative, the global priority of each alternative is summed up. For example, the weight of the alternative WTE82 (0.14,0.28,0.79,1.44;1,1)(0.17,0.31,0.71,1.26;0.8,0.8) is obtained as follows:

$$\sum \tilde{\tilde{r}} = (0.031,0.070,0.24,0.47;1.1)(0.03,0.080,0.21,0.40;0.8,0.8)$$

$$\oplus (0.00,0.007,0.022,0.043;1,1)(0.005,0.008,0.019,0.037;0.8,0.8)$$

$$\oplus (0.008,0.017,0.055,0.11;1.1)(0.010,0.019,0.049,0.099;0.8,0.8)$$

$$\oplus (0.004,0.007,0.021,0.043;1.1)(0.004,0.008,0.019,0.036;0.8,0.8)$$

$$\oplus (0.010,0.018,0.05,0.10;1.1)(0.011,0.021,0.048,0.09;0.8,0.8)$$

$$\oplus (0.002,0.004,0.01,0.02;1.1)(0.002,0.004,0.011,0.022,0.8,0.8)$$

$$\oplus (0.004,0.007,0.02,0.055;1.1)(0.004,0.008,0.022,0.046;0.8,0.8)$$

$$\oplus (0.083,0.14,0.36,0.57;1.1)(0.09,0.16,0.33,0.51,0.8,0.8)$$

$$= (0.14,0.28,0.79,1.44;1.1)(0.17,0.31,0.71,1.26;0.8,0.8)$$

Fuzzy and crisp weights of each criterion are presented in Table 9.8. The type-2 fuzzy weights are defuzzified using Equation 9.13. For the alternative WTE82, it is calculated as follows:

$$\frac{(1.44-0.14)+(1.\ 0.28-0.14)+(1.\ 0.79-0.14)}{4}+0.14$$

$$\frac{+\left[\dfrac{(0.46-0.17,)+(0.8.\ 0.10,-0.17)+(0.8.\ 0.25-0.17)}{4}+0.17\right]}{2}=0.617$$

TABLE 9.6

Type-2 Fuzzy Weights of the Alternatives with Respect to the Criteria

Alternatives	Reliability	Technical Characteristics
WTE82	(0.23,0.38,0.77,1.14;1,1)	(0.18,0.25,0.44,0.63;1,1)
	(0.26,0.41,0.72,1.05;0.8,0.8)	(0.20,0.27,0.41,0.57;0.8,0.8)
WTS101	(0.051,0.075,0.15,0.24;1,1)	(0.29,0.41,0.67,0.88;1,1)
	(0.056,0.081,0.14,0.22;0.8,0.8)	(0.31,0.43,0.64,0.83;0.8,0.8)
WTV100	(0.13,0.19,0.41,0.66;1,1)	(0.041,0.053,0.092,0.13;1,1)
	(0.14,0.21,0.38,0.59;0.8,0.8)	(0.044,0.056,0.086, 0.12;0.8,0.8)
WTG87	(0.031,0.042,0.087,0.15;1,1)	(0.038,0.049,0.081,0.11;1,1)
	(0.033,0.044,0.07,0.13;0.8,0.8)	(0.041,0.052,0.076, 0.10;0.8,0.8)
Alternatives	**Performance**	**Cost**
WTE 82	(0.14,0.21,0.40,0.58;1,1)	(0.049,0.065,0.10,0.14;1,1)
	(0.16,0.23,0.38,0.53;0.8,0.8)	(0.052,0.068,0.10,0.13;0.8,0.8)
WTS101	(0.28,0.40,0.70,0.94;1,1)	(0.033,0.042,0.069,0.099;1,1)
	(0.31,0.43,0.66,0.88;0.8,0.8)	(0.035,0.04,0.065, 0.091;0.8,0.8)
WTV100	(0.063,0.082,0.15,0.22;1,1)	(0.27,0.36,0.59,0.78;1,1)
	(0.066,0.087,0.14,0.20;0.8,0.8)	(0.28,0.38,0.56,0.73;0.8,0.8)
WTG87	(0.034,0.04,0.073,0.10;1,1)	(0.21,0.30,0.52,0.72;1,1)
	(0.036,0.045,0.069, 0.09;0.8,0.8)	(0.23,0.32,0.49,0.67;0.8,0.8)
Alternatives	**Availability**	**Maintenance**
WTE82	(0.24,0.34,0.57,0.77;1,1)	(0.12,0.16,0.29,0.43;1,1)
	(0.26,0.36,0.54,0.71;0.8,0.8)	(0.13,0.17,0.27,0.39;0.8,0.8)
WTS101	(0.10,0.13,0.21,0.29;1,1)	(0.054,0.073,0.12,0.17;1,1)
	(0.10,0.14,0.20,0.27;0.8,0.8)	(0.058,0.077,0.11,0.15;0.8,0.8)
WTV100	(0.15,0.22,0.37,0.52;1,1)	(0.29,0.42,0.75,1.03;1,1)
	(0.17,0.23,0.35,0.48;0.8,0.8)	(0.32,0.45,0.70,0.96;0.8,0.8)
WTG87	(0.053,0.07,0.12,0.19;1,1)	(0.060,0.086,0.15,0.23;1,1)
	(0.056,0.074,0.11,0.17;0.8,0.8)	(0.066,0.09,0.14,0.21;0.8,0.8)
Alternatives	**Cooperation**	**Domesticity**
WTE82	(0.25,0.40,0.78,1.13;1,1)	(0.44,0.56,0.85,1.04;1,1)
	(0.28,0.43,0.73,1.04;0.8,0.8)	(0.47,0.59,0.81,1.00;0.8,0.8)
WTS101	(0.030,0.040,0.079,0.13;1,1)	(0.068,0.088,0.13,0.18;1,1)
	(0.032,0.043,0.073, 0.11;0.8,0.8)	(0.073,0.093,0.13,0.17;0.8,0.8)
WTV100	(0.11,0.17,0.33,0.50;1,1)	(0.058,0.077,0.12,0.17;1,1)
	(0.12,0.18,0.31,0.45;0.8,0.8)	(0.06,0.081,0.12,0.16;0.8,0.8)
WTG87	(0.072,0.10,0.19,0.31;1,1)	(0.060,0.07,0.11,0.16;1,1)
	(0.077,0.10,0.18,0.28;0.8,0.8)	(0.063,0.078,0.11,0.15;0.8,0.8)

TABLE 9.7

Type-2 Fuzzy Priorities of the Alternatives with Respect to the Criteria

Criteria	WTE82	WTS101	WTV100	WTG87
Reliability	(0.031,0.070,0.24,0.47;1,1) (0.03,0.080,0.21,0.40;0.8,0.8)	(0.006,0.013,0.048,0.10;1,1) (0.008,0.015,0.042,0.086;0.8,0.8)	(0.017,0.036,0.12,0.27;1,1) (0.020,0.042,0.11,0.22;0.8,0.8)	(0.004,0.007,0.026,0.065;1,1) (0.004,0.008,0.023,0.052;0.8,0.8)
Technical characteristics	(0.00,0.007,0.022,0.043;1,1)(0.005,0.008,0.019,0.037;0.8,0.8)	(0.006,0.012,0.034,0.060;1,1) (0.007,0.014,0.030,0.053;0.8,0.8)	(0.009,0.001,0.004,0.009;1,1) (0.001,0.001,0.004,0.008;0.8,0.8)	(0.009,0.001,0.004,0.007;1,1) (0.001,0.001,0.003,0.006;0.8,0.8)
Performance	(0.008,0.017,0.055,0.11;1,1) (0.010,0.019,0.049,0.099;0.8,0.8)	(0.016,0.031,0.097,0.19;1,1) (0.019,0.035,0.086,0.16;0.8,0.8)	(0.003,0.006,0.020,0.046;1,1) (0.004,0.007,0.018,0.038;0.8,0.8)	(0.002,0.00,0.010,0.021;1,1) (0.002,0.003,0.008,0.01;0.8,0.8)
Cost factors	(0.004,0.007,0.021,0.043;1,1) (0.004,0.008,0.019,0.036;0.8,0.8)	(0.002,0.004,0.013,0.029;1,1) (0.00,0.005,0.012,0.02;0.8,0.8)	(0.022,0.040,0.11,0.23;1,1) (0.025,0.045,0.10,0.19;0.8,0.8)	(0.017,0.034,0.10,0.21;1,1) (0.020,0.038,0.09,0.18;0.8,0.8)
Availability	(0.010,0.018,0.05,0.10;1,1) (0.011,0.021,0.048,0.09;0.8,0.8)	(0.004,0.007,0.020,0.040;1,1) (0.004,0.008,0.018,0.034;0.8,0.8)	(0.006,0.012,0.035,0.072;1,1) (0.007,0.013,0.031,0.061;0.8,0.8)	(0.002,0.003,0.011,0.026;1,1) (0.002,0.004,0.010,0.021;0.8,0.8)
Maintenance	(0.002,0.004,0.01,0.026;1,1) (0.002,0.004,0.011,0.022;0.8,0.8)	(0.001,0.001,0.005,0.010;1,1) (0.001,0.002,0.004,0.008;0.8,0.8)	(0.005,0.010,0.032,0.063;1,1) (0.006,0.012,0.028,0.054;0.8,0.8)	(0.001,0.002,0.006,0.014;1,1) (0.001,0.002,0.005,0.011;0.8,0.8)
Cooperation	(0.004,0.007,0.02,0.055;1,1) (0.004,0.008,0.022,0.046;0.8,0.8)	(0.004,0.008,0.002,0.006;1,1) (0.005,0.008,0.002,0.005;0.8,0.8)	(0.001,0.003,0.010,0.024;1,1) (0.002,0.003,0.009,0.020;0.8,0.8)	(0.001,0.001,0.006,0.01;1,1) (0.001,0.002,0.005,0.012;0.8,0.8)
Domesticity	(0.083,0.14,0.36,0.57;1,1) (0.096,0.16,0.33,0.51;0.8,0.8)	(0.012,0.023,0.060,0.10;1,1) (0.014,0.026,0.054,0.090;0.8,0.8)	(0.011,0.020,0.054,0.098;1,1) (0.012,0.022,0.048,0.085;0.8,0.8)	(0.011,0.019,0.050,0.089;1,1) (0.012,0.021,0.04,0.078;0.8,0.8)
Total	(0.14,0.28,0.79,1.44;1,1) (0.17,0.31,0.71,1.26;0.8,0.8)	(0.05,0.096,0.28,0.54;1,1) (0.060,0.10,0.25,0.46;0.8,0.8)	(0.069,0.13,0.40,0.82;1,1) (0.080,0.14,0.35,0.69;0.8,0.8)	(0.040,0.07,0.22,0.45;1,1) (0.046,0.083,0.19,0.38;0.8,0.8)

TABLE 9.8

Fuzzy and Crisp Weights

	Type-2 Fuzzy Weights	Crisp Weights	Normalized Crisp Weights
WTE82	(0.14,0.28,0.79,1.44;1,1) (0.17,0.31,0.71,1.26;0.8,0.8)	0.617	0.457
WTS101	(0.05,0.096,0.28,0.54;1,1) (0.060,0.10,0.25,0.46;0.8,0.8)	0.224	0.166
WTV100	(0.069,0.13,0.40,0.82;1,1) (0.080,0.14,0.35,0.69;0.8,0.8)	0.327	0.242
WTG87	(0.040,0.07,0.22,0.45;1,1) (0.046,0.083,0.19,0.38;0.8,0.8)	0.181	0.134

According to the results, WTE82 is the best alternative with a weight of 0.457, which is followed by WTV100 (0.242), WTS101 (0.166), and WTG87 (0.134).

9.6 Conclusions

The superiority of renewable energies to others is well known in today's society. Wind energy systems have high investment costs but low operating expenses. Wind energy is plentiful and readily available and capturing its power does not deplete the natural resources. Electricity generated by wind turbines does not pollute water and air. Hence, wind energy causes less smog, less acid rain, and fewer greenhouse gas emissions. In this chapter, we focus on the wind energy technology selection decision, which is one of the essential decisions requiring evaluation of large number of criteria. The study contributes to the literature by proposing a multicriteria multiexpert decision-making model for wind energy technology selection problems with IT2 fuzzy sets.

There are a large number of alternatives with regard to wind technologies, which increase the complexity of the problem. In future studies, we suggest more wind energy technologies be considered in the fuzzy multicriteria analysis. In addition, subcriteria may be considered in the problem hierarchy. Other fuzzy MCDM methods such as TOPSIS, VIseKriterijumska Optimizacija I Kompromisno Resenje (VIKOR), AHP, and ELECTRE using hesitant or intuitionistic fuzzy sets for the same problem can also be applied and the obtained results can be compared with this study.

References

Abdullah, L., Najib, L. (June 1, 2014) A new type-2 fuzzy set of linguistic variables for the fuzzy analytic hierarchy process, *Expert Systems with Applications*, 41(7), 3297–3305.

Acikgoz, C. (2011) Renewable energy education in Turkey, *Renewable Energy*, 36(2), 608–611.

Balezentis, T., Zeng, S.Z. (2013) Group multi-criteria decision making based upon interval-valued fuzzy numbers: An extension of the MULTIMOORA method, *Expert Systems with Applications*, 40(2), 543–550.

Buckley, J.J. (1985) Fuzzy hierarchical analysis, *Fuzzy Sets and Systems*, 17, 233–247.

Castillo, O., Melin, P. (2008) *Type-2 Fuzzy Logic: Theory and Applications, Studies in Fuzziness and Soft Computing*. Berlin/Heidelberg: Springer-Verlag.

Çelik, E., Aydin, N., Gumus, A.T. (2014) A multiattribute customer satisfaction evaluation approach for railtransit network: A real case study for Istanbul, Turkey, *Transport Policy*, 36, 283–293.

Cevik Onar, S., Kılavuz, T. (2015) Risk analysis of wind energy investments in Turkey, *Human and Ecological Risk Assessment: An International Journal*, 21(5), 1230–1245.

Cevik Onar, S., Oztaysi, B., Otay, İ., Kahraman, C. (2015) Multi-expert wind energy technology selection using interval-valued intuitionistic fuzzy sets, *Energy*, 90(1), 274–285.

Chen, T.-Y. (2013) A linear assignment method for multiple-criteria decision analysis with interval type-2 fuzzy sets, *Applied Soft Computing*, 13(5), 2735–2748.

Chen, T.-Y. (2014) An interactive signed distance approach for multiple criteria group decision-making based on simple additive weighting method with incomplete preference information defined by interval type-2 fuzzy sets, *International Journal of Information Technology and Decision Making*, 13(5), 979–1012.

Chen, S.-M., Lee, L.-W. (2010a) Fuzzy multiple attributes group decision-making based on the interval type-2 TOPSIS method, *Expert Systems with Applications: An International Journal*, 37(4), 2790–2798.

Chen, S.-M., Lee, L.-W. (2010b) Fuzzy multiple criteria hierarchical group decision-making based on interval type-2 fuzzy sets, *IEEE Transaction on Systems, Man, and Cybernetics—Part A: Systems and Humans*, 40(5), 1120–1128.

Chen, T.-Y., Chang, C.-H., Lu, J.R. (2013) The extended QUALIFLEX method for multiple criteria decision analysis based on interval type-2 fuzzy sets and applications to medical decision making, *European Journal of Operational Research*, 226, 615–625.

Chowdhury, S., Zhang, J., Messac, A., Castillo, L. (2013) Optimizing the arrangement and the selection of turbines for wind farms subject to varying wind conditions, *Renewable Energy*, 52, 273–282.

Dursun, B., Gokcol, C. (2014) Impacts of the renewable energy law on the developments of wind energy in Turkey, *Renewable and Sustainable Energy Reviews*, 40, 318–325.

EWEA Report. (2014) The European Wind Energy Association, Wind in Power 2014 European Statistics, http://www.ewea.org/fileadmin/files/library/publications/statistics/EWEA-Annual-Statistics-2014.pdf (Access date May 10, 2015).

International Energy Agency (IEA) Report. (2014) http://www.iea.org/ (Accessed on May 10, 2015).

Kahraman, C., Cevik Onar, S., Oztaysi, B. (2016a) A comparison of wind energy investment alternatives using interval-valued intuitionistic fuzzy benefit/cost analysis, *Sustainability*, 8, 118.

Kahraman, C., Onar, S.C., Oztaysi, B. (2015) Fuzzy multi-criteria decision-making: A literature review, *International Journal of Computational Intelligence Systems*, 8(4), 637–666.

Kahraman, C., Onar, S.Ç., Öztaysi, B. (2016b) Fuzzy decision making: Its pioneers and supportive environment, *Studies in Fuzziness and Soft Computing*, 341, 21–58.

Kahraman, C., Öztayşi, B., Çevik Onar, S. (2016c) A comprehensive literature review of 50 years of fuzzy set theory, *International Journal of Computational Intelligence Systems*, 9, 3–24.

Kahraman, C., Öztayşi, B., Uçal Sarı, İ., Turanoğlu, E. (2014) Fuzzy analytic hierarchy process with interval type-2 fuzzy sets, *Knowledge-Based Systems*, 59, 48–57.

Karnik, N.N., Mendel, J.M. (2001) Centroid of a type-2 fuzzy set, *Information Sciences*, 132, 195–220.

Kaya, T., Kahraman, C. (2010) Multi-criteria renewable energy planning using an integrated fuzzy VIKOR & AHP methodology: The case of Istanbul, *Energy*, 35(6), 2517–2527.

Kayal, P., Chanda, C.K. (2015) A multi-objective approach to integrate solar and wind energy sources with electrical distribution network, *Solar Energy*, 112, 397–410.

Kiliç, M., Kaya, İ. (2015) Investment project evaluation by a decision making methodology based on type-2 fuzzy sets, *Applied Soft Computing*, 27, 399–410.

Lee, A.H.I., Chen H.H., Kang, H.Y. (2009) Multi-criteria decision making on strategic selection of wind farms, *Renew Energy*, 34, 120–126.

Mendel, J.M., John, R.I., Liu, F. (2006) Interval type-2 fuzzy logic systems made simple, *IEEE Transactions on Fuzzy Systems*, 14(6), 808–821.

Montoya, F.G., Manzano-Agugliaro, F., López-Márquez, S., Hernández-Escobedo, Q., Gil, C. (2014) Wind turbine selection for wind farm layout using multi-objective evolutionary, *Expert Systems with Applications*, 41(15), 6585–6595.

Oztaysi, B., Onar, S.C., Bolturk, E., Kahraman, C. (2015) Hesitant fuzzy analytic hierarchy process, IEEE International Conference on Fuzzy Systems, 2015-November, art. no. 7337948.

Perkin, S., Garrett, D., Jensson, P. (2015) Optimal wind turbine selection methodology: A case-study for Búrfell, *Iceland Renewable Energy*, 75, 165–172.

Sahu, B.K., Hiloidhari, M., Baruah, D.C. (2013) Global trend in wind power with special focus on the top five wind power producing countries, *Renewable and Sustainable Energy Reviews*, 19, 348–359.

Sanchez-Lozano, J.M., García-Cascales, M.S., Lamata, M.T. (2014) Identification and selection of potential sites for onshore wind farms development in Region of Murcia, Spain, *Energy*, 73, 311–324.

Sarja, J., Halonen, V. (2013) Wind turbine selection criteria: A customer perspective, *Journal of Energy and Power Engineering*, 7, 1795–1802.

Turkish Wind Energy Statistics Report. (2015) http://www.tureb.com.tr/files/bilgi_bankasi/turkiye_res_durumu/3.turkiye_ruzgar_enerjisi_istatistik_raporu_2015.pdf (Accessed date July 26, 2015).

Wang, W., Liu, X., Qin, Y. (2012) Multi-attribute group decision making models under interval type-2 fuzzy environment, *Knowledge-Based Systems*, 30, 121–128.

Web 1. www.windustry.org (Accessed date May 10, 2014).

Web 2. Global Wind Energy Council (GWEC), Windustry, http://www.gwec.net/ (Accessed date May 10, 2014).

Wu, Y., Geng, S. (2014) Multi-criteria decision making on selection of solar–wind hybrid powerstation location: A case of China, *Energy Conversion and Management*, 81, 527–533.

Wu, Y., Geng, S., Xu, H., Zhang, H. (2014) Study of decision framework of wind farm project plan selection under intuitionistic fuzzy set and fuzzy measure environment, *Energy Conversion and Management*, 87, 274–284.

Zadeh, L.A. (1965) Fuzzy sets, *Information and Control*, 8, 338–353.

Zadeh, L.A. (1975) The concept of a linguistic variable and its application to approximate reasoning-I, *Information Sciences*, 8, 199–249.

10

A Decision Support for Prioritizing Process Sustainability Tools Using FAHP

Vinodh S and Vimal KEK

CONTENTS

10.1 Introduction

The contribution of manufacturing organizations to the gross domestic product of developing countries like India is vital (Tripathi and Ahmad, 2015). Sustainability is regarded as a powerful strategy by manufacturing organizations in the competitive scenario (Gunasekaran and Spalanzani, 2012). The core dimensions/orientations of sustainability include material, product, and process (Vinodh, 2011). However, when considering the sustainability of an organization, the manufacturing process can have an adverse effect on human health and the environment because of emissions, particulates, and other wastes (Vimal et al., 2015a). Thus, the process dimension has gained vital importance in recent days (Haapala et al., 2013). Previously, sustainability research was focused on selecting the sustainable process. However, the selected process may not be economically competitive. Thus, the focus has been shifted toward improving sustainability of the current process with the application of various tools.

Tools and methods are being developed in the viewpoint of the process dimension of sustainability (Rashid et al., 2008; Vimal et al., 2015a). The identified set of tools needs to be evaluated with the set of governing sustainability criteria to understand their importance and benefits. Selection of appropriate tools in the context of process sustainability is a typical decision-making problem with several criteria. This chapter reports a study in which the process of selecting sustainability tools is formulated as a multicriteria decision-making (MCDM) problem. Ten governing criteria and five tools are considered in the decision-making problem. The analytic hierarchy process (AHP) has been used as the solution methodology. Since the inputs for the alternative solutions involve vagueness and impreciseness, Fuzzy Analytic Hierarchy Process (FAHP) has been used. The five alternative tools include energy modeling, waste modeling, carbon footprint analysis, parametric optimization, and water foot print analysis. Based on computation, it has been found that the carbon footprint analysis tool ranks first followed by other strategies.

The subsequent sections of the chapter include a critical review on process sustainability tools, process sustainability indicators, and sustainability applications of FAHP, a description of FAHP methodology, and a case analysis of an injection molding process, followed by results and conclusions.

10.2 Literature Review

A literature review is performed for the following perspectives: process sustainability tools, sustainable manufacturing criteria, and AHP applications in the sustainability domain.

10.2.1 Review of Process Sustainability Tools

Seidel et al. (2007) identified waste minimization and energy efficiency improvements as the important sustainable manufacturing strategies under the process improvement loop. From a product perspective, end-of-life (EoL) strategies are important because of material scarcity, which was not a concern in earlier days. On the other hand, depletion of natural resources, globalization, and other economic revolutions are the reasons for the shift toward resource conservation in addition to human efficiency. Furthermore, Rashid et al. (2008) benchmarked four sustainable manufacturing strategies: waste minimization, material efficiency, resource efficiency, and eco-efficiency based on definition, scope, practicality, and compatibility. Other than these strategies, Garetti and Taisch (2012) illustrated the significance of facilitating newer technology for successfully achieving sustainability in the manufacturing process.

From the above review, we can understand that important challenges or obstacles toward achieving sustainability include resource depletion, energy consumption, and emissions. Based on the earlier understanding, the following tools were identified from the literature. These tools were considered from the perspective of contribution toward improvement in process sustainability. The tools identified include energy modeling (Kellens et al., 2012; Seow and Rahimfard, 2011), waste minimization (Rashid et al., 2008), material efficiency (Rashid et al., 2008; Young et al., 1994), resource efficiency (Rashid et al., 2008; Pearce, 2001), carbon footprint analysis (Čuček et al., 2012; Dormer et al., 2013), water footprint analysis (Čuček et al., 2012; Pfister et al., 2011), particulate studies (Vimal et al., 2015a,b), EoL disposal studies (Vimal et al., 2015a), parametric optimization (Sivapirakasam et al., 2011), and education and training (Muduli et al., 2013). Based on these studies, the most suitable tools for the process will be identified in line with the expert team's opinion.

10.2.2 Review of Sustainable Manufacturing Indicators

Sustainable production is described with five components based on the definition given by the Lowell Center for Sustainable Production (1998). The sustainable criteria or measures are identified considering the earlier definition, which addresses the triple bottom line (TBL) aspects of sustainability. During recent times, the number of sustainable manufacturing measures has increased, as the demand for sustainable manufacturing increases due to green awareness as a result of globalization (Tseng, 2013).

In the literature, many studies have focused on sustainability assessment using measures/indicators. Veleva et al. (2001) proposed a set of indicators considering the environmental, health, and safety aspects of production. Tseng et al. (2009a) applied indicators and the analytic network process (ANP) to assess the environmental performance of a multinational original equipment manufacturer. Bhanot et al. (2015) discussed various drivers and barriers of sustainable manufacturing. The responses were collected from researchers usings a questionnaire and validated using statistical methods. Dubey et al. (2015) developed a framework for sustainable manufacturing performance assessment aimed at world class sustainable practices.

However, Veleva and Ellenbecker (2001) disagreed with the possibility of a consensus set of sustainable production indicators applicable to all types of manufacturing processes. In addition, many studies address performance evaluation using indicators; however, social and economic perspectives were mostly neglected. Thus, 10 criteria suitable for the current study have been identified from various research studies from the TBL perspective and are listed in Table 10.1.

TABLE 10.1

Sustainable Manufacturing Governing Criteria

Sustainability Orientation	Criteria	References
Environment	Greenhouse gas (GhG) emissions (C_1)	Veleva and Ellenbecker (2001), Veleva et al. (2001), Tseng et al. (2009a), Tseng et al. (2013), Ocampo and Clark (2015), and Vimal and Vinodh (2015)
	Water consumption (C_2)	Veleva et al. (2001), Veleva and Ellenbecker (2001), Moneim et al. (2013), Tseng et al. (2009a), and Ocampo and Clark (2015)
	Land utilization (C_3)	Evans et al. (2009), Ocampo and Clark (2015), and Vimal and Vinodh (2015)
	Resource consumption (C_4)	Veleva et al. (2001), Tseng et al. (2009a), Lin et al. (2010), Ocampo and Clark (2015), and Vimal and Vinodh (2015)
	Energy consumption (C_5)	Veleva et al. (2001), Veleva and Ellenbecker (2001), Ocampo and Clark (2015), and Vimal and Vinodh (2015)
Economic	Productivity (C_6)	Veleva and Ellenbecker (2001) and Ocampo and Clark (2015)
	Economic feasibility (C_7)	Veleva et al. (2001), Veleva and Ellenbecker (2001), and Ocampo and Clark (2015)
Social	Green awareness (C_8)	Rusinko (2007), Garbie (2015), and Vimal and Vinodh (2015)
	Well-being and job satisfaction (C_9)	Veleva et al. (2001), Veleva and Ellenbecker (2001), Moneim et al. (2013), Tseng et al. (2009a), and Ocampo and Clark (2015)
	Health and safety (C_{10})	Tseng et al. (2009a), Tseng et al. (2013), Ocampo and Clark (2015), and Vimal and Vinodh (2015)

10.2.3 Review of Applications of FAHP in the Sustainability Domain

The AHP methodology has a wide range of applications in the domain of sustainable development ranging from supplier selection to manufacturing of products.

Huang et al. (2011) conducted a literature analysis to understand MCDM application in the field of environmental management and concluded that around 48% of the studies use the AHP process. Pohekar and Ramachandran (2004) further validated the above conclusion in the field of sustainable energy planning. The important studies on AHP applications are detailed here. Gupta et al. (2011) identified a large set of product sustainability metrics. The usage of a large number of metrics is difficult to handle; thus, AHP methodology was applied to prioritize the important metrics. With the help of a case study from the electronics industry, the proposed methodology was demonstrated. Gumus (2009) evaluated hazardous waste transportation with the help of an AHP-based hybrid methodology. Kahraman et al. (2009)

compared renewable energy alternatives using an AHP-based axiomatic design methodology. Ghadimi et al. (2012) proposed a FAHP-based product sustainability assessment methodology and validated it with a case study conducted in the automobile industry. Biju et al. (2015) developed a customer and sustainability requirements evaluation matrix to enable sustainable product development during the early design stage. The authors used FAHP to determine the weightings of the customer and sustainability requirements. Calabrese et al. (2016) conducted a materiality assessment for reporting sustainability through a quantitative approach. The method ensured reliable reporting of sustainable indicators. Other applications of AHP include energy planning (Kaya and Kahraman, 2010), sustainable supplier assessment (Dai and Blackhurst, 2012), and cleaner production implementation (Tseng et al., 2009b). These studies validate the practical usability of AHP methodology in the field of sustainable manufacturing.

10.3 Methodology for Ranking Process Sustainability Tools

An MCDM problem has been formulated for ranking sustainability tools using governing criteria. AHP is widely used for handling MCDM problems in real situations (Dai and Blackhurst, 2012). From the literature, it was found that AHP has a wide range of applications in decision-making in sustainable manufacturing domain. AHP works on the principle of considering independence rather than other MCDM techniques like graph theory and ANP considers both independence and interdependence. However, the consideration of interdependence is not required because of the nature of problem. For example, governing criteria and tools selected are independent. But, from the literature, it was found that in decision-making problem, expert judgment with crisp number is relatively difficult because of uncertainty and associated vagueness. In order to deal with uncertainty in human judgment, fuzzy logic was found to be deployed very successfully. Fuzzy logic is based on fuzzy set theory to deal with imprecise information using membership functions. Fuzzy set theory was emphasized by Zadeh (1965), which is a set with a membership value in the interval [0, 1]. The steps in FAHP methodology for ranking sustainability tools are listed below (Ayağ and Özdemir, 2006).

Step 1: Selection of governing criteria for evaluating the identified tools.

Step 2: Constitution of an expert team comprising "industrial experts" with knowledge of manufacturing processes and process planning in the field of new product development and "sustainable manufacturing experts" with experience in sustainable manufacturing implementation across various companies. Brainstorming sessions are organized among experts to generate ideas. Environmental impact assessment concepts, difficulties in deploying sustainability tools, and procedures to deploy various practices are discussed. The objective is to arrive at a consensus opinion among experts.

Step 3: Identification of case process, development of process model toward understanding various inputs and outputs.

Step 4: Selection of suitable tools for a candidate process from the tools listed in Table 10.2.

Step 5: Determination of rank using FAHP methodology (development of hierarchy structure).

> **Step 5.1:** Construction of pairwise comparison for three levels (refer to Figure 10.2) namely, alternatives, criteria, and criteria group.
>
> **Step 5.2:** Construction of α–cut fuzzy comparison matrices.
>
> **Step 5.3:** Fuzzy-based synthesis to obtain importance rating and λ_{max} (for consistency test).
>
> **Step 5.4:** Consistency verification for the matrices obtained from experts.
>
> **Step 5.5:** Computation of priority score for each alternative.

10.4 Case Study

The case study has been conducted to improve sustainable performance of a manufacturing process. To understand the model, process modeling has to be done. The process model of the manufacturing process consists of inputs and outputs of the manufacturing process. With the help of this model, the process can be completely visualized. Thus, for demonstration purposes, the process model for injection molding has been developed as shown in Figure 10.1. After rigorous analysis of the injection molding process model, the expert team shortlisted five tools from the list of sustainability tools. The five tools considered by the experts include energy modeling, waste minimization, carbon footprint analysis, parametric optimization, and water foot print analysis. The details about the five tools and their deployment are discussed below.

10.4.1 Description of Process Sustainability Tools

The tools identified based on expert opinion are detailed below.

10.4.1.1 Energy Modeling

In the context of the metal removal process, 90% of the environmental impact is due to electrical energy consumed (Vimal et al., 2015b). As far as manufacturing processes are concerned, electrical energy consumption has been

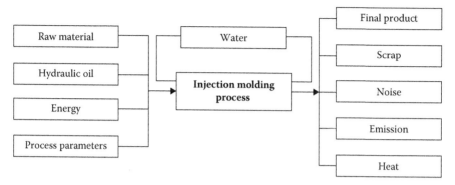

FIGURE 10.1
Model pertaining to injection molding process.

studied extensively in the past. Newman et al. (2012) identified possible ways to improve the energy efficiency of manufacturing process through energy consumption mathematical modeling and energy-conscious metal cutting. Gutowski et al. (2006) further studied various electrical energy requirements for manufacturing processes through analysis of specific energy requirements. The energy studies are viewed from four levels, namely, process, machine, line, and factory. Many researchers adopted different methods to study energy consumption and the efficiency of manufacturing processes such as energy efficiency of discrete manufacturing, energy-based modeling, energy profile creation, thermodynamic analysis, power study, energy optimization, modeling embodied product energy, and axiomatic modeling.

Still, the literature suggests that scope for energy studies toward improving energy consumption in various stages of manufacturing process. After modeling energy consumption in various phases, steps have been taken toward optimizing energy consumption through adoption of various methods. Finally, the improvement is quantified with the help of life cycle assessment (LCA) methods.

10.4.1.2 Waste Minimization

Environmental waste has been referred to as the ninth waste in recent times (Vinodh et al., 2011). Waste minimization aims to reduce wastes from raw material and ingredient use, product loss, water consumption and effluent generation, packaging, factory and office consumables, energy consumption, and other solids. However, reduction of wastes at the source point conserves natural resources; hence, the focus has been shifted toward waste control rather than minimization. Waste minimization is deployed with the help of detailed flow analysis of the identified process. Based on material flow analysis, various wastes throughout the process have been mapped. Then, suitable strategies have been formulated for waste reduction. On the other hand,

EoL disposal strategies have been decided based on environmental impact assessment.

10.4.1.3 Carbon Footprint Analysis

Increased CO_2 emissions result in an unambiguous warming on earth. This results in increased air and sea temperature resulting in rapid increase in seawater level. Thus, minimization of the carbon footprint is a goal of the modern world to protect people. For this, ISO 14067 (Finkbeiner, 2009) outlined standards for carbon footprint analysis of products and requirements and guidelines for quantification and communication were detailed. This forms the methodology for systematic activities toward minimization of greenhouse gase (GhG) level. Based on this, effort has been made to deploy this tool toward improving process sustainability performance.

10.4.1.4 Parametric Optimization

In general, optimization will have a trade-off effect on various parameters to achieve improved sustainable performance (Vimal et al., 2015a). However, optimization has a positive correlation with GhG emission, energy consumption, and waste (Yan and Li, 2013). In the past, most parametric optimization studies were conducted with the objective of improving energy efficiency.

Considering the impact of optimization on environmental performance, optimization is considered an important tool. Through the development of a process model, important parameters have been identified. The significance of the identified parameters with regard to environmental responses is evaluated by the expert team or with a trial study, based on which experiment designed have been planned (Vimal et al., 2015b).

10.4.1.5 Water Footprint Analysis

The main problem with excessive usage of water is scarcity of water. Due to overpopulation, the availability of water is becoming limited. Apart from population growth, other reasons for increased water stress level are economic growth and climate change. This includes water used for operations and water used for production of raw materials required by plants. Thus, it is necessary to optimize water consumption to sustain this world. Unlike carbon footprint analysis, water footprint analysis methodology is still evolving. Thus, the methodology to assess the impacts created by water use was not practically applied. On the other hand, Pfister et al. (2011) developed a water stress index to assess the impacts of blue water at the national level. Also, ISO 14046 (ISO, 2012) outlined various operational steps of water footprint analysis. Overall, the optimization of water consumption has a great influence on sustaining the operations of a manufacturing process.

For the identified five tools, the steps shown in the methodology section have been applied to derive the rank. The steps are detailed as follows: After the identification of governing criteria (Table 10.1), a discussion session is arranged among experts to shortlist the process sustainability tools. After defining the criteria and alternatives, a hierarchial structure is formed as shown in Figure 10.2.

1. Formation of pairwise comparison matrix: The pairwise comparison matrix is obtained from experts for three levels, namely, the TBL perspective, criteria, and alternatives. Triangular fuzzy numbers $(\tilde{1}, \tilde{3}, \tilde{5}, \tilde{7}, \tilde{9})$ were used to express the relationship between factors. The fuzzy pairwise comparison matrix is formed as shown in Equation 10.1. Using this definition, these numbers are described in Table 10.2, where $\tilde{a}_{ij} = 1$, if $i = j$ and $\tilde{a}_{ij} = 1,3,5,7,9$. When $i \neq j$, a reciprocal

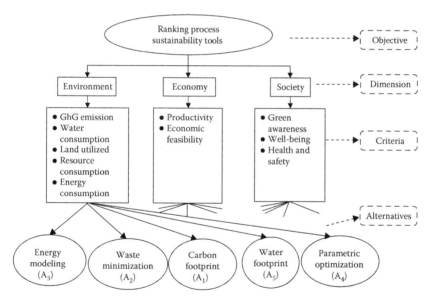

FIGURE 10.2
Hierarchial structure for process sustainability tools selection.

TABLE 10.2

Triangular Fuzzy Numbers Used

Crisp Value	Fuzzy Number	Definition	Membership Function
1	$\tilde{1}$	Equally important/preferred	(1,1,3)
3	$\tilde{3}$	Moderately more important/preferred	(2,3,4)
5	$\tilde{5}$	Strongly more important/preferred	(4,5,6)
7	$\tilde{7}$	Very strongly more important/preferred	(6,7,8)
9	$\tilde{9}$	Extremely more important/preferred	(8,9,10)

value is automatically assigned to the reverse comparison within the matrix. That is, if $\tilde{a}_{ij}$ is a matrix value assigned to the relationship of component i to component j, then $\tilde{a}_{ji}$ is equal to $1/\tilde{a}_{ij}$.

$$\tilde{A} = \begin{pmatrix} 1 & \tilde{a}_{12}\ldots\ldots\tilde{a}_{1n} \\ \tilde{a}_{21} & 1 & \ldots\ldots\tilde{a}_{2n} \\ & \vdots\vdots\vdots\vdots\vdots \\ \tilde{a}_{1n}\tilde{a}_{n2}\ldots\ldots & 1 \end{pmatrix} \tag{10.1}$$

As an example, the pairwise comparison matrix of the TBL perspective is shown in Table 10.3. After obtaining fuzzy numbers from experts, $\propto$-cut matrix ($\propto$ denotes confidence level) is derived using Equation 10.2 and is shown in Table 10.3. The average of values in the $\propto$-cut matrix is used to obtain crisp numbers; this shown in Table 10.4.

$$\tilde{1}_{\propto} = [1, 3 - 2\propto]; \tilde{3}_{\propto} = [1 + 2\propto, 5 - 2\propto]; \tilde{5}_{\propto} = [3 + 2\propto, 7 - 2\propto];$$

$$\tilde{7}_{\propto} = [5 + 2\propto, 9 - 2\propto]; \tilde{9}_{\propto} = [7 + 2\propto, 11 - 2\propto] \tag{10.2}$$

TABLE 10.3

Pairwise Comparison Matrix for Triple Bottom Line Perspective

	Environment	Economy	Society
Environment	$\tilde{1}$	$\tilde{3}$	$\tilde{5}$
Economy	$\tilde{3}^{-1}$	$\tilde{1}$	$\tilde{3}$
Society	$\tilde{5}^{-1}$	$\tilde{3}^{-1}$	$\tilde{1}$

TABLE 10.4

$\propto$ - Cut and Final Comparison Matrix for Triple Bottom Line Perspective

	$\propto$ Cut			Final Comparison Matrix		
	Environment	Economy	Society	Environment	Economy	Society
Environment	1	[2,4]	[4,6]	1	3	5
Economy	[1/4, 1/2]	1	[2,4]	0.37	1	3
Society	[1/6, 1/4]	[1/4, 1/2]	1	0.20	0.37	1

TABLE 10.5

Eigenvalue and Consistency Check for Triple Bottom Line Perspective

	Environment	Economy	Society	λ_{max}	Eigenvector	Normalized Eigenvector
Environment	1	3	5		0.912	0.63
Economy	0.37	1	3	3.099	0.381	0.26
Society	0.20	0.37	1		0.154	0.11

Similarly, 14 pairwise comparison matrices were collected (one for the TBL perspective, three for criteria comparison under three perspectives, and 10 for alternative comparison with respect to 10 criteria).

2. Eigenvector (e-vector) is calculated using Equation 10.3. The eigenvalue computed for the TBL perspective is shown in Table 10.5.

$$Aw = \lambda_{max}w \qquad (10.3)$$

3. The consistency of the obtained pairwise comparison matrix is checked using Equation 10.4:

$$CR = \frac{CI}{RI} \qquad (10.4)$$

where

$$CI = \frac{\lambda_{max} - n}{n - 1}$$

Sample calculation:

$$CI = \frac{3.099 - 3}{2} = 0.0495$$

$$CR = \frac{0.0495}{0.58} = 0.085, \text{ which is less than } 0.1$$

As a sample, the matrix obtained and its computation of weights at the criteria level and attribute level are shown in Tables 10.6 and 10.7. Table 10.6 shows the pairwise comparison matrix obtained for criteria under the economic perspective and Table 10.7 shows the pairwise comparison matrix obtained for comparison of criteria with respect to the "energy modeling" tool. Similarly, a consistency check has been done for all 14 matrices.

TABLE 10.6

Pairwise Comparison Matrix of Criteria for the under Economy Perspective

	C_6	C_7	λ_{max}	Eigenvector	Normalized Eigenvector
C_6	$\tilde{1}$	$\tilde{5}$	2	0.98	0.83
C_7	$\tilde{5}$	$\tilde{1}$	CR = 0	0.19	0.17

TABLE 10.7

Pairwise Comparison Matrix of Alternatives with Respect to "Productivity" Criterion

C_6	A_1	A_2	A_3	A_4	A_5	A_1	A_2	A_3	A_4	A_5	λ_{max}	Eigenvector	Normalized Eigenvector
A_1	1	1	5	7	9	1	1.5	5	7	9		0.82	0.46
A_2	$\tilde{1}$	1	1	5	7	.75	1	1.5	5	7	5.286	0.482	0.27
A_3	$\tilde{5}$	$\tilde{1}$	1	3	5	.208	.75	1	3	5	CR =	0.28	0.15
A_4	$\tilde{7}$	$\tilde{5}$	$\tilde{3}$	1	3	.146	.208	.375	1	3	0.06<0.1	0.119	0.07
A_5	$\tilde{9}$	$\tilde{7}$	$\tilde{5}$	$\tilde{3}$	1	.113	.146	.208	.375	1		0.062	0.03

10.5 Results and Discussion

For ranking the alternatives, a cumulative score has been computed using Equation 10.5. As an example, the cumulative score computed for the "energy modeling" tool is shown here:

$$\text{Cumulative score}_j = \sum_{i=1}^{5} W_k \times w_i \times S_i \tag{10.5}$$

Notations

w_i—Weights of TBL perspective

S_i—Alternative scores with respect to criteria

W_k—Weights of criteria

Cumulative score of energy modeling tool:

$$= (0.637 \times 0.428 \times 0.129) + (0.637 \times 0.315 \times 0.154) + (0.637 \times 0.132 \times 0.135)$$
$$+ (0.637 \times 0.076 \times 0.125) + (0.637 \times 0.047 \times 0.146) + (0.257 \times 0.833 \times 0.156)$$
$$+ (0.257 \times 0.166 \times 0.141) + (0.104 \times 0.65 \times 0.154) + (0.104 \times 0.278 \times 0.146)$$
$$+ (0.104 \times 0.071 \times 0.131) = 0.143$$

TABLE 10.8

Cumulative Scores

Criterion	Weights of Triple Bottom Line	Criterion Weights	Alternative Performance					Cumulative Score Computation				
			A_1	A_2	A_3	A_4	A_5	A_1	A_2	A_3	A_4	A_5
C_1	Environment 0.637	0.428	0.49	0.28	0.12	0.06	0.03	0.134	0.076	0.035	0.017	0.009
C_2		0.31	0.46	0.27	0.15	0.06	0.03	0.093	0.055	0.031	0.013	0.007
C_3		0.132	0.50	0.26	0.13	0.05	0.03	0.043	0.022	0.011	0.004	0.002
C_4		0.076	0.46	0.29	0.12	0.07	0.03	0.022	0.014	0.006	0.003	0.0017
C_5		0.047	0.51	0.24	0.14	0.06	0.03	0.015	0.007	0.004	0.001	0.001
C_6	Economy 0.257	0.833	0.45	0.27	0.15	0.07	0.03	0.098	0.059	0.033	0.015	0.007
C_7		0.166	0.44	0.3	0.14	0.07	0.03	0.019	0.012	0.006	0.003	0.001
C_8	Society 0.104	0.65	0.51	0.23	0.15	0.06	0.03	0.035	0.016	0.010	0.004	0.002
C_9		0.278	0.51	0.23	0.14	0.06	0.04	0.014	0.006	0.004	0.001	0.001
C_{10}		0.071	0.46	0.29	0.13	0.07	0.03	0.003	0.002	0.0009	0.0005	0.0002
								0.480	**0.273**	**0.1438**	**0.066**	**0.035**

Similarly, scores were computed for other alternatives and are shown in Table 10.8.

Based on this computaion (Table 10.4), it is found that carbon footprint analysis (cumulative score = 0.480) is a significant tool for sustainable performance improvement of the injection molding (IM) process. Subsequently, other tools will be subjected to implementation. The procedure demonstrated using the IM process is generic in nature and can be applied for any process with minor modifications based on the process model.

10.6 Conclusions

The modern manufacturing scenario realizes the importance of sustainability from a process viewpoint. Selection of an apropriate process sustainability tool is essential to attain a competitive advantage. In this chapter, 10 criteria from the TBL perspective of sustainability are considered with five alternatives. The formulated decision-making problem is solved using FAHP. The ranking of alternatives based on FAHP is carbon footprint analysis > waste minimization > energy modeling > optimization > water footprint analysis. The study provides appropriate guidelines for engineers to select the best tool for sustainable performance improvement of the manufacturing process.

References

Ayağ, Z., and Özdemir, R. G. (2006). A fuzzy AHP approach to evaluating machine tool alternatives. *Journal of Intelligent Manufacturing, 17*(2), 179–190.

Bhanot, N., Rao, P. V., and Deshmukh, S. G. (2015). Enablers and barriers of sustainable manufacturing: Results from a survey of researchers and industry professionals. *Procedia CIRP, 29*, 562–567.

Biju, P. L., Shalij, P. R., and Prabhushankar, G. V. (2015). Evaluation of customer requirements and sustainability requirements through the application of fuzzy analytic hierarchy process. *Journal of Cleaner Production, 108*, 808–817.

Calabrese, A., Costa, R., Levialdi, N., and Menichini, T. (2016). A fuzzy analytic hierarchy process method to support materiality assessment in sustainability reporting. *Journal of Cleaner Production, 121*, 248–264.

Čuček, L., Klemeš, J. J., and Kravanja, Z. (2012). A review of footprint analysis tools for monitoring impacts on sustainability. *Journal of Cleaner Production, 34*, 9–20.

Dai, J., and Blackhurst, J. (2012). A four-phase AHP–QFD approach for supplier assessment: A sustainability perspective. *International Journal of Production Research, 50*(19), 5474–5490.

Dormer, A., Finn, D. P., Ward, P., and Cullen, J. (2013). Carbon footprint analysis in plastics manufacturing. *Journal of Cleaner Production, 51*, 133–141.

Dubey, R., Gunasekaran, A., and Chakrabarty, A. (2015). World-class sustainable manufacturing: Framework and a performance measurement system. *International Journal of Production Research, 53*(17), 5207–5223.

Evans, A., Strezov, V., and Evans, T. J. (2009). Assessment of sustainability indicators for renewable energy technologies. *Renewable and Sustainable Energy Reviews, 13*(5), 1082–1088.

Finkbeiner, M. (2009). Carbon footprinting—Opportunities and threats. *International Journal of Life Cycle Assessment, 14*(2), 91–94.

Garbie, I. H. (2015). Sustainability awareness in industrial organizations. *Procedia CIRP, 26*, 64–69.

Garetti, M., and Taisch, M. (2012). Sustainable manufacturing: Trends and research challenges. *Production Planning Control, 23*(2–3), 83–104.

Ghadimi, P., Azadnia, A. H., Yusof, N. M., and Saman, M. Z. M. (2012). A weighted fuzzy approach for product sustainability assessment: A case study in automotive industry. *Journal of Cleaner Production, 33*, 10–21.

Gumus, A. T. (2009). Evaluation of hazardous waste transportation firms by using a two step fuzzy-AHP and TOPSIS methodology. *Expert Systems with Applications, 36*(2), 4067–4074.

Gunasekaran, A., and Spalanzani, A. (2012). Sustainability of manufacturing and services: Investigations for research and applications. *International Journal of Production Economics, 140*(1), 35–47.

Gupta, A., Vangari, R., Jayal, A. D., and Jawahir, I. S. (2011). Priority evaluation of product metrics for sustainable manufacturing. In *Global Product Development* (pp. 631–641). Germany: Springer Berlin Heidelberg.

Gutowski, T., Dahmus, J., and Thiriez, A. (2006). Electrical energy requirements for manufacturing processes. In 13th CIRP International Conference on Life Cycle Engineering, pp. 623–627.

Haapala, K. R., Zhao, F., Camelio, J., Sutherland, J. W., Skerlos, S. J., Dornfeld, D. A., Jawahir, I. S., Clarens, A. F., and Rickli, J. L. (2013). A review of engineering research in sustainable manufacturing. *Journal of Manufacturing Science and Engineering, 135*(4), 041013.

Huang, I. B., Keisler, J., and Linkov, I. (2011). Multi-criteria decision analysis in environmental sciences: Ten years of applications and trends. *Science of the Total Environment, 409*(19), 3578–3594.

ISO 14046.CD.1 (2012). Water footprint—Requirements and guidelines. Geneva, Switzerland

Kahraman, C., Kaya, İ., and Cebi, S. (2009). A comparative analysis for multiattribute selection among renewable energy alternatives using fuzzy axiomatic design and fuzzy analytic hierarchy process. *Energy, 34*(10), 1603–1616.

Kaya, T., and Kahraman, C. (2010). Multicriteria renewable energy planning using an integrated fuzzy VIKOR & AHP methodology: The case of Istanbul. *Energy, 35*(6), 2517–2527.

Kellens, K., Dewulf, W., Overcash, M., Hauschild, M., and Duflou, J.R. (2012). Methodology for systematic analysis and improvement of manufacturing unit process life cycle inventory, part 1: Methodology description. *International Journal of the Life Cycle Assessment, 17*(1), 69–78.

Lowell Center for Sustainable Production. (1998). Sustainable production: A working definition. In: Informal Meeting of the Committee Members.

Moneim, A. F. A., Galal, N. M., and Shakwy, M. E. (2013). Sustainable manufacturing indicators. Proceedings of the Global Climate Change: Biodiversity and Sustainability, Alexandria, Egypt, pp. 15–18.

Muduli, K., Govindan, K., Barve, A., Kannan, D., and Geng, Y. (2013). Role of behavioural factors in green supply chain management implementation in Indian mining industries. *Resources, Conservation and Recycling*, 76, 50–60.

Newman, S. T., Nassehi, A., Imani-Asrai, R., and Dhokia, V. (2012). Energy efficient process planning for CNC machining. *CIRP Journal of Manufacturing Science Technology*, 5(2), 127–136.

Ocampo, L. A., and Clark, E. E. (2015). An analytic hierarchy process (AHP) approach in the selection of sustainable manufacturing initiatives: A case of a semiconductor manufacturing firm in the Philippines. *International Journal of the Analytic Hierarchy Process*, 7(1), 32–49.

Pearce, D. (2001). Measuring resource productivity, Paper to DTI/Green Alliance Conference, February 2001, London.

Pfister, S., Bayer, P., Koehler, A., and Hellweg, S. (2011). Environmental impacts of water use in global crop production: Hotspots and trade-offs with land use. *Environmental Science and Technology*, 45(13), 5761–5768.

Pohekar, S. D., and Ramachandran, M. (2004). Application of multi-criteria decision making to sustainable energy planning—A review. *Renewable and Sustainable Energy Reviews*, 8(4), 365–381.

Rashid, S. H., Evans, S., and Longhurst, P. (2008). A comparison of four sustainable manufacturing strategies. *International Journal of Sustainable Engineering*, 1(3), 214–229.

Rusinko, C. (2007). Green manufacturing: An evaluation of environmentally sustainable manufacturing practices and their impact on competitive outcomes. *IEEE Transactions on Engineering Management*, 54(3), 445–454.

Seidel, R. H. A., Shahbazpour, M., and Seidel, M. C. (2007). Establishing sustainable manufacturing practices in SMEs. In *2nd International Conference on Sustainability Engineering and Science, Talking and Walking Sustainability*.

Seow, Y., and Rahimifard, S. (2011). A framework for modelling energy consumption within manufacturing systems. *CIRP Journal of Manufacturing Science and Technology*, 4(3), 258–264.

Sivapirakasam, S. P., Mathew, J., and Surianarayanan, M. (2011). Multi-attribute decision making for green electrical discharge machining. *Expert Systems with Applications*, 38(7), 8370–8374.

Tripathi, R., and Ahmad, I. (2015). Prospect for unorganised manufacturing sector in India: A comparative study with respect to China. *International Journal of Applied Research*, 1(6), 170–172.

Tseng, M. L. (2013). Modeling sustainable production indicators with linguistic preferences. Journal of Cleaner Production, 40, 46–56.

Tseng, M. L., Divinagracia, L., and Divinagracia, R. (2009a). Evaluating firm's sustainable production indicators in uncertainty. *Computers and Industrial Engineering*, 57(4), 1393–1403.

Tseng, M. L., Lin, Y. H., and Chiu, A. S. (2009b). Fuzzy AHP-based study of cleaner production implementation in Taiwan PWB manufacturer. *Journal of Cleaner Production*, 17(14), 1249–1256.

Veleva, V., and Ellenbecker, M. (2001). Indicators of sustainable production: Framework and methodology. *Journal of Cleaner Production*, 9(6), 519–549.

Veleva, V., Hart, M., Greiner, T., and Crumbley, C. (2001). Indicators of sustainable production. *Journal of Cleaner Production*, *9*(5), 447–452.

Vimal, K. E. K., and Vinodh, S. (2015). LCA integrated ANP framework for selection of sustainable manufacturing processes. *Environmental Modeling and Assessment*, *21*(4), 507–516.

Vimal, K. E. K., Vinodh, S., and Raja, A. (2015a). Modelling, assessment and deployment of strategies for ensuring sustainable shielded metal arc welding process—A case study. Journal of Cleaner Production, 93, 364–377.

Vimal, K. E. K., Vinodh, S., and Raja, A. (2015b). Optimization of process parameters of SMAW process using NN-FGRA from the sustainability view point. Journal of Intelligent Manufacturing, 1–22. doi:10.1007/s10845-015-1061-5.

Vinodh, S. (2011). Assessment of sustainability using multi-grade fuzzy approach. *Clean Technologies and Environmental Policy*, *13*(3), 509–515.

Vinodh, S., Arvind, K. R., and Somanaathan, M. (2011). Tools and techniques for enabling sustainability through lean initiatives. *Clean Technologies and Environmental Policy*, *13*(3), 469–479.

Yan, J., and Li, L. (2013). Multi-objective optimization of milling parameters—The trade-offs between energy, production rate and cutting quality. *Journal of Cleaner Production*, *52*, 462–471.

Young, J. E., Sachs, A., and Ayres, E. (1994). *The Next Efficiency Revolution: Creating a Sustainable Materials Economy*. Washington, DC: Worldwatch Institute.

Zadeh, L. A. (1965). Fuzzy sets. *Information and Control*, *8*(3), 338–353.

11

Use of FAHP for Occupational Safety Risk Assessment: An Application in the Aluminum Extrusion Industry

Muhammet Gul and Ali Fuat Guneri

CONTENTS

11.1 Introduction

Risk is defined as relating to a technical or sociotechnical *system* in which *events* may take place in the *future* that have *unwanted* consequences to *assets* that we want to protect (Rausand, 2011). The first stage in controlling risks is to perform a risk assessment in order to see what is required to be done (Health and Safety Executive, 1998). This includes identifying hazards, deciding the harmfulness of them, evaluating all possible risks, reducing them, and recording the results, respectively (Health and Safety Executive, 1998; Main, 2004; Gul and Guneri, 2016). The plentifulness in risk assessment methods points out the importance of selection of the appropriate one (Guneri et al., 2015). Several methods have been developed to assess risks up to now. These methods are mostly examined under two main groups: qualitative and quantitative (Tixier et al., 2002; Reniers et al., 2005; Marhavilas et al., 2011). Methods integrated with multicriteria decision-making (MCDM) are included in quantitative risk assessment methods and have been applied to occupational health and safety (OHS) risk assessment.

After the new OHS Law (number 6331) came into force, carrying of risk assessment became mandatory for public and private sector stakeholders in Turkey. Employers have faced a problem about which risk assessment method they should apply. Constraints such as scope, practicality, cost, and sensitivity may affect the selection process (Guneri et al., 2015; Guneri and Gul, 2013). In line with the growth of the country, the aluminum industry in Turkey has also grown (Günay, 2006). However, companies in the sector have faced new hazards. The International Aluminum Institute (IAI) carries out comprehensive projects to control measures against identified hazards and risks. İt is suggested that enhanced process control, positive technological improvement, and better planning can reduce potential risks (Wesdock and Arnold, 2014). In order to manage risks, new and effective risk assessment methods are required for companies to eliminate occupational hazards or reduce them to an acceptable level in work-places systematically, benefit from the participation of employees at all levels with a better teamwork (Gul and Guneri, 2016).

MCDM is a discipline of operations research that makes the decision maker's preferences explicit in decision-making environments of multiple criteria (Gul and Guneri, 2016). In MCDM methods, it is often difficult for experts to give a precise rating to an alternative with respect to the criteria. Therefore, the application of probabilistic risk analysis methods may not give satisfactory results due to the incomplete risk data or high level of uncertainty. For this reason, fuzzy MCDM (FMCDM) is used to model the uncertainty and vagueness. Calculating the relative importance of criteria using fuzzy numbers instead of crisp numbers is one of the significant advantages of FMCDM methods. So, in this chapter, we use FMCDM methods (fuzzy analytic hierarchy process [FAHP] and fuzzy technique for order preference by similarity to ideal solution [FTOPSIS]) in assessment of potential hazards.

This chapter combines FAHP and FTOPSIS for occupational safety risk assessment. Buckley's FAHP is used to rank the risk parameters of the proportional risk assessment (PRA) method: *probability, severity,* and *frequency.* In evaluating the ranking order of hazard groups, FTOPSIS is utilized. The combined method considers linguistic variables in evaluation of criteria and alternatives has the advantages of managing uncertainties, simultaneous consideration of positive and the negative ideal solutions, simple computations, and a logical concept (Gul and Guneri, 2016; Mahdevari et al., 2014). The proposed method aims to identify the potential hazards and presents control measures. In the chapter, a case study showing the application of the proposed approach is provided through a look at the aluminum extrusion industry in Turkey.

The rest of the chapter is organized as follows: Section 11.2 presents the previous research in the related literature. In Section 11.3, methods comprising the proposed method are described. Section 11.4 presents the case study of the proposed methodology in the aluminum extrusion industry. The conclusions of the study and some future directions are given in Section 11.5.

11.2 Literature Review

Risk assessment is a crucial aspect for almost all sectors. The aluminum industry contains high risk operations and much attention is paid to the risk assessment process. Many studies are available in the OHS risk management literature with different risk assessment approaches. Marhavilas and Koulouriotis (2008) and Supciller et al. (2015) focused on the PRA method. The first study presented two risk assessment techniques named the *proportional technique* and *decision matrix technique* and presented a case study of these techniques in the aluminum extrusion industry in Greece. Real data over a 5.5-year time period of 1999–2004 was used to analyze risks. Upon comparison of the results, the two methods were found compatible. The second study integrated PRA method with fuzzy sets. It proposed a fuzzy PRA method to overcome the drawbacks of the classical PRA method. A case study was carried out in a textile firm manufacturing towels and bathrobes. Classical PRA and fuzzy PRA methods were compared and risk assessment based on fuzzy sets provided more precise measurements than the classical PRA method.

In the literature, many MCDM- and FMCDM-based risk assessment methods have been examined in order to find causes and characteristics of accidents and evaluate the workplace safety conditions of various areas such as energy (Mohsen and Fereshteh, 2017; Mahdevari et al., 2014; Kang et al., 2014), manufacturing (Kokangül et al., 2017; Gul and Guneri, 2016; Djapan et al., 2015; Grassi et al., 2009; Hu et al., 2009), transportation and supply chain (Akyuz, 2017; Akyuz and Celik, 2015; John et al., 2014), construction (Efe et al., 2016; Liu and Tsai, 2012; Ebrahimnejad et al., 2010), chemistry and biochemistry (Othman et al., 2016), health, safety, and medicine (Gul et al., 2017), and education (Hassanain et al., 2017).

Regarding the energy area, Mohsen and Fereshteh (2017) proposed a failure mode and effect analysis (FMEA) based on fuzzy VIseKriterijumska Optimizacija I Kompromisno Resenje (VIKOR)-FAHP-entropy methods for the failure modes risk assessment of a geothermal power plant. Mahdevari et al. (2014) identified and ranked hazards of a coal deposit in Iran using FTOPSIS. They distinguished the risks into 12 different groups according to the index value from a fuzzy method. They also showed the control measures taken within the scope of their study. Kang et al. (2014) proposed a risk evaluation model for oil storage tank zones based on the theory of inherent and controllable hazards. Inherent hazards were evaluated by the major hazards method, which is based on the probability and severity of events. The risk factors of controllable hazards were identified by fault tree analysis (FTA). These weights of factors were calculated by AHP. After that, a fuzzy comprehensive evaluation mode for controllable hazards was established. A risk-matrix method was applied to order the risk rank of the oil storage tank zone. The proposed model combines four different methods on risk assessment.

A number of research studies were carried out in the area of manufacturing safety management by various researchers. Recently, a study by Kokangül et al. (2017) was performed regarding the usability of the class intervals in the Fine–Kinney risk assessment method for the results obtained with the AHP method. A case was analyzed for a large manufacturing company. Gul and Guneri (2016) used the FAHP method in order to enable OHS experts to use linguistic variables for weighting two parameters derived from the L-type matrix method. They presented a case study in an aluminum plate manufacturing factory and determined orders of priority of hazard groups using FTOPSIS. They expanded the case study considering the application of the proposed risk assessment methodology for hazard types in each department of the factory. On conclusion of their study, control measures were taken for the hazards that placed first in the intradepartment rankings. Djapan et al. (2015) proposed FAHP for determining risk levels at the workplaces in a central Serbian manufacturing small and medium enterprise. The results showed that ergonomic risk exists more often in the observed small- and medium-sized enterprise (SME). The developed model contributed to real-life practice by decreasing risk factor values and increasing the quality and efficiency of work processes at the workplaces. Grassi et al. (2009) developed a MCDM risk assessment model for a country-specific sausage production company in Italy. FTOPSIS was used to determine the risk index of hazardous activities. They proposed to add three parameters named undetectability, sensitivity to nonexecution of maintenance, and sensitivity to nonutilization of personal protective equipment (PPE) as well as severity and probability. Hu et al. (2009) carried out a risk assessment of green components to hazardous substance using FMEA and FAHP. They used FMEA parameters as criteria and weighted them with FAHP. Then, a risk priority value was calculated for each of the components by multiplying the weights and FMEA scores of two managers.

Akyuz (2017) applied the analytic network process (ANP) with human factors analysis and a classification system to evaluate potential operational causes in cargo ships. They provided novel knowledge by presenting a different perspective during marine accident analysis in which priority weights of accident causes related to human error were calculated by ANP. Also, a fuzzy DEMATEL (The Decision Making Trial and Evaluation Laboratory) -based approach was developed in a marine environment by Akyuz and Celik (2015) to evaluate critical operational hazards in a gas freeing process. John et al. (2014) used FAHP as a risk assessment methodology in seaport operations in addition to the methods of the evidential reasoning (ER) approach, fuzzy set theory, and expected utility. They applied FAHP to rank the risk factors while using ER to synthesize them.

Regarding the construction industry, Efe et al. (2016) applied FMEA and the fuzzy preference ranking organization method for enrichment of evaluations (FPROMETHEE) for failures in the construction industry. Linguistic variables were used to assess occurrence, severity, and detection factors related to weights. For selecting the most serious failure modes, FPROMETHEE was used.

Liu and Tsai (2012) developed a risk assessment model using quality function deployment (QFD), fuzzy ANP (FANP), and FMEA methods in a construction company in Taiwan. QFD was used in order to represent the relationships among construction items, hazard types, and hazard causes. FANP was used to identify important hazard types and hazard causes while FMEA was used to assess the risk value of hazards. Ebrahimnejad et al. (2010) proposed a project risk assessment model using FTOPSIS and fuzzy linear programming technique for multidimensional analysis of preference (FLINMAP) methods for an Iran build–operate–transfer (BOT) power plant project.

Othman et al. (2016) presented a structured methodology for incorporating prioritization in hazard and operability analysis using AHP. The hazards of a process identified using hazard and operability studies (HAZOP) were quantitatively weighted and ranked by AHP. The method was applied to a simple reactor unit and a more complex system of a dividing wall column pilot plant as case studies. Gul et al. (2017) suggested a risk assessment model for hospitals using FAHP-fuzzy VIseKriterijumska Optimizacija I Kompromisno Resenje (FVIKOR). FAHP was used to determine the weights of the five risk factors used by Grassi et al. (2009). FVIKOR was then applied for determination of risk groups in each department of the case study hospital. Hassanain et al. (2017) proposed a fire safety ranking system for student housing facilities. The weights of fire safety attributes were determined using AHP.

From the above-mentioned review of the literature, it is concluded that this chapter contributes to the literature on OHS risk assessment by FMCDM methods in some aspects: (1) a hybridization of two FMCDM methods is proposed to resolve the shortcomings of a precise risk score calculation and decrease the inconsistency in decision-making; (2) the evaluations of three risk parameters of the PRA method and hazard prioritizations are executed by OHS experts using and a full consensus; and (3) unlike the conventional PRA method, experts determine criteria weights using the pairwise comparison feature of Buckley's FAHP.

11.3 Methodology

In this section, the risk assessment process is presented step by step along with a flowchart from Main (2012). Hereafter, the risk assessment methods are introduced. The PRA method and its limitations are emphasized in Section 11.3.1. Finally, Buckley's FAHP and FTOPSIS methodologies are explained.

11.3.1 Risk Assessment Process

The risk assessment process involves certain steps (Main, 2012). The steps are identifying hazards, assessing risks, reducing the identified risks, and documenting the results (also shown in Figure 11.1). Risk assessment uses the available data to determine the magnitude, severity, and frequency of specific events systematically. The risk assessment is the central part of the risk management process, the purpose of which is to establish a proactive safety strategy by investigating potential risks (Rausand, 2011; Mahdevari et al., 2014). The first step is establishing project parameters and identifying the assessment scope. In the second step, hazards are identified through different approaches. Main (2012) mentioned many methods for identifying hazards such as intuitive operational and engineering sense, examination of system specifications and expectations, review of relevant codes, regulations and consensus standards, interviews with system users or operators, checklists, brainstorming, review of historical data, and so on. The third step is about assessing initial risks. The main focus of this chapter is within this step.

   Main (2012) generated four substeps to assess initial risks. (1) A risk scoring system is selected. Three-parameter (probability, severity, and frequency) scoring systems are frequently used in the literature for the PRA method. (2) For each hazard, the severity rating is assessed. Severity is assessed according to the personal injury, the value of property or equipment damaged, the loss of working time, and so on. (3) The probability and frequency of hazards are assessed. This is related to the duration and extent of exposure, training, and awareness, and the characteristics of the hazard. (4) This process is about deriving an initial risk level from the selected risk scoring system. The fourth of main steps in the risk assessment process is reducing risks. This step enables the process to become more efficient so that significant risks are eliminated quickly by using a hazard control hierarchy (Main, 2012). After risk reduction is performed, a second assessment is carried out to validate that the selected measures reduce the risks effectively. This is the step of assessing residual risks. As shown in Figure 11.1, the process follows a decision step hereafter. The risk assessment team decides that the risks are reduced to an acceptable level. The last step includes documentation of the results (Marhavilas et al., 2011).

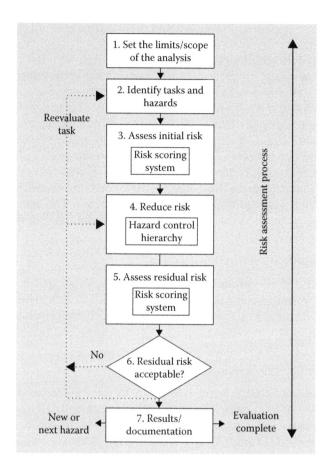

FIGURE 11.1
Risk assessment process (From Main, B. W., *Professional Safety*, 49, 37–47, 2012.)

11.3.2 Risk Assessment Methods

In the risk assessment process, using an apparent technique has several benefits. Initially, it identifies occupational hazards and proposes improvement precautions more efficiently than conventional safety processes. It is required to use a risk assessment technique to reveal the main occupational hazards in workplaces systematically, benefiting from the experiences of employees through better teamwork, and obtaining the same results at the end of the review in each department. Selecting the appropriate risk assessment method among several methods that have different outputs, steps, and applications is of vital importance. Outputs of the risk assessment vary with respect to the selected method type. Currently, many risk assessment methods are available in the literature. They are generally classified under two common groups in terms of estimating the risks, probability of the risks, and possible effects. These are qualitative and quantitative methods (Ceylan and Bashelvaci, 2011).

Saat (2009) indicates the existence of more than 150 risk assessment methods in his study. Pinto et al. (2011) mention the most commonly used occupational risk assessment (ORA) methods in industry. Some of them are as follows: HAZOP, FTA, FMEA, human reliability assessment (HRA), preliminary hazard analysis (PHA), checklists, and what-if (coarse analysis). Tixier et al. (2002) divide the risk assessment methods into two main groups: qualitative and quantitative. They identify 62 methodologies in their study.

Marhavilas et al. (2011) aimed to determine and study, analyze, and elaborate, classify, and categorize the main risk assessment methods by reviewing the scientific literature from 2000 to 2009 in their work. The main difference between these methods is the specific ways to find the value of the risk.

11.3.3 PRA Method

Marhavilas and Koulouriotis (2008) and Supciller et al. (2015) proposed a proportional formula for calculating the quantified risk using the PRA technique. Calculation of the risk value is proportionally achieved by the following equation:

$$R = P * S * F$$

where R is the risk value, P is the probability factor, S is the severity of harm factor, and F is the frequency (or the exposure) factor. Initially, measurement of the probability, severity, and frequency ratings is determined with this method (Tables 11.1 through 11.3). Then, the risk values are constructed. The acceptability level of the risks is also interpreted according to Table 11.4.

TABLE 11.1

Probability Ratings

By Marhavilas and Koulouriotis (2008)		By Supciller et al. (2015)	
P	Description of the Event	P	Description of the Event
10	Unavoidable	10	Can be expected, almost surely
9	Almost assured	6	Very possible
8	Frequent	3	Unusual, but possible
7	Probable	1	Only possible over a longer duration
6	Probability slightly >50%	0.5	Very unlikely
5	Probability 50%	0.2	Virtually impossible
4	Probability slightly <50%	–	–
3	Almost improbable	–	–
2	Improbable	–	–
1	Impossible	–	–

TABLE 11.2

Severity Ratings

By Marhavilas and Koulouriotis (2008)		By Supciller et al. (2015)	
S	Description of the Event	S	Description of the Event
10	Death	100	Disastrous, many fatalities
9	Permanent total inefficiency	40	Disaster, several fatalities
8	Permanent serious inefficiency	15	Seriously, on fatality
7	Permanent slight inefficiency	7	Significant, life-threatening injuries
6	Absence from the work >3 weeks, and return with health problems	3	Important
		1	First aid needed
5	Absence from the work >3 weeks, and return after full recovery	–	–
4	Absence from the work >3 days and <3 weeks, and return after full recovery	–	–
3	Absence from the work >3 days, and return after full recovery	–	–
2	Slight injury without absence from the work, and with full recovery	–	–
1	No human injury	–	–

TABLE 11.3

Frequency Ratings

By Marhavilas and Koulouriotis (2008)		By Supciller et al. (2015)	
F	Description of the Event	F	Description of the Event
10	Permanent presence of damage	10	Constantly
9	Presence of damage every 30 s	6	Every day, during working hours
8	Presence of damage every 1 min	3	Weekly or occasionally
7	Presence of damage every 30 min	2	Monthly
6	Presence of damage every 1 h	1	A few times per year
5	Presence of damage every 8 h	0.5	Very rare
4	Presence of damage every 1 week	–	–
3	Presence of damage every 1 month	–	–
2	Presence of damage every 1 year	–	–
1	Presence of damage every 5 years	–	–

In this chapter, a fuzzy approach enabling OHS experts to use linguistic terms for evaluating three risk parameters derived from PRA is proposed in order (1) to deal with limitation in crisp risk score calculation and (2) decrease the inconsistency in the decision-making process. The PRA method has some shortcomings. It is based on an equal criteria weight for probability, severity, and frequency. Different evaluations of the criteria may

TABLE 11.4

Risk Ratings

By Marhavilas and Koulouriotis (2008)		By Supciller et al. (2015)	
R	Urgency Level of Required Actions	R	Urgency Level of Required Actions
700–1000	Precaution required immediately	>400	Stop work and ensure that the risk is eliminated immediately
500–700	Precaution required earlier than 1 day	200–400	Immediate precaution required. Ensure that the risk is eliminated immediately
300–500	Precaution required earlier than 1 month	70–200	Correction is required. Ensure that the risk is reduced. Report the risk to manager immediately
200–300	Precaution required earlier than 1 year	20–70	Risk requires attention
<200	Immediate precaution is not necessary but event surveillance is required	<20	No attention required

lead to different meanings (Grassi et al., 2009). For example, hazards with high probability, high frequency, and low severity can be classified at the same level as ones with low probability, high frequency, and high severity. The new developed PRA technique-based fuzzy method has some advantages. (1) The use of FAHP in the proposed fuzzy MCDA approach avoids the shortcomings of a crisp risk score calculation. (2) The evaluations of three risk parameters and hazard rankings are made by experienced OHS experts with a full consensus. (3) Unlike a classic PRA method, experts assign criteria weights using the pairwise comparison technique of FAHP. (4) A more accurate risk calculation is provided to the stakeholders of aluminum industry with the new approach.

11.3.4 FAHP

FAHP is one of the most commonly used MCDM methods based on fuzzy sets. Because AHP cannot provide a subjective thinking approach, FAHP is proposed in order to solve hierarchical fuzzy problems. There is more than one FAHP method developed in the FMCDM literature (Ho, 2008; Kubler et al., 2016). Buckley (1985) proposed fuzzy priorities of comparison ratios whose membership functions were trapezoidal. Chang (1996) introduced a new FAHP approach with the use of triangular fuzzy numbers for pairwise comparison and the use of the extent analysis method for the synthetic extent values of the pairwise comparisons. Buckley's (1985) method was used in the case application of this chapter. However, other methods have some limitations. For instance, the extent analysis method could not make full use of all the information of the fuzzy comparison matrices and might cause an irrational zero weight within the selection criteria (Chan and Wang, 2013). The steps of Buckley's FAHP method followed in this study are given below (Tzeng and Huang, 2011; Gül et al., 2012):

Step 1: Pairwise comparisons are constructed among all the criteria in the hierarchy system. Linguistic terms are assigned by asking which is more important of two elements/criteria, such as

$$\tilde{M} = \begin{pmatrix} 1 & \tilde{a}_{12} & \cdots & \tilde{a}_{1n} \\ \tilde{a}_{21} & 1 & \cdots & \tilde{a}_{2n} \\ \vdots & \vdots & \ddots & \vdots \\ \tilde{a}_{n1} & \tilde{a}_{n2} & \cdots & 1 \end{pmatrix} = \begin{pmatrix} 1 & \tilde{a}_{12} & \cdots & \tilde{a}_{1n} \\ 1/\tilde{a}_{21} & 1 & \cdots & \tilde{a}_{2n} \\ \vdots & \vdots & \ddots & \vdots \\ 1/\tilde{a}_{n1} & \tilde{a}_{n2} & \cdots & 1 \end{pmatrix} \qquad (11.1)$$

$$\tilde{a}_{ij} = \begin{cases} \tilde{1}, \tilde{3}, \tilde{5}, \tilde{7}, \tilde{9} & \text{criterion } i \text{ is of relative importance to criterion } j \\ 1 & i = j \\ \tilde{1}^{-1}, \tilde{3}^{-1}, \tilde{5}^{-1}, \tilde{7}^{-1}, \tilde{9}^{-1} & \text{criterion } j \text{ is of relative importance to criterion } i \end{cases} \qquad (11.2)$$

Step 2: A fuzzy geometric mean matrix is defined using the geometric mean technique:

$$\tilde{r}_i = \left(\tilde{a}_{i1} \otimes \tilde{a}_{i2} \otimes \cdots \otimes \tilde{a}_{in} \right)^{1/n} \qquad (11.3)$$

Step 3: The fuzzy weights of each criterion are obtained by Equation 11.4:

$$\tilde{w}_i = \tilde{r}_i \otimes \left(\tilde{r}_1 \oplus \tilde{r}_2 \oplus \cdots \oplus \tilde{r}_n \right)^{-1} \qquad (11.4)$$

Here, $\tilde{w}_i$ is the fuzzy weight of criterion i and $\tilde{w}_i = (lw_i, mw_i, uw_i)$. lw_i, mw_i, uw_i are the lower, middle, and upper value of the fuzzy weight of criterion i.

Step 4: The center of area (CoA) method is used to find the best non-fuzzy performance (BNP), as in Equation 11.5:

$$w_i = [(uw_i - lw_i) + (mw_i - lw_i)] / 3 + lw_i \qquad (11.5)$$

11.3.5 Fuzzy TOPSIS

TOPSIS was developed by Hwang and Yoon (1981) to find the best alternative based on the compromise solution concept. The compromise solution concept can be regarded as selecting the solution with the shortest distance from the ideal solution and the farthest distance from the negative ideal solution. Since the ratings while evaluating alternatives against criteria usually refer to the subjective uncertainty, TOPSIS has been extended to consider the situation of fuzzy numbers (Tzeng and Huang, 2011; Celik et al., 2012). We followed the procedure of Chen's (2000) FTOPSIS method in the case study of this chapter. The steps are listed in the following (Tzeng and Huang, 2011; Kutlu and Ekmekçioğlu, 2012):

Step 1: The scores of alternatives with respect to each criterion are obtained considering a decision-making group with K experts using the following formula $\tilde{x}_{ij} = \frac{1}{K}[\tilde{x}_{ij}^1(+)\tilde{x}_{ij}^1(+)....(+)\tilde{x}_{ij}^K]$. $A = \{A_i | i = 1,....,m\}$ shows the set of alternatives, $C = \{C_j | j = 1,....,n\}$ represent the criteria set, where $X = \{X_{ij} | i = 1,....,m; j = 1,....,n\}$ denotes the set of fuzzy ratings and $\tilde{w} = \{\tilde{w}_j | j = 1,....,n\}$ is the set of fuzzy weights. The linguistic variables are described by triangular fuzzy numbers as follows: $\tilde{x}_{ij} = (a_{ij}, b_{ij}, c_{ij})$.

Step 2: Normalized ratings are determined by Equation 11.6:

$$\tilde{r}_{ij} = \begin{cases} (\dfrac{a_{ij}}{c_j^*}, \dfrac{b_{ij}}{c_j^*}, \dfrac{c_{ij}}{c_j^*}), & \text{where } c_j^* = \max_i c_{ij} \text{ if } j \in \text{benefit criteria} \\[2ex] (\dfrac{a_j^-}{c_{ij}}, \dfrac{b_j^-}{b_{ij}}, \dfrac{c_j^-}{a_{ij}}), & \text{where } a_j^- = \min_i a_{ij} \text{ if } j \in \text{cost criteria} \end{cases} \tag{11.6}$$

Step 3: Weighted normalized ratings are obtained by Equation 11.7:

$$\tilde{v}_{ij} = \tilde{w}_j(x)\tilde{r}_{ij}, \quad i = 1,....,m; j = 1,....,n \tag{11.7}$$

Step 4: The fuzzy positive ideal point (FPIS, A') and the negative ideal point (FNIS, A⁻) are derived as in Equations 11.8 and 11.9, where J_1 and J_2 are the benefit and cost attributes, respectively:

$$\text{FPIS} = A^* = \{\tilde{v}_1^*, \tilde{v}_2^*,...., \tilde{v}_n^*\} \text{ where } \tilde{v}_j^* = (1,1,1) \tag{11.8}$$

$$\text{FNIS} = A^- = \{\tilde{v}_1^-, \tilde{v}_2^-,...., \tilde{v}_n^-\} \text{ where } \tilde{v}_j^- = (0,0,0) \tag{11.9}$$

Step 5: The next step is about calculating the separation from the FPIS and the FNIS between the alternatives. The separation values can also be obtained using the Euclidean distance as in Equations 11.10 and 11.11:

$$\tilde{S}_i^* = \sqrt{\sum_{j=1}^{n}[\tilde{v}_{ij} - \tilde{v}_j^*]^2}, \quad i = 1,....,m \tag{11.10}$$

$$\tilde{S}_i^- = \sqrt{\sum_{j=1}^{n} [\tilde{v}_{ij} - \tilde{v}_j^-]^2}, \quad i = 1, \ldots, m \tag{11.11}$$

Step 6: Then, the defuzzified separation values are derived using the CoA defuzzification method, to calculate the similarities to the ideal solution. Next, the similarities to the ideal solution are given as Equation 11.12:

$$C_i^* = \tilde{S}_j^- / (\tilde{S}_j^* + \tilde{S}_j^-), \quad i = 1, \ldots, m \tag{11.12}$$

The preferred orders are ranked according to C_i^* in descending order to finally select the best alternatives.

11.3.6 The Proposed Approach

Fuzziness enables transformation of the uncertainty and vagueness of human judgments into a mathematical equation (Kutlu and Ekmekçioğlu, 2012; Gul and Guneri, 2016). In this chapter, a fuzzy approach enabling OHS experts to use linguistic terms for evaluating three parameters of the PRA technique is proposed to consider the disadvantages of a crisp risk score calculation and decrease the inconsistency in the decision-making process for the aluminum industry. Initially, a team consisting of OHS experts identifies the potential main hazards. Second, a pairwise comparison matrix for three risk parameters is constructed. Hereafter, Buckley's FAHP is applied to determine the weights of these risk factors. Then, experts' linguistic evaluations of each hazard with respect to risk parameters are aggregated to get a mean value. With the aid of the obtained fuzzy decision matrix a, implementation of FTOPSIS is carried out. In the FTOPSIS process, by using the weights of risk parameters and the fuzzy decision matrix, a weighted normalized fuzzy decision matrix is constructed. Subsequently, FPIS and FNIS and the distance of each hazard from FPIS and FNIS are calculated, respectively. In the final step, FTOPSIS closeness coefficients of processes are obtained. According to the closeness coefficients, the ranking order of all main hazard groups is determined. Figure 11.2 shows the proposed FAHP-FTOPSIS combined risk assessment model based on PRA.

11.4 A Case Application for the Aluminum Extrusion Industry

11.4.1 Aluminum Extrusion and Description of the Observed Plant

Extrusion is defined as a plastic deformation process in which a block of metal (billet) is forced to flow by compression through the die opening of a cross-sectional area smaller than that of the original billet (Saha, 2000). There are two main

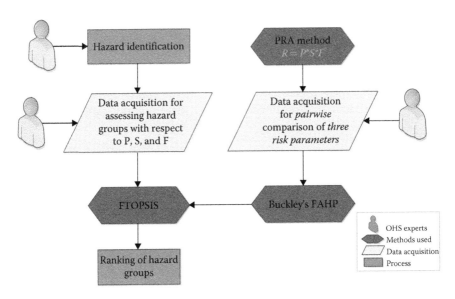

FIGURE 11.2
The proposed approach.

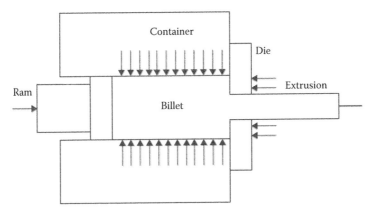

FIGURE 11.3
Principles and definition of direct extrusion process. (From Saha, P. K., *Aluminum Extrusion Technology*, ASM International, 2000.)

types of extrusion, namely direct (Figure 11.3) and indirect; they are both used commonly in aluminum industries. The producible cross-sections vary from solid round and rectangular shapes to L-shapes, T-shapes, tubes, and many other different types (Marhavilas and Koulouriotis, 2008).

Extrusion might be cold or hot based on the alloy and the method used. In hot extrusion, the billet is preheated to assist plastic deformation (Saha, 2000). Aluminum extrusions are used in various areas such as the construction industry, window, and door frame systems, prefabricated houses/building structures, roofing, and exterior cladding, curtain walling, road and rail vehicles, airframes, and marine applications (Marhavilas and Koulouriotis, 2008).

The observed plant in this study is located in Turkey. The firm a leading aluminum extrusion manufacturer in Turkey. It has a large billet casting area, new and replacement mold manufacturing. It produces various aluminum profiles for various sectors and purposes according to world quality standards.

11.4.2 Risk Evaluation Using the Proposed Approach

A group of OHS experts identified the main hazard items such as falling of materials (MH1), lack of annual periodic controls of grounding (MH2), lack of machine guards for moving and rotating parts (MH3), poor stacking of materials and the possibility of falling from height (MH4), negatives that may occur in the extraction process (MH5), lack of periodic maintenance and written instructions for use of compressors (MH6), overthrow of the tubes and lack of use and storage instructions (MH7), negatives experienced in cutting profiles (MH8), negatives in the molding processes (MH9), being open of electrical panel cover, hazards caused by safety signs and leakage relay (MH10), negatives in the chemical coating process (MH11), incompatibility of chemicals used (MH12), misuse of appropriate personal protective equipment (PPE) in the painting process (MH13), fall of pending profiles (MH14), storing of oil drums besides oxygen tubes (MH15), nonavailability of adequate safety measures against dust (MH16), failure to get education for appropriate vocational training of employees (MH17), failure to get education for basic OSH training of employees (MH18), failiure to renew and update the risk analysis (MH19), failure to reach electrical panels due to placing materials in front of them (MH20), diseases that may occur due to lack of hygiene in the dressing area (MH21), failure to use PPEs (MH22), and sinking of aluminum shavings (MH23).

After the identification of the main hazards, evaluations of three OHS experts in linguistic variables are used to determine the importance of three risk parameters (P, S, and F) using the pairwise comparison provided by Buckley's FAHP method. In this chapter, the OHS experts use the linguistic terms from Gul and Guneri (2016) to evaluate the risk parameters' weights. The evaluation in linguistic form is presented in Table 11.5. In pairwise comparison of the risk parameters probability and severity, the answers of three experts are very weak (VW), absolutely weak (AW), and absolutely weak (AW), respectively. When comparing probability and frequency, they respond fairly weak (FW), fairly weak (FW), and equal (E). Finally, when comparing the parameters severity and frequency, the responses are fairly strong (FS), very strong (VS), and very strong (VS). Upon review of all the pairwise comparisons, the weights of risk parameters are obtained as (0.225, 0.488, 0.288) for P, S, and F, respectively.

TABLE 11.5

Evaluations of Occupational Health and Safety Experts in Linguistic Variables and Weights of the Risk Parameters

OHS Experts 1-2-3	Probability	Severity	Frequency	Weight
Probability	E,E,E	VW,AW,AW	FW,FW,E	0.225
Severity	–	E,E,E	FS,VS,VS	0.488
Frequency	–	–	E,E,E	0.288

Note: The consistency index CI = 0.043, the random consistency index RI = 0.58, and the consistency ratio CR = CI/RI = 0.0742. Since CR is 0.07, which is less than 10%, the consistency test is satisfied.

The advantages of using FAHP for this application are:

1. FAHP is a common MCDM method and generally used in weighting the risk parameters of traditional OHS risk assessment methods such as L-matrix, X-matrix, PRA, Fine–Kinney, and FMEA. Using FAHP in weighting the risk parameters of these methods enables the expert to provide an accurate hazard score. That is, FAHP calculates weight values of the risk parameters of PRA apart from traditional OHS risk assessment methods, which mostly consider equal weights.

2. Moreover, different combinations of judgments on the parameters may lead to a completely different meaning. For example, hazards with high probability and severity and low frequency could be classified at the same level as hazards with low probability and severity and high frequency. These limitations are expressed in the literature (Grassi et al., 2009). So, this chapter considers weighting of the three parameters of PRA using Buckley's FAHP calculations.

3. This enables the experts to utilize group decision-making in assessing hazards, give relative importance among the risk parameters by pairwise comparison, and use linguistic terms for risk parameter evaluation.

Fuzzy linguistic evaluations of each risk parameter with respect to the main hazard groups following the FTOPSIS procedure are applied by using the weights of three PRA parameters obtained from Buckley's FAHP. In this study, the OHS experts evaluate the main hazard groups using linguistic variables as provided by Gul and Guneri (2016). The evaluations of the OHS experts in linguistic variables for the three risk parameters with respect to hazards are carried out. For instance, the three occupational safety (OS) experts evaluated the main hazard (MH1) as medium good (MG), fair (F), and medium good (MG), respectively, for probability (P), medium poor (MP), medium poor (MP), and poor (P), respectively, for severity (S), and medium poor (MP), medium poor (MP), and poor (P), respectively, for frequency (F).

TABLE 11.6

Fuzzy Decision Matrix and Weighted Fuzzy Decision Matrix

Hazard Group	Fuzzy Decision Matrix			Weighted Fuzzy Decision Matrix		
	Probability	Severity	Frequency	Probability	Severity	Frequency
MH1	(0.5;0.7;0.9)	(0.83;0.97;1)	(0.07;0.23;0.43)	(0.08;0.16;0.29)	(0.29;0.46;0.69)	(0.01;0.07;0.17)
MH2	(0.7;0.9;1)	(0.9;1;1)	(0.83;0.97;1)	(0.12;0.2;0.32)	(0.31;0.48;0.69)	(0.17;0.29;0.4)
MH3	(0.9;1;1)	(0.5;0.7;0.9)	(0.1;0.3;0.5)	(0.15;0.22;0.32)	(0.17;0.33;0.62)	(0.02;0.09;0.2)
MH4	(0.7;0.9;1)	(0.37;0.57;0.77)	(0.1;0.3;0.5)	(0.12;0.2;0.32)	(0.13;0.27;0.53)	(0.02;0.09;0.2)
MH5	(0.77;0.93;1)	(0.9;1;1)	(0.3;0.5;0.7)	(0.13;0.21;0.32)	(0.31;0.48;0.69)	(0.06;0.15;0.28)
MH6	(0.37;0.57;0.77)	(0.9;1;1)	(0.13;0.27;0.43)	(0.06;0.13;0.24)	(0.31;0.48;0.69)	(0.03;0.08;0.17)
MH7	(0.37;0.57;0.77)	(0.9;1;1)	(0.07;0.2;0.37)	(0.06;0.13;0.24)	(0.31;0.48;0.69)	(0.01;0.06;0.15)
MH8	(0.5;0.7;0.9)	(0.5;0.7;0.9)	(0.1;0.3;0.5)	(0.08;0.16;0.29)	(0.17;0.33;0.62)	(0.02;0.09;0.2)
MH9	(0.5;0.7;0.9)	(0.5;0.7;0.9)	(0.07;0.23;0.43)	(0.08;0.16;0.29)	(0.17;0.33;0.62)	(0.01;0.07;0.17)
MH10	(0.5;0.7;0.9)	(0.9;1;1)	(0.07;0.2;0.37)	(0.08;0.16;0.29)	(0.31;0.48;0.69)	(0.01;0.06;0.15)
MH11	(0.5;0.7;0.9)	(0.37;0.57;0.77)	(0.1;0.3;0.5)	(0.08;0.16;0.29)	(0.13;0.27;0.53)	(0.02;0.09;0.2)
MH12	(0.5;0.7;0.9)	(0.37;0.57;0.77)	(0.1;0.3;0.5)	(0.08;0.16;0.29)	(0.13;0.27;0.53)	(0.02;0.09;0.2)
MH13	(0.9;1;1)	(0.77;0.93;1)	(0.07;0.2;0.37)	(0.15;0.22;0.32)	(0.26;0.45;0.69)	(0.01;0.06;0.15)
MH14	(0.9;1;1)	(0.7;0.9;1)	(0.07;0.2;0.37)	(0.15;0.22;0.32)	(0.24;0.43;0.69)	(0.01;0.06;0.15)
MH15	(0.3;0.5;0.7)	(0.9;1;1)	(0.07;0.2;0.37)	(0.05;0.11;0.22)	(0.31;0.48;0.69)	(0.01;0.06;0.15)
MH16	(0.7;0.9;1)	(0.7;0.9;1)	(0.07;0.2;0.37)	(0.12;0.2;0.32)	(0.24;0.43;0.69)	(0.01;0.06;0.15)
MH17	(0.37;0.57;0.77)	(0.5;0.7;0.9)	(0.3;0.5;0.7)	(0.06;0.13;0.24)	(0.17;0.33;0.62)	(0.06;0.15;0.28)
MH18	(0.37;0.57;0.77)	(0.5;0.7;0.9)	(0.3;0.5;0.7)	(0.06;0.13;0.24)	(0.17;0.33;0.62)	(0.06;0.15;0.28)
MH19	(0.5;0.7;0.9)	(0.3;0.5;0.7)	(0.13;0.3;0.5)	(0.08;0.16;0.29)	(0.1;0.24;0.48)	(0.03;0.09;0.2)
MH20	(0.57;0.77;0.93)	(0.9;1;1)	(0.07;0.2;0.37)	(0.09;0.17;0.3)	(0.31;0.48;0.69)	(0.01;0.06;0.15)
MH21	(0.57;0.77;0.93)	(0.1;0.3;0.5)	(0.07;0.23;0.43)	(0.09;0.17;0.3)	(0.03;0.14;0.34)	(0.01;0.07;0.17)
MH22	(0.5;0.7;0.9)	(0.7;0.9;1)	(0.07;0.2;0.37)	(0.08;0.16;0.29)	(0.24;0.43;0.69)	(0.01;0.06;0.15)
MH23	(0.5;0.7;0.9)	(0.7;0.9;1)	(0.07;0.2;0.37)	(0.08;0.16;0.29)	(0.24;0.43;0.69)	(0.01;0.06;0.15)

Note: The fuzzy weight vector is (0.165;0.224;0.317) for the probability parameter, (0.345;0.478;0.687) for the severity parameter, and (0.202;0.298;0.4) for the frequency parameter.

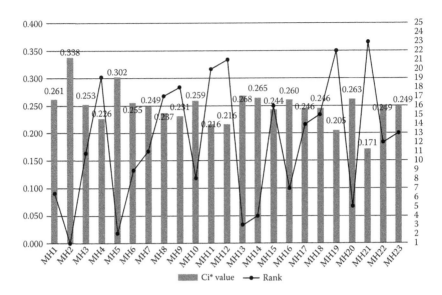

FIGURE 11.4
Fuzzy technique for order preference by similarity to ideal solution (FTOPSIS) final ranking.

The fuzzy linguistic variables are then transformed into fuzzy triangular values as shown in Table 11.6. This is the first stage of the FTOPSIS analysis. By the addition of the fuzzy risk parameter weights into the FTOPSIS calculation, the weighted fuzzy decision matrix is generated using Equation 11.7. We consider the FPIS and FNIS values as (1, 1, 1) and (0, 0, 0) as in Section 11.3.5. The distance from the FPIS and FNIS of each alternative is calculated using Equations 11.10 and 11.11. The last step presents the similarities to an ideal solution using Equation 11.12. The FTOPSIS analyses results are provided in Figure 11.4. According to the results, the ranking of the hazards is ordered by giving C_i^* values. The closest value to 1 is ranked the highest risk, whereas risks having a C_i^* value farthest from 1 are ranked the lowest risk (Mahdevari et al., 2014; Kutlu and Ekmekçioğlu, 2012; Gul and Guneri, 2016). Results show that the most important hazards are from the main electrical panel and fuses (MH2), flying objects due to the mold squeezing or breaking during pressing in extrusion process area (MH5), failure to use appropriate PPE in the painting process (MH13), and fall of pending profiles (MH14). Overall risk management cannot eliminate risks altogether but can only suggest suitable measures to control. Therefore, after completing each stages of risk assessment, each risk should be controlled or eliminated. If this is not possible, they should be reduced to the acceptable level (Mahdevari et al., 2014).

The advantages of using FTOPSIS for this application can be explained as follows. Given the rise in model complexity, a FTOPSIS method is also attached to the proposed approach so as to guarantee risk prioritization coherence even when a large number of hazards must be considered. The method allows experts to assign judgments to the hazard groups with respect to three parameters by means of linguistic terms, which are better interpreted by humans and fuzzy in nature and are then transferred into fuzzy numbers. In this study, FTOPSIS is applied to analyze the health and safety the hazards of the aluminum extrusion industry because it offers more capability in handling uncertainties, simultaneous consideration of the positive and the negative ideal points, simple computations, and a logical concept.

11.4.3 Potential Control Measures of the Hazards

This chapter applied a FAHP-FTOPSIS combined approach to rank the hazards. Fuzzy numbers can be adapted to the real cases instead of crisp numbers (Mahdevari et al., 2014). In order to reduce risks, a hazard control hierarchy is required. The order of control hierarchy as in Barnes (2009) helps to determine potential control measures in the observed plant. This order is the most efficient way to control measures of the main hazards. Unless the best control measure is possible, the plant executives may use other measures to reduce risks. Some control measures for the main hazard groups are suggested below.

For hazard MH2, an emergency action plan should be activated and several control measures should be taken. First, direct or alternative current fuses must be in a covered box. Second, without interrupting the current using a switch or circuit breaker, the table should not be opened. Third, with regard to dispatch tables, tables where employees in the worksite can reach must be inside a covered box made from sheet metal and kept locked. Switches, fuses, and cables to the tables exceeding 1000 volts should be easily accessible and the way of immediately responding.

To reduce the risks related to MH5, a required action plan earlier than a 1-year period should be activated and some control measures should be taken as follows. In the extrusion lines of the observed plant there is no machine protective equipment available in case of flying objects due to mold break during pressing processes. Therefore, the environment of the press should be protected with a shatterproof panel in order to protect both the press operator and other employees against any flying objects. The machine operator should see the profile jamming during the profile pressing process without being close to the press material outlet. Extra lighting in this area should be provided since the ambient lighting is insufficient. Since the press is a machine with a high noise level, engineering measures must be taken to prevent occupational diseases with hearing loss.

The ventilation where the electrostatic powder coating process is carried out is inadequate. The existing fan used for ventilation is insufficient. In order to protect workers from occupational diseases that may arise in the future, a more powerful fan must be placed. It must be designed to provide enough fresh air. Also, employees must be trained about the use of PPEs.

As discussed by plant management, the storage area is insufficient. Materials and chemicals are stacked poorly. The worksite and its environment should be arranged ergonomically, ensuring the proper functioning of the processes. Materials should be placed in a tidy way and materials in danger of falling must be placed in appropriate areas. They should not be placed on the stairs. Flammable materials should be kept in separate areas of the plant. Risk assessment and management is clearly a continuing process and the adequacy of control measures should also be subject to continuous improvement, review, and revision if necessary (Mahdevari et al., 2014).

11.5 Conclusions and Future Directions

In this chapter, a new occupational safety risk assessment approach that combines FAHP and FTOPSIS is proposed. The proposed method is applied through a case study in an aluminum extrusion plant. First, Buckley's FAHP is used to rank the *probability, severity,* and *frequency* risk parameters of the PRA method. Then in evaluating the ranking orders of the main hazard groups, FTOPSIS is utilized. The proposed fuzzy-based approach allows the interpretation of the risks more realistically by giving pairwise comparisons among the probability, severity, and frequency parameters. The proposed method identifies the potential hazards and presents control measures for early warning. Results show that the most important main hazards come from the main electrical panel and fuses (MH2), flying objects due to a mold squeezing or breaking during pressing in the extrusion process area (MH5), failure to use appropriate PPE in the painting process (MH13), and fall of pending profiles (MH14).

However, risk assessment and management are continuing processes, and criteria that may affect the risk score may change over time. Therefore, the OHS experts and top management should track risks and controls over certain periods. For future studies, some of the other MCDM methods such as ELECTRE (ELimination Et Choix Traduisant la REalite'), GRA (grey relational analysis), MOORA (The multi-objective optimization on the basis of ratio analysis), DEMATEL (The decision-making trial and evaluation laboratory), and so on could be used together with the ones used in this study to assess the risks of the aluminum industry.

References

Akyuz, E. (2017). A marine accident analysing model to evaluate potential operational causes in cargo ships. *Safety Science, 92,* 17–25.

Akyuz, E. and Celik, E. (2015). A fuzzy DEMATEL method to evaluate critical operational hazards during gas freeing process in crude oil tankers. *Journal of Loss Prevention in the Process Industries, 38,* 243–253.

Barnes, M. (2009). Risk assessment workbook for mines. Metalliferous, extractive and opal mines, and quarries, Mine Safety Operations, [IGA-019 (TRIM: OUT09/16488)].

Buckley, J. J. (1985). Fuzzy hierarchical analysis. *Fuzzy Sets and Systems, 17*(3), 233–247.

Celik, E., Gul, M., Gumus, A. T., and Guneri, A. F. (2012). A fuzzy TOPSIS approach based on trapezoidal numbers to material selection problem. Journal of Information Technology Applications and Management, 19(3), 19–30.

Ceylan, H. and Başhelvacı, V. S. (2011). Risk analysis with risk assessment matrix method: An application. International Journal of Engineering Research and Development, 3(2), 25–33 (In Turkish).

Chan, H. K. and Wang, X. (2013). Fuzzy extent analysis for food risk assessment. In *Fuzzy Hierarchical Model for Risk Assessment* (pp. 89–114). London: Springer.

Chang, D. Y. (1996). Applications of the extent analysis method on fuzzy AHP. European Journal of Operational Research, 95(3), 649–655.

Chen, C. (2000). Extensions of the TOPSIS for group decision-making under fuzzy environment. Fuzzy Sets *and Systems, 114*, 1–9.

Djapan, M. J., Tadic, D. P., Macuzic, I. D., and Dragojovic, P. D. (2015). A new fuzzy model for determining risk level on the workplaces in manufacturing small and medium enterprises. *Proceedings of the Institution of Mechanical Engineers, Part O: Journal of Risk and Reliability, 229*(5), 456–468.

Ebrahimnejad, S., Mousavi, S. M., and Seyrafianpour, H. (2010). Risk identification and assessment for build–operate–transfer projects: A fuzzy multi attribute decision making model. *Expert Systems with Applications, 37*(1), 575–586.

Efe, B., Yerlikaya, M. A., and Efe, Ö. F. (2016). İş Güvenliğinde Bulanık Promethee Yöntemiyle Hata Türleri ve Etkilerinin Analizi: Bir İnşaat Firmasında Uygulama (Failure mode and effects analysis with fuzzy Promethee method in occupational accidents: An application in a construction firm). Gümüşhane Üniversitesi Fen Bilimleri Enstitüsü Dergisi, 6(2), 126–137. (In Turkish)

Grassi, A., Gamberini, R., Mora, C., and Rimini, B. (2009). A fuzzy multi-attribute model for risk evaluation in workplaces. Safety Science, 47(5), 707–716.

Gul, M., Ak, M. F., and Guneri, A. F. (2017). Occupational health and safety risk assessment in hospitals: A case study using two-stage fuzzy multi criteria approach. *Human and Ecological Risk Assessment: An International Journal, 23*(2), 187–202..

Gül, M., Çelik, E., Güneri, A. F., and Gümüş, A. T. (2012). Simülasyon ile bütünleşik çok kriterli karar verme: Bir hastane acil departmanı için senaryo seçimi uygulaması. *Istanbul Commerce University Journal of Science, 11*(22), 1–18 (In Turkish).

Gul, M. and Guneri, A. F. (2017). A fuzzy multi criteria risk assessment based on decision matrix technique: A case study for aluminum industry. *Journal of Loss Prevention in the Process Industries, 40*, 89–100.

Günay, D. (2006). *Alüminyum Sektörü Hakkında Bir Değerlendirme*. Türkiye Kalkınma Bankası A.Ş. Ekonomik ve Sosyal Araştırmalar Müdürlüğü, Ankara.

Guneri, A. F. and Gul, M. (2013). Prioritization of risk evaluation methods for occupational safety with fuzzy multi criteria decision making. In 26th European Conference on Operational Research, 1–4 July, Rome, Italy.

Guneri, A. F., Gul, M., and Ozgurler, S. (2015). A fuzzy AHP methodology for selection of risk assessment methods in occupational safety. International Journal of Risk Assessment and Management, 18(3–4), 319–335.

Hassanain, M. A., Hafeez, M. A., and Sanni-Anibire, M. O. (2017). A ranking system for fire safety performance of student housing facilities. Safety science, 92, 116–127.

Health and Safety Executive. (1998). Electrical Safety on Construction Sites. INDG231 HSE Books ISBN 0 7176 1207 4, Suffolk, United Kingdom: Health and Safety Executive

Ho, W. (2008). Integrated analytic hierarchy process and its applications—A literature review. *European Journal of Operational Research, 186*(1), 211–228.

Hu, A. H., Hsu, C. W., Kuo, T. C., and Wu, W. C. (2009). Risk evaluation of green components to hazardous substance using FMEA and FAHP. *Expert Systems with Applications, 36*(3), 7142–7147.

Hwang, C. L. and Yoon, K. (1981). *Multiple Attribute Decision Making: Methods and Applications: A State of the Art Survey*. New York, NY: Springer-Verlag.

John, A., Paraskevadakis, D., Bury, A., Yang, Z., Riahi, R., and Wang, J. (2014). An integrated fuzzy risk assessment for seaport operations. *Safety Science, 68,* 180–194.

Kang, J., Liang, W., Zhang, L., Lu, Z., Liu, D., Yin, W., and Zhang, G. (2014). A new risk evaluation method for oil storage tank zones based on the theory of two types of hazards. *Journal of Loss Prevention in the Process Industries, 29*, 267–276.

Kokangül, A., Polat, U., and Dağsuyu, C. (2017). A new approximation for risk assessment using the AHP and Fine Kinney methodologies. *Safety Science, 91*, 24–32.

Kubler, S., Robert, J., Derigent, W., Voisin, A., and Le Traon, Y. (2016). A state-of the-art survey & testbed of fuzzy AHP (FAHP) applications. *Expert Systems with Applications, 65*, 398–422.

Kutlu, A. C. and Ekmekçioğlu, M. (2012). Fuzzy failure modes and effects analysis by using fuzzy TOPSIS-based fuzzy AHP. *Expert Systems with Applications, 39*(1), 61–67.

Liu, H. T. and Tsai, Y. L. (2012). A fuzzy risk assessment approach for occupational hazards in the construction industry. *Safety Science, 50*(4), 1067–1078.

Mahdevari, S., Shahriar, K., and Esfahanipour, A. (2014). Human health and safety risks management in underground coal mines using fuzzy TOPSIS. *Science of the Total Environment, 488*, 85–99.

Main, B. W. (2012). Risk assessment: A review of the fundamental principles. *Professional Safety, 49*(12), 37–47.

Marhavilas, P. K. and Koulouriotis, D. E. (2008). A risk-estimation methodological framework using quantitative assessment techniques and real accidents' data: Application in an aluminum extrusion industry. *Journal of Loss Prevention in the Process Industries, 21*(6), 596–603.

Marhavilas, P. K., Koulouriotis, D., and Gemeni, V. (2011). Risk analysis and assessment methodologies in the work sites: On a review, classification and comparative study of the scientific literature of the period 2000–2009. *Journal of Loss Prevention in the Process Industries, 24*(5), 477–523.

Mohsen, O. and Fereshteh, N. (2017). An extended VIKOR method based on entropy measure for the failure modes risk assessment—A case study of the geothermal power plant (GPP). *Safety Science, 92*, 160–172.

Othman, M. R., Idris, R., Hassim, M. H., and Ibrahim, W. H. W. (2016). Prioritizing HAZOP analysis using analytic hierarchy process (AHP). *Clean Technologies and Environmental Policy, 18*(5), 1345–1360.

Pinto, A., Nunes, I. L., and Ribeiro, R. A. (2011). Occupational risk assessment in construction industry—Overview and reflection. Safety Science, 49(5), 616–624.

Rausand, M. (2011). *Risk Assessment: Theory, Methods, and Applications* (Vol. 115). Hoboken, NJ: John Wiley & Sons.

Reniers, G. L. L., Dullaert, W., Ale, B. J. M., and Soudan, K. (2005). Developing an external domino accident prevention framework: Hazwim. *Journal of Loss Prevention in the Process Industries, 18*(3), 127–138.

Saat, M. B. (2009). *Implementation of integrated occupational health and safety risk assessment methods, checklist and matrix methods, to a construction site*, MS Thesis, Ankara: Gazi University Institute of Science and Technology (In Turkish).

Saha, P. K. (2000). Aluminum Extrusion Technology. Ohio, USA: ASM International.

Supciller, A. A. and Abali, N. (2015). Occupational health and safety within the scope of risk analysis with fuzzy proportional risk assessment technique (Fuzzy Prat). *Quality and Reliability Engineering International, 31*(7), 1137–1150.

Tixier, J., Dusserre, G., Salvi, O., and Gaston, D. (2002). Review of 62 risk analysis methodologies of industrial plants. *Journal of Loss Prevention in the Process Industries, 15*(4), 291–303.

Tzeng, G. H. and Huang, J. J. (2011). *Multiple Attribute Decision Making: Methods and Applications*. Boca Raton, FL: CRC Press.

Wesdock, J. C. and Arnold, I. M. (2014). Occupational and environmental health in the aluminum industry: Key points for health practitioners. *Journal of Occupational and Environmental Medicine, 56*(5 Suppl), S5.

12

Assessing the Management of Electronic Scientific Journals Using Hybrid FDELPHI and FAHP Methodology

Nara Medianeira Stefano, Raul Otto Laux, Nelson Casarotto Filho, Lizandra Lupi Garcia Vergara, and Izabel Cristina Zattar

CONTENTS

12.1 Introduction

Unlike tangible assets (Castilla-Polo and Gallardo-Vázquez, 2016), intangible assets have the major strategic characteristic of singularity. That is what makes them unique assets, difficult to acquire, develop, and even copy as well as maybe even legally protected. This feature has provided a source of competitive advantage for intangible assets to allow them to face competition and excel in their markets. The intangible asset is interpreted as something that does not have a tangible reality.

In the literature related to the economy, the knowledge asset is often used as the equivalent to the intangible asset, whereas in the literature related to management it is equivalent to *intellectual capital* (IC) (Hsu et al., 2014; Verma and Dhar, 2016). Knowledge can present itself in different ways. Codified or explicit knowledge is formalized in some kind of document, regardless of the informational support selected to register it. Tacit knowledge is that which the individual has acquired throughout life, with experience. It is usually difficult to formalize or explain to another person because it is subjective and inherent to the abilities of a person. Thus, similar to products, services (Pak et al., 2015; Talley and Ng, 2016) are a set of coded and uncoded knowledge; however, services are more strongly dependent on intangible knowledge.

IC management (ICM) is an area that has increasingly been highlighted in several research areas. Nevertheless, it is important to emphasize that it is not about technology (it is a facilitator). However, the information environments require agility, modernization, and people that are generators of knowledge. Despite the spreading of literature regarding the identification, evaluation, and ICM in various sectors of the economy, the importance and study of intangibles for the management of scientific journals (Corera-Álvarez and Molina-Molina, 2016) is quite limited. The scientific journals (Bomfá, 2009) fall into the services category because they provide for the user (reader, author, and reviewer) the service of receiving, processing, evaluating, and communicating/disseminating scientific information. Bomfá (2009) points out that in the case of electronic scientific journals, service is defined as a process, because the management of the editorial process covers from when the scientist idealizes the research idea (input), financial support, human resources, technology used, thematic approach to research, among others, to the dissemination phase of the research (outputs): review, publication, distribution channels targeted to users, forms of storage, and retrieval of information.

However, because the introduction of electronic journals, many have been conducted to determine its behavior researches (Buela-Casal, 2004; Bar-Ilan and Fink, 2005; Raza and Upadhyay, 2006; Moghaddam and Moballeghi, 2008; Khan and Ahmad, 2009; Nicholas et al., 2010; Gresty and Edwards-Jones, 2012; Vasishta, 2013; Homol, 2014; Bravo, Díez, and Merino, 2015; Taquette and Minayo, 2016). Nowadays, investments related to electronic, digital, and other intangible collaboration services have become more important than ever. In fact, the information services are facing significant challenges to identify and assess the value of their intangible assets. So, in this context, intangible assets of electronic journals can also be classified according to human capital (HC), organizational (or structural) capital, and relational capital, that is, from the perspective of IC. Therefore, the identification of intangible assets into three distinct categories of assets is the first important step to assess the management of an electronic scientific journal. The activity of promoting science involves not only the effort of making the results of research studies public but also the task of managing the flows prior to those processes. Technical support is needed to operationalize equipment, technology, media and communication resources, and publishing. Thus, the measurement of knowledge involving IC is very important for people involved with the business so they can judge their procedures or get help improving their decision-making.

This chapter proposes criteria for assessing the management of electronic journals using the IC approach based on a hybrid mathematical methodology, integrating the *fuzzy Delphi* (FDELPHI) and *Fuzzy Analytic Hierarchy Process* (FAHP) methods. The FDELPHI method was used to raise the critical factors (criteria/subcriteria) present in the management of electronic scientific journals. The FAHP method was applied to calculate the relative weights of the selected criteria/subcriteria that affect the management.

Therefore, the use of the FAHP approach results in some advantages, such as:

- Fuzzy numbers are preferable to understand human judgment because it is diffuse in nature.
- The adoption of fuzzy numbers allows the decision maker(s) to have the freedom to estimate what they want.
- The approach handles uncertain data well because human feelings are unpredictable and uncertain. To model this kind of uncertainty, the fuzzy set can be incorporated in the pairwise comparison as an extension of the analytic hierarchy process (AHP).

This chapter is organized as follows: Section 12.2 deals with the evaluation and management of scientific journals; Section 12.3 describes the hybrid FDELPHI and FAHP methodology; and Section 12.4 presents the analysis and discussion of data.

12.2 Evaluation and Management of Scientific Journals

Scientific journals are one of the most important segments of the academic and scientific environment and one of the most valuable resources in the scholarly communication chain. Researchers have been trying to evaluate the use of scientific journals in various ways, such as through questionnaires, interviews, and quotes. Despite the advances of information technology and the migration of the printed newsletter to electronic media, there are no fundamental changes in the nature of methodologies of management of scientific journals.

The first attempts at journal review, considering the examination of quality indicators, date from the early 1960s from a model developed by the United Nations Education Science and Culture Organization (UNESCO). In Brazil, some studies have been conducted to develop an efficient system of periodic assessment. The study by Krzyzanowski and Ferreira (1998, 57), which considers scientific journals under the categories of form and substance (content performance), today serves as a reference for the Coordination for the Improvement of Higher Education Personnel (CAPES), of the Ministry of Education (MEC), for the classification of publications.

According to Yamamoto et al. (2002), these systems are characterized by the establishment of a set of parameters that consider intrinsic and extrinsic aspects, formal and of merits and translated by indicators that allow the score and consequent ranking of journals. A review of scientific and technical journals arises from the need to define measurable parameters, which may reflect the quality of the recorded information. In Brazil, these evaluations are held by committees of CAPES, the institution responsible for the evaluation of postgraduate courses and for the scientific production of universities (Fachin, 2002; Duarte and Rodrigues, 2012).

The Qualis criteria of evaluation are used by CAPES to stratify the quality of the intellectual production of postgraduation programs. This process was designed to meet the specific needs of the evaluation system and it is based on the information given by the collected data. As a result, a list of the classification of the vehicles used by graduate programs for the dissemination of their production is available.

Qualis measures the quality of articles and other types of production from an analysis of the quality of dissemination vehicles, that is, journals (CAPES, 2012). The classification is performed by assessment areas and undergoes an annual update process. These vehicles are grouped in strata indicative of quality, A1 (the highest), A2, B1, B2, B3, B4, B5, C (with weight zero), within the national and international areas.

In international journals, the impact factor (IF) of the Journal Citation Reports (JCR) database of the Institute for Scientific Information (ISI) is used for levels A and B, with A-level journals having an IF equal or greater to 0.5 and B-level journals having an IF inferior to 0.5. In the case of periodicals edited abroad that are not part of the JCR, the Area Committee will examine other available information, for example, on the site of the journal, to define its classification. IFs are calculated by dividing the number of citations in the current year of articles published in the two previous years by the total number of articles published in the two previous years. The IF is thus a measure of the frequency with which a given journal is cited in a particular year.

Besides providing the citation information, JCR also indicates the speed at which these citations (of a given periodic) appear in the literature by calculating the average number of times an article is cited during the year it was published (Kieling and Gonçalves, 2007). The citation system was invented primarily as a way to understand how scientific discoveries and innovations are communicated and how they work (Szymanski et al., 2012). Initially, it was not seen as a tool to individually evaluate scientists, universities, or academic systems. The citation system is useful for monitoring how scientific ideas are propagated among researchers and scientists (individual) and how they communicate the results of their research.

The IF of journals is increasingly becoming an important parameter for evaluating journals worldwide. Besides this method, Web Impact Factor (WIF) is also available. Ingwersen (1998) proposed the WIF as the equivalent of the ISI's IF. The WIF is based on the number of connections made on a website, in comparison to the size (usually the number of pages) of the site. It also provides quantitative tools for ranking, evaluating, categorizing, and comparing websites, domains, and subdomains.

Although IF is the most used tool in the literature (Fachin, 2002; Buela-Casal, 2004; Dong et al., 2005; Kieling and Gonçalves, 2007; Moghaddam and Moballeghi, 2008; Bornmann et al., 2011; Szymanski et al., 2012; Machado and Santos, 2016), other proposals for evaluating scientific journals are found in the literature. The evaluation criteria presented by databases such as Scientific Electronic Librarym Online (SciELO), Scopus, ISI Platform/Web of Knowledge, and *Periódicos de Acesso Livre* (OASIS.Br) are noteworthy. SciELO and *Qualis CAPES* establish their own criteria, policies, and procedures for admission and continuation of scientific journals in their collection.

Concerning scientific journal management, electronic or printed, open access or restricted, the most important aspect is sustainability. This can refer to the economic aspect or article submissions because a scientific journal with no articles does not exist.

Dubini and Giglia (2009) comment that sustainability has the following foundations: effectiveness (achieving the proposed objectives), efficiency (minimizing the resources used to achieve the proposed objectives), and durability (the possibility to operate over time and also the introduction of innovative solutions).

The sustainability of a journal involves categories such as (Guanes and Guimarães, 2012; Sandes-Guimarães, 2013) costs, indexing, accessibility, navigability, quality of articles, editorial body, and referees, among other items. In the case of sustainability of a scientific journal, this not only refers to the economic aspects of the process and online publishing. Along with the costs, it also refers to the accessibility, the retrieval of information, navigability, and interactivity, that is, characteristics of Information and Communication Technology (ICT), especially the Internet.

As the management of a scientific journal involves many activities, when this is managed erroneously, it compromises its "health." The FDELPHI method enables "filtering" of the main critical factors (criteria/subcriteria) present in the management of electronic scientific journals.

Therefore, choosing a suitable model for the management of electronic scientific journals will not only depend on financial expenses but also on the mission objectives, publisher values, resources, risk tolerance, and institutional or corporate affiliation.

12.3 Proposed Hybrid Fuzzy Methodology

In this chapter, the mathematical modeling was based on FDELPHI and FAHP (hybrid methodology) to assess the management criteria for electronic scientific journals, using the approach of IC to improve their performance. In this work, an Excel spreadsheet was used for the development of all calculations. The use of FDELPHI and FAHP in this chapter is due to:

- The FDELPHI (Mikaeil et al., 2013; Wang et al., 2014; Parameshwaran et al., 2015; Lin et al., 2016) method can solve the fuzziness of common understanding of expert opinions. The fuzziness of common understanding of experts could be solved by using the fuzzy theory and evaluated on a more flexible scale. The efficiency and quality of questionnaires could be improved. It is a methodology in which the subjective data of experts are transformed into quasiobjective data using statistical analysis and fuzzy operations. The main advantages of this method are that it can reduce the numbers of surveys to save time and cost and it also includes the individual attributes of all experts.

- In the conventional AHP, the pairwise comparison is made by using a ratio scale. Even though the discrete scale has the advantages of simplicity and ease of use, it does not take into account the uncertainty associated with the mapping of one's perception (or judgment) to a number. To deal with the uncertainty and vagueness from the subjective perception and the human experience in the decision-making process, many FAHP (Büyüközkan et al., 2008; Paksoy et al., 2012; Chamoli, 2015; Singh and Dasgupta, 2016) methods have been proposed by various authors. Decision makers usually find that they are more confident to give interval judgments than fixed-value judgments. This is because usually he/she is unable to be explicit about his/her preferences due to the fuzzy nature of the comparison process.

A hybrid mathematical modeling was based on two phases: in the first place (Phase I), decision makers identify the problem, categories, criteria, and subcriteria from the perspective of the IC. This initial stage was conducted through the FDELPHI method. Figure 12.1 shows the proposed mathematical modeling to evaluate management criteria for electronic journals using the FAHP and FDELPHI tools.

In Phase II, a matrix of pairwise comparisons was built. Through the FAHP method, proposed by Chang (1996), the vector of weights of a paired array was determined.

12.3.1 Description of the Mathematical Modeling

In this work, a hybrid fuzzy methodology is used (Figure 12.2), that is, FDELPHI and FAHP.

The following subsection discusses the detailed phases of the mathematical modeling.

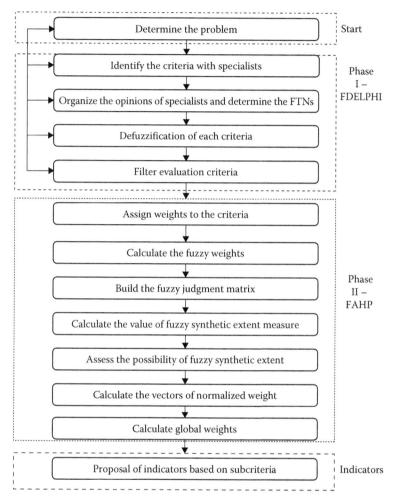

FIGURE 12.1
Proposed mathematical model used in this study.

12.3.2 Phase I—FDELPHI

In the first stage, the FDM (Multicriteria Method Delphi), the steps adopted are modified based on Chang and Wang (2006), Kuo and Chen (2008), Hsu et al. (2010), Wang and Durugbo (2013), and Stefano et al. (2015).

Step 1: Organize a panel of experts and a questionnaire to allow specialists to express their choices on the importance of each criterion, along with a set of S criteria, in a range from 1 to 5 or 1 to 7. A score is then referred to as R_i, $i \in S$, where the index of i criteria is evaluated by k experts.

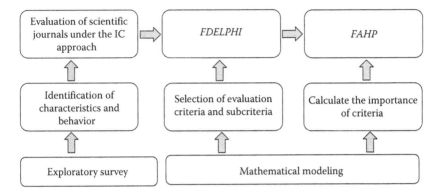

FIGURE 12.2
Scheme to evaluate the management criteria for scientific journals in approach using the concept of intellectual capital (IC).

Step 2: Arrange the opinions of experts from a questionnaire and determine the triangular fuzzy numbers (TFNs) for the index $O_i = (L_i, M_i, U_i)$ for each criterion i. L_i indicates the minimum value of the classification of all the experts, that is (Equation 12.1):

$$L_i = \text{Min}(L_{ik}) \tag{12.1}$$

M_i is the geometric mean of the rating of all specialists for the criterion i. And it can be obtained through Equation 12.2:

$$M_i = (R_{i1} \times R_{i2} \times \dots R_{ik})^{\frac{1}{k}} \tag{12.2}$$

U_i indicates the maximum rating of the experts and is calculated by Equation 12.3:

$$M_i = \text{Max}(L_{ik}) \tag{12.3}$$

Step 3: Once the TFNs are determined for all criteria, the area center (AC) approach (Hsieh et al., 2004) is used to defuzzify the TFNs of each evaluation criterion, that is (Equation 12.4):

$$G_i = \frac{(U_i - L_i) + (M_i - L_i)}{3} + L_i \tag{12.4}$$

Step 4: Filter the evaluation criteria, defining a limit α. That is, if $G_i \geq \alpha$ then the i evaluation criterion is selected; if $G_i < \alpha$ then the i evaluation criterion is eliminated.

12.3.3 Phase II—FAHP

The purpose of this phase is to determine the weight of the criteria/subcriteria that influence the management of electronic journals the most. In this study, the FAHP method proposed by Chang (1996) and described below was used to calculate the weights of the criteria.

The fuzzy triangular scale of preference used in this study is given in Table 12.1, and is based on the approach of Chang (1996), who developed the application of TFNs for the linguistic variables of the comparison scale paired to the FAHP and the extended analysis method (analytical measurement) to the values of the paired comparison. This method is used because the steps of this approach are similar to the conventional AHP and relatively easier than other FAHP approaches.

A comparison of pairs is performed using a ratio scale. The scale used is a five-point scale that uses TFNs. These numbers are used to indicate the relative strength of each pair of elements in the same hierarchy. The scores from the pairwise comparisons are transformed into linguistic variables, which, according to Buckley (1985), are represented as TFNs.

The following shows the detailed method proposed by Chang (1996), and used by Kahraman et al. (2004), Kutlu and Ekmekçioğlu (2012), Cho and Lee (2013), and Stefano (2014) among many other researchers. The steps to be followed for the application of FAHP are as follows:

Step 1: To form comparisons of pairs of attributes using the fuzzy numbers, which consist of low, medium and higher values at the same level of the hierarchical structure.

Step 2: The value of the fuzzy synthetic extent with respect to the ith object is defined by Equations 12.5 through 12.8:

$$S_i = \sum_{j=1}^{m} M_{gi}^j \otimes \left[\sum_{i=1}^{n} \sum_{j=1}^{m} M_{gi}^j \right]^{-1} \tag{12.5}$$

TABLE 12.1

Relationship between Linguistic Variables and Their Relevance Functions

Linguistic Scale of Importance	Fuzzy Numbers for FAHP	Fuzzy Triangular Scale	Fuzzy Triangular Reciprocal Scale
Precisely equal	–	(1, 1, 1)	(1, 1, 1)
Equally important	1	(0.66, 1, 1.5)	(0.66, 1, 1.5)
Less important	2	(1, 1.5, 2)	(0.5, 0.66, 1)
Moderately important	3	(1.5, 2, 2.5)	(0.4, 0.5, 0.66)
Very important	4	(2, 2.5, 3)	(0.33, 0.4, 0.5)
Extremely important	5	(2.5, 3, 3.5)	(0.29, 0.33, 0.4)

Source: Adapted from Chang, D.Y., *European Journal of Operational Research*, 95, 649–655, 1996 and Calabrese, A. et al., *Expert Systems with Applications*, 40, 3747–3755, 2013.

$$\sum_{j=1}^{n} M_{ij} = \left(\sum_{j=1}^{n} l_{ij}, \sum_{j=1}^{n} m_{ij}, \sum_{j=1}^{n} u_{ij} \right), i = 1;2;3\dots,n \qquad (12.6)$$

$$\sum_{i=1}^{m}\sum_{j=1}^{n} M_{gi}^{j} = \left(\sum_{i=1}^{m}\sum_{j=1}^{n} l_{ij}, \sum_{i=1}^{m}\sum_{j=1}^{n} m_{ij}, \sum_{i=1}^{m}\sum_{j=1}^{n} u_{ij} \right) \qquad (12.7)$$

$$\sum_{i=1}^{m}\sum_{j=1}^{n} M_{ij} = \left(\frac{1}{\displaystyle\sum_{i=1}^{m}\sum_{j=1}^{n} u_{ij}}, \frac{1}{\displaystyle\sum_{i=1}^{m}\sum_{j=1}^{n} m_{ij}}, \frac{1}{\displaystyle\sum_{i=1}^{m}\sum_{j=1}^{n} l_{ij}} \right) \qquad (12.8)$$

Step 3: The degree of possibility of $M_2 = (l_2, m, u_2) \geq M_1 = (l_1, m_1, u_1)$ is set (Equation 12.9) as

$$V(M_2 \geq M_1) = \sup_{y \geq x} \left[\min(\mu_{M_2}(x), \mu_{M_2}(y)) \right] \qquad (12.9)$$

And, it may be equivalent to Equation 12.10:

$$V(M_2 \geq M_1) = \mathrm{hgt}(M_1 \cap M_2) = \mu_{M_2}(d) \begin{cases} 1; if\ m_2 \geq m_1 \\ 0; if\ l_1 \geq l_2 \\ \dfrac{l_1 - u_2}{(m_2 - u_2) - (m_1 - l_1)}, cc \end{cases} \qquad (12.10)$$

where d is the ordinate of greater intersection point D, between μ_{M_1} and μ_{M_2}. To compare M_1 and M_2, the values $V(M_1 \geq M_2)$ and $V(M_2 \geq M_1)$ are needed.

Step 4: The level of possibility for a convex fuzzy number to be greater than k convex fuzzy numbers can be defined by Equation 12.11:

$$(M \geq M_1; M_2\dots; M_k) = V\left[(M \geq M_1)e(M \geq M_2) and\dots and(M \geq M_k) \right. \\ = \min V(M \geq M_i), i = 1,2,3\dots,k \left. \right] \qquad (12.11)$$

Equation 12.11 takes the form of Equation 12.12:

$$d'^{(A_i)} = \min V(S_j \geq S_i)$$ (12.12)

For $k = 1, 2..., n, k \neq i$. Following on this, the weight vector (Equation 12.13) is given by

$$W' = \delta\left(d'^{(A_1)}, d'^{(A_2)}..., d'^{(A_n)}\right)^T$$ (12.13)

where $A_i (i = 1, 2..., n)$ has n elements.

Step 5: Through standardization, the weight vectors are normalized by Equation 12.14:

$$W = \left(d(A_1); d(A_2)...; d(A_n)\right)^T$$ (12.14)

Step 6: This step involves the calculation of overall weights (Huang et al., 2008; Ju et al., 2012) for the subcriteria. They are calculated by multiplying the weight of the subcriteria by the weight of the criteria to which it belongs. The overall weights are denoted by $w^i_{sub} = (w_{i1}, w_{i2}..., w_{ini})$, where n_i is the number of subcriteria with respect to the i_n criterion.

12.3.4 Sampling

The selection of journals for this research is based on a criterion, the *Qualis* (we used journals with *Qualis* A1, A2, B1, and B2), which is stipulated by Coordination for the Improvement of Higher Education Personnel (CAPES). The sample is characterized as simple and random. This sample provides accuracy and efficiency as well as being an easy procedure to apply (because all elements of the population have the same probability of belonging to the sample). Journals were selected from diverse areas (administration, engineering, environmental science, physical education, education, urban and regional planning/demography). A total of 27 questionnaires were sent out and 14 were returned. The electronic journals that were part of the sample were *Journal of Cleaner Production, Semina, Ciências Agrárias, Transinformação, Iberoamerican Journal of Industrial Engineering (IJIE), Ciências Sociais Unisinos, Custos e @gronegócio Online, Estudos Feministas, Revista em Agronegócios e Meio Ambiente, Revista Colabor@, Revista da Educação física da UEM, Brazilian Journal of Chemical Engineering, Revista Gestão & Planejamento, Revista Ambiência,* and *Brazilian Journal of Poultry Science.*

12.4 Analysis and Results

12.4.1 Application of FDELPHI

The following shows the first stage of the methodology used for evaluating the criteria for electronic science journals management, in other words the FDELPHI method.

12.4.1.1 Selection of Criteria by Using the FDELPHI Method

To start the survey of criteria, a prequestionnaire was prepared (based on the literature on scientific journals, and interviews with experts, i.e., editors of scientific journals) to assess the importance of each item. Table 12.2 shows the maximum, minimum, and geometric mean values with respect to each proposed criterion.

Through this review of the questionnaire, carried out through the FDELPHI method, seven criteria that reported a G_i smaller or equal than 3 (α) were eliminated.

TABLE 12.2

Selection and Evaluation of Criteria through the FDELPHI Method (Second Round)

Criteria	Subcriteria	Geometric Mean	Max	Min	G_i
Tangibles	Physical space where the journal is located.	3.07	5.00	2.00	3.36
	Equipment and technological resources used to make it feasible to work with the journal.	4.84	5.00	4.00	4.61
	Support materials, such as papers, to view the printed article.	2.63	4.00	2.00	2.88
Website's interface	Displaying search systems.	4.22	5.00	3.00	3.07
	Showing the navigation system.	4.69	5.00	4.00	4.56
	Providing usability for the readers of the journal.	4.69	5.00	4.00	4.56
	Providing accessibility.	4.65	5.00	3.00	4.22
	Organizing the information content.	4.54	5.00	4.00	4.51
	Handling different types of documents, such as Word documents, PDFs, RTFs, and others.	3.26	4.00	3.00	3.42
	Having a labeling system	4.40	5.00	4.00	4.47
	Having DOI, ISSN, e-ISSN licenses, creative commons attribution, among others.	4.26	5.00	4.00	4.42
	Making the periodicity explicit.	3.26	4.00	3.00	3.42
	Displaying contact information.	4.54	5.00	4.00	4.51
	Ownership of the copyright.	3.84	4.00	3.00	3.61
	Presenting permission for reproduction.	3.84	4.00	3.00	3.61

(Continued)

TABLE 12.2 (*Continued*)

Selection and Evaluation of Criteria through the FDELPHI Method (Second Round)

Criteria	Subcriteria	Geometric Mean	Max	Min	G_i
Standardization of the specification	Having an abstract.	4.84	5.00	4.00	4.61
	Displaying bibliographic caption.	4.84	5.00	4.00	4.61
	Showing the history of the editorial management.	4.69	5.00	4.00	4.56
	Presenting the mission and editorial policy.	3.84	4.00	3.00	3.61
Standardization of articles	Showing authorship, authors' affiliation, and contacts.	4.84	5.00	4.00	4.61
	Having a summary and an abstract.	5.00	5.00	5.00	5.00
	Standardization of bibliographic references.	4.40	5.00	4.00	4.47
	Having keywords.	5.00	5.00	5.00	5.00
	Having content division (abstract, introduction, methodology, results, conclusion, and references).	4.84	5.00	4.00	4.61
	Having an electronic update of the documents.	3.26	4.00	3.00	3.42
	Having page numbering.	2.83	3.00	2.00	2.61
	Having standardization of headings, subheadings, charts, figures, tables, and equations.	4.84	5.00	4.00	4.61
	Presenting good definitions of figures and items.	4.13	5.00	4.00	4.38
	Observing the size (length) of the articles proposed by the journal.	3.68	4.00	3.00	3.56
	Showing article dates of receipt and acceptance.	3.84	4.00	3.00	3.61
	Showing the correct page number order.	5.00	5.00	5.00	5.00
Guidelines and standards	Ensuring standards that avoid endogeny (referring to authors, institutions, geographic regions).	4.54	5.00	4.00	4.51
	Displaying information about previous editions.	4.13	5.00	4.00	4.38
	Providing a template with instructions for authors to use on their articles.	4.84	5.00	4.00	4.61
	Clarifying the arbitration by peers system.	4.26	5.00	4.00	4.42
	Making the areas of publication of the journal explicit.	4.26	5.00	4.00	4.42

(*Continued*)

TABLE 12.2 (*Continued*)

Selection and Evaluation of Criteria through the FDELPHI Method (Second Round)

Criteria	Subcriteria	Geometric Mean	Max	Min	G_i
Guidelines and standards (*continued*)	Making clear the minimal number of articles or editions to be published in the National Counsel of Technological and Scientific Development (CNPQ) classification for large areas.	3.96	5.00	3.00	3.99
	Working with CAPES criteria directed to journals.	4.84	5.00	4.00	4.61
	Making the copyright accessible for authors (intellectual property).	3.84	4.00	3.00	3.61
Deadlines	Meeting the deadline for launching the edition.	4.26	5.00	4.00	4.42
	Meeting the deadline of evaluations.	4.13	5.00	4.00	4.38
	Meeting the deadline, when there is a need for modifications by the authors.	4.13	5.00	4.00	4.38
Quality of articles	Having a screening process before sending the article for a review.	4.50	5.00	3.00	4.17
	Having articles with an innovative character.	3.39	4.00	3.00	3.46
	Attracting authors with tradition in their fields.	3.84	4.00	3.00	3.61
	Properly designating a referee within the subject area to evaluate the product.	4.13	5.00	4.00	4.38
	Publishing current articles.	4.54	5.00	4.00	4.51
	Publishing articles aligned with the areas proposed by the journal.	4.26	5.00	4.00	4.42
	Publishing original articles.	4.40	5.00	4.00	4.47
Financial resources	Finding financial sources for the journal.	4.69	5.00	4.00	4.56
	Conducting research projects for financing agencies in the area to acquire resources.	3.96	5.00	3.00	3.99
Training	Seeking training for publishers.	4.40	5.00	4.00	4.47
	Seeking training for scholarship students.	4.40	5.00	4.00	4.47
Competence	Ability of publishers to manage processes and resolve any problems.	4.84	5.00	4.00	4.61
	Ability of scholarship students to perform their activities.	4.84	5.00	4.00	4.61

(*Continued*)

TABLE 12.2 (*Continued*)

Selection and Evaluation of Criteria through the FDELPHI Method (Second Round)

Criteria	Subcriteria	Geometric Mean	Max	Min	G_i
Competence (*continued*)	Ability of referees to conduct the evaluations.	4.13	5.00	4.00	4.38
	Ability of the editorial staff to carry out their activities.	4.13	5.00	4.00	4.38
	Ability of the technical staff to carry out their activities.	3.84	4.00	3.00	3.61
Knowledge	The knowledge of editors to accomplish their tasks.	5.00	5.00	5.00	5.00
	The knowledge of scholarship students to accomplish their tasks.	4.84	5.00	4.00	4.61
	The knowledge of the referees to do their jobs.	4.54	5.00	4.00	4.51
	The knowledge of the technical staff to perform their tasks.	3.84	4.00	3.00	3.61
Turnover	Consecutive exchanges of editors.	2.16	4.00	1.00	2.39
	Consecutive exchanges of scholarship students.	2.12	4.00	1.00	2.37
	Consecutive exchanges of the technical staff.	2.03	3.00	1.00	2.01
Attracting new members	Attracting new readers and authors for the journal.	3.54	4.00	3.00	3.51
	Attracting qualified referees for the journal.	3.96	5.00	3.00	3.99
Recognition	Thanking for evaluations carried out by referees.	4.26	5.00	4.00	4.42
	Thanking and rewarding the work(s) of scholarship student(s).	4.13	5.00	4.00	4.38
	Thanking for reader's preference for the journal.	3.26	4.00	3.00	3.42
Teamwork	Having synergy between those responsible for assembling the edition (scholarship students and editors).	4.69	5.00	4.00	4.56
Partnerships	Eventually partner up with events, symposiums, and other magazines.	2.90	4.00	2.00	2.97
Accuracy of the information	Having accuracy and consistency of information provided to readers, authors, and referees.	4.69	5.00	4.00	4.56
Visibility	Perform disclosure of the periodic.	5.00	5.00	5.00	5.00

(*Continued*)

TABLE 12.2 (*Continued*)

Selection and Evaluation of Criteria through the FDELPHI Method (Second Round)

Criteria	Subcriteria	Geometric Mean	Max	Min	G_i
Benchmarking	Use best practices of top journals in the area.	4.40	5.00	4.00	4.47
Indexations	Registering in databases, repositories, and national and international virtual libraries.	3.20	4.00	2.00	3.07
Communication	Warning readers about the launching of new editions and news.	4.26	5.00	4.00	4.42
Prestige	Indexing the journal in databases of high recognition.	5.00	5.00	5.00	5.00
	Working toward the impact factor.	5.00	5.00	5.00	5.00
Satisfaction survey	Conducting satisfaction surveys to see how the journal is seen by readers.	2.25	3.00	2.00	2.42
Category					
Organizational capital		3.65	5.00	3.00	3.88
Human capital		4.69	5.00	4.00	4.56
Relational capital		3.50	5.00	3.00	3.83

12.4.2.2 Application of FAHP

First, the categories, criteria, and subcriteria of the questionnaire have been simplified with acronyms, for example, criteria (C) and subcriteria (SC). Table 12.3 shows the abbreviations used and their meanings.

The AHP model is formed by the main objective categories, criteria, and subcriteria determined in the first step. The AHP model is composed of four levels. The purpose of the evaluation is located on the first level. Categories are on the second level, the criteria are located on the third level, and the subcriteria are located on the fourth level.

Step 1: The formation of the comparisons of pairs of attributes for the criteria was carried out using fuzzy numbers. The same was done for the categories and subcriteria.

Step 2: The value of the *fuzzy* synthetic measure is calculated. Initially, this was performed for the following categories: organizational capital (OC), HC, and relational capital (RC):

$$\sum_{j=1}^{3} OC_{g1}^{j} = (1,1,1) \oplus (0.29,0.33,0.4) \oplus (0.66,1,1.5) = (1.95,2.3,2.9)$$

TABLE 12.3

Categories and Acronyms Used in the Questionnaire Survey

Category	Criteria	Subcriteria
Organizational capital (OC)	Tangibles (C_1)	Equipment and technological resources used to make it feasible to work with the journal. (SC_1)
		Physical space where the journal is located. (SC_2)
	Website interface quality (C_2)	Providing usability (user interaction with the interface) to readers of the journal. (SC_3)
		Providing accessibility (make the interface usable for anyone). (SC_4)
		Providing functionality (refers to clarity of links, easily switching between pages and processes, easily finding information, appropriate content presentation, users performing tasks with no errors and problems). (SC_5)
		Friendliness of the colors of the website of the journal. (SC_6)
		Configuring the system to accommodate different types of documents, such as Word, PDF, RTF, and others. (SC_7)
		Showing ISSN, e-ISSN license, creative commons attribution (intellectual property). (SC_8)
	Standardization of the specification (C_3)	Having an abstract. (SC_9)
		Displaying a bibliographic caption. (SC_{10})
		Showing the history of the editorial management. (SC_{11})
		Presenting the mission and editorial policy. (SC_{12})
	Normalization of articles (C_4)	Showing authorship, affiliation, and contact(s) of author(s). (SC_{13})
		Having summary and abstract. (SC_{14})
		Having standardization of references. (SC_{15})
		Having keywords. (SC_{16})
		Having content division (abstract, introduction, methodology, results, conclusion, and references). (SC_{17})
		Having an electronic update of submission of documents. (SC_{18})
		Having standardization of headings, subheadings, charts, figures, tables, equations. (SC_{19})
		Presenting good definitions of figures and items(s). (SC_{20})
		Observing the size (length) of the articles proposed by the journal. (SC_{21})
		Showing article date of receipt and acceptance. (SC_{22})
		Showing the correct page number order. (SC_{23})

(Continued)

TABLE 12.3 (*Continued*)

Categories and Acronyms Used in the Questionnaire Survey

Category	Criteria	Subcriteria
Organizational capital (OC)	Policies, guidelines, and standards (C_5)	Ensuring standards that avoid endogeny (referring to authors, institutions, and geographic regions) (SC_{24})
		Displaying information about previous editions (archiving). (SC_{25})
		Providing a template with instructions for authors to use on their articles. (SC_{26})
		Clarifying the arbitration by peers system. (SC_{27})
		Making the areas of publication of the journal explicit. (SC_{28})
		Making clear the minimal number of articles or editions to be published in the CNPQ classification for large areas. (SC_{29})
		Working with CAPES criteria directed to journals. (SC_{30})
		Making the copyright accessible for authors (intellectual property). (SC_{31})
		Making the periodicity explicit. (SC_{32})
		Making the editorial staff explicit (SC_{33})
		Presenting permission for reproduction. (SC_{34})
		Displaying contact information. (SC_{35})
	Deadlines (C_6)	Meeting the deadline for launching the edition. (SC_{36})
		Meeting the deadline of evaluation. (SC_{37})
		Meeting the deadline, when there is a need for modifications by the authors. (SC_{38})
	Quality of articles (C_7)	Having a screening process before sending the article for a review. (SC_{39})
		Having articles with an innovative character. (SC_{40})
		Attracting authors with tradition in their fields. (SC_{41})
		Properly designating a referee within the subject area to evaluate the product. (SC_{42})
		Publishing current articles. (SC_{43})
		Publishing articles aligned with the areas proposed by the journal. (SC_{44})
		Publishing original articles. (SC_{45})
		Having a prior review as to the writing of the articles. (SC_{46})
	Financial resources (C_8)	Finding financial sources for the article. (SC_{47})
		Conduct research projects for financing agencies in the area to acquire resources. (SC_{48})
Human capital (HC)	Training (C_9)	Seeking training for publishers. (SC_{49})
		Seeking training for involved editors. (SC_{50})
	Competence (C_{10})	Ability of publishers to manage processes and resolve problems. (SC_{51})
		Ability of referees to conduct the evaluations. (SC_{52})

(Continued)

TABLE 12.3 (*Continued*)

Categories and Acronyms Used in the Questionnaire Survey

Category	Criteria	Subcriteria
Human capital (HC) (*continued*)		Ability of the editorial staff to carry out their activities. (SC$_{53}$)
		Ability of the technical staff to carry out their activities. (SC$_{54}$)
	Knowledge (C$_{11}$)	The knowledge of editors to accomplish their tasks. (SC$_{55}$)
		The knowledge of referees to do their jobs. (SC$_{56}$)
		The knowledge of technical staff to perform their tasks. (SC$_{57}$)
	Attracting new members (C$_{12}$)	Attracting new readers and authors for the journal. (SC$_{58}$)
		Attracting qualified referees for the journal. (SC$_{59}$)
	Recognition (C$_{13}$)	Thanking for evaluations carried out by referees. (SC$_{60}$)
		Thanking for reader's preference for the journal. (SC$_{61}$)
	Teamwork (C$_{14}$)	Having synergy between those responsible for assembling the edition. (SC$_{62}$)
		Delegating functions to other employees. (SC$_{63}$)
Relational capital (RC)	Accuracy of the information (C$_{15}$)	Having accuracy and consistency of information provided to readers, authors, and referees. (SC$_{64}$)
		Warning readers about the launching of new editions and news. (SC$_{65}$)
	Visibility (C$_{16}$)	Perform disclosure of the periodic. (SC$_{66}$)
		Use best practices of top journals in the area. (SC$_{67}$)
		Registering in databases, repositories, and national and international virtual libraries. (SC$_{68}$)
		Indexing the journal in databases of high recognition. (SC$_{69}$)
		Working towards the impact factor. (SC$_{70}$)

$$\sum_{j=1}^{3} \mathrm{HC}_{g2}^{j} = (2.5, 3, 3.5) \oplus (1, 1, 1) \oplus (2.5, 3, 3.5) = (6, 7, 8)$$

$$\sum_{j=1}^{3} \mathrm{RC}_{g3}^{j} = (0.66, 1, 1.5) \oplus (0.29, 0.33, 0.4) \oplus (1, 1, 1) = (1.95, 2.3, 2.9)$$

$$\sum_{i=1}^{3} \sum_{j=1}^{3} M_{gi}^{j} = (1.95, 2.3, 2.9) \oplus (6, 7, 8) \oplus (1.95, 2.3, 2.9) = (10, 11.6, 13.8)$$

$$\left[\sum_{i=1}^{m}\sum_{j=1}^{n}M_{ij}\right]^{-1} = \left(\frac{1}{10}, \frac{1}{11,6}, \frac{1}{13,8}\right) = (0.0725, 0.0862, 0.1000)$$

$$S_1(OC) = (1.95, 2.3, 2.9) \otimes (0.0725, 0.0862, 0.1000) = (0.145, 0.198, 0.290)$$

$$S_2(HC) = (6, 7, 8) \otimes (0.0725, 0.086,; 0.1000) = (0.435, 0.603, 0.800)$$

$$S_3 RC = (1.95, 2.3, 2.9) \otimes (0.0725, 0.0862, 0.1000) = (0.145, 0.198, 0.290)$$

After performing step 2 for the "Categories," the same was done for the "Criteria (C)," as shown below. The synthetic values for each criterion were calculated using Equations 12.5 through 12.8.

$$\sum_{j=1}^{16} C_{g1}^j = (1, 1, 1) \oplus (0.33, 0, 4, 0, 5) \oplus (0.33, 0.4, 0.5) \oplus \ldots \oplus (0.2, 0.33, 0.4)$$
$$= (3, 6.5 =, 7.8)$$

$$\sum_{j=1}^{16} C_{g2}^j = (11.1, 15.5, 21.7)$$

$$\sum_{j=1}^{16} C_{g3}^j = (11.1, 15.5, 21.7)$$

$$\sum_{j=1}^{16} C_{g16}^j = (27.5, 34, 41.5)$$

$$\sum_{i=1}^{16}\sum_{j=1}^{16} M_{gi}^j = (11.1, 15.5, 21.7) \oplus (11.1, 15.5, 21.7) \oplus \cdots \oplus (27.5, 34, 41.5)$$
$$= (284.3, 368.5, 471.9)$$

$$\left[\sum_{i=1}^{m}\sum_{j=1}^{n}M_{ij}\right]^{-1} = \left(\frac{1}{284,3}, \frac{1}{368,5}, \frac{1}{471,9}\right) = (0.000212, 0.00271, 0.00352)$$

$$S_1(C_1) = (3, 6.5, 7.8) \otimes (0.000212, 0.00271, 0.00352) = (0.00064, 0.0176, 0.027)$$

$$S_2(C_2) = (11.1, 15.5, 21.7) \otimes (0.000212, 0.00271, 0.00352) = (0.0024, 0.1420, 0.076)$$

$$S_3(C_3) = (11.1, 15.5, 21.7) \otimes (0.000212, 0.00271, 0.00352) = (0.0024, 0.1420, 0.076)$$

...

$$S_{16}(C_{16}) = (27.5, 34, 41.5) \otimes (0.000212, 0.00271, 0.00352) = (0.58, 0.0921, 0.146)$$

The same procedure was performed for the subcriteria.

Step 3: Using the results of the previous step, the weight vectors are calculated. According to Equations 12.11 and 12.12, the degree of possibility of $S_i = (l_j; m_j; u_j) \geq S_i = (l_i; m_i; u_i)$ can be calculated by comparing the values of S_i. The minimum degree of possibility is calculated by $d'(i)$ of $V(S_j \geq S_i)$ for $i; j = 1, 2, 3 ..., k$:

$$S_4(C_4) = (0, 58, 0.0921, 0.146)$$

$$S_1(C_1) = (0.00064, 0.0176, 0, 027)$$

$$V(S1 \geq S4) = \frac{0.58 - 0.027}{(0.0176 - 0.027) - (0.0921 - 0.058)}$$

The results for the degree of possibility for the criteria are shown in Table 12.4.

Step 4: In this step, the minimum degree of possibilities is calculated $< d'(i)V_j \geq S_i$, for $i, j = 1, 2, 3..., k$. As a result we obtained the following weight vector: $(0.22, 0.58, 0.58, 0.58, 1, 0.58, 0.58, 1, 0.58, 0.58, 0.58, 1, 0.58, 1, 1, 1, 1)^T$; for the aspects, the results were as follows: $(0.80, 0.80, 1, 1)^T$. The same procedure was done for the subcriteria.

TABLE 12.4

Degree of Possibility of $V(S_1 \geq S_i)$ for the Criteria

$V(S_1 \geq S_i)$	Value	...	$V(S_j \geq S_i)$	Value
$V(S_1 \geq S_2)$	0.50	...	$V(S_{16} \geq S_1)$	1
$V(S_1 \geq S_3)$	0.50	...	$V(S_{16} \geq S_2)$	1
$V(S_1 \geq S_4)$	0.22	...	$V(S_{16} \geq S_{15})$	1

Step 5: Normalization of weights. After the normalization of the value of these weights with respect to the main goal, the results are as follows: $W = (0.018, 0.042, 0.042, 0.092, \ldots, 0.092)$ Following, the weights of the subcriteria are calculated similarly. Table 12.5 shows the final weight of each criterion.

From the foregoing in Table 12.5, it can be seen that the HC category had the highest weight, that is, the most influence over the management of a scientific e-journal. In fact, HC is the "engine" in regard to the management of a scientific e-journal; without this factor activities cannot be carried out. In addition, it is also the main source of sustainable competitive advantage. As electronic journals require changes in social structures, these changes will be appropriated by individuals. This shows that internal structures are dependent on control and on human behavior.

It can be asserted that the *Human Capital* (HC) in an electronic journal stands out for having three main characteristics, namely

Competences—the knowledge and ability that editors have to manage the activities of a journal, that is, publishing, communication with readers and authors, raising funds, and attracting new members (authors, referees), among many others involving the management of an electronic journal. The competencies of those involved, especially the editors, include the ability to find flaws in submitted manuscripts and the ability to work together with authors constructively to correct these flaws and improve the publication. Therefore, a requirement of editors is that they are up to date regarding the area of knowledge (ANPAD, 2010) and the research methods used.

Attitudes—the behavioral dimension of individuals. It is expected that a major characteristic of a scientific journal editor is to motivate the search for improvements, delegate certain functions to those involved, and therefore, create a group that works with synergy. Also the question of cordiality is highlighted (ANPAD, 2010), which is important as reviews/opinions are prepared. It is desirable that the editor moderates criticism from reviewers/referees, has sensitivity when communicating ideas, is consistent with comments and suggestions with regard to the the rules declared by the journal, and is receptive to new ideas. Thus, the editor should review the reviewers' comments to the author so that they in turn receive a constructive evaluation, even if the article does not remain in the review process for publication.

Intellectual agility—the ability of those involved to apply their knowledge for leverage of the performance of the electronic journal. It is not unheard of for editors to not know the present activities and processes of their own publication. In addition, usually in a journal, there are not a large number of employees and as a consequence,

TABLE 12.5

Weights of the Subcriteria

Categories	Weight (W)	W (%)
Organizational capital (OC)	0.20	0.20
Human capital (HC)	0.60	0.60
Relational capital (RC)	0.20	0.20
Total	**1.00**	**100**
Criteria	**Weight (W)**	**W (%)**
Tangibles (C_1)	0.018	1.8
Website interface quality (C_2)	0.042	4.2
Standardization of the specification (C_3)	0.042	4.2
Normalization of articles (C_4)	0.092	9.2
Policies, guidelines, and standards (C_5)	0.042	4.2
Deadlines (C_6)	0.042	4.2
Quality articles (C_7)	0.092	9.2
Financial resources (C_8)	0.042	4.2
Training (C_9)	0.042	4.2
Competence (C_{10})	0.042	4.2
Knowledge (C_{11})	0.092	9.2
Attracting new members (C_{12})	0.042	4.2
Recognition (C_{13})	0.092	9.2
Teamwork (C_{14})	0.092	9.2
Accuracy of the information (C_{15})	0.092	9.2
Visibility (C_{16})	0.092	9.2
Total	**1.00**	**100**

there can be an excess demand of services. Consequently, this will cause problems in its management and cause a delay in the launch of an issue. Therefore, the editors are the ones who coordinates the editorial process and promote development of the journal. Their commitment is that the journal offers what may be the best in terms of new knowledge on a particular subject area, and they work on this throughout the process of selection of papers while adhering to the highest ethical standards. This is the ability to transfer knowledge from one context to another, whether in innovation processes, imitation, adaptation, or conversion.

It is emphasized that these three characteristics are commonly found throughout the publishing process of a scientific journal. In addition, it should also be recognized that all of these elements are equally important. For example, teamwork is essential to an electronic or printed journal. The publication of a scientific journal is a complex and demanding job. Editors, editorial staff, and all teams involved are sure to face many challenges (Vrana, 2012) such as reviewers who do not have the reviews/opinions in on time, authors who submit low–quality manuscripts, authors who submit manuscripts with incorrect formatting, insufficient financial support, insufficient number of members on the editorial board, a chief editor with excessive obligations outside of the journal, lack of volunteers to help the editor(s), and in the case of printed journals, the high cost of newspaper printing, among many other problems and difficulties of scientific journals.

It can be considered, from the many factors mentioned, that even with a small team, synergy is essential to the progress of a scientific journal and it will affect the quality of the final product.

> **Step 6:** Calculation of the global weight of the subcriteria. The was calculated as follows: for example, the global level of subcriteria with respect to "Financial Resources (REFIN)" is $w^1_{Subc} = 0.042 \otimes (0.0082, 0.0042) = (0.0003437, 0.000177)$. Table 12.6 shows the results for all subcriteria.

As shown in Table 12.6, the subcriterion with the greatest weight was the SC8 (0.4032) (show the International Standard Serial Number—ISSN, e-ISSN license, creative commons attribution); this is a very important factor when considering the management of electronic journals as well as printed journals. The participation of editors and others involved in the management and publishing of journals process also helps in polishing and discovering new knowledge because in this process one can learn about different areas of knowledge (covered by the journal) and about the procedures of editing.

In contrast, an editorial team with failures will lead to poor management, making the journal or the journal lose its credibility and reputation, thereby causing its closure. The results indicate the following:

1. For "Categories," the HC was appointed as the one that most likely (60%) influences the management of electronic scientific journals. Indeed, HC is the key for a journal, electronic or printed. In a knowledge-based economy, intangible resources and especially competencies (derived from the HC) are crucial for the organizations to survive. In this sense, that is, HC, the most notable criterion was the "knowledge of those involved in the journal (KNOW)" with 9.2%. Knowledge of individuals in an organization is a very valuable feature because with it you may learn new techniques, resolve problems, create essential skills, and obtain competitive advantages. In the case of journals, editors can be considered the "guardians" of scientific knowledge and also have the central role in the development of science because it is up to them to accept or reject (of course with the participation of the referees) the manuscripts. The OC and RC "Categories" are also important; both got 20% on the viewpoint of the editors as to their influence in the management of journal. In addition, the rest of the criteria

TABLE 12.6

Global Weight of Subcriteria

Criteria	Subcriteria	Local Weight	Global Weight	Global Weight (%)
C_1	SC_1	0.7100	0.0129	1.29
	SC_2	0.2900	0.0051	0.51
C_2	SC_3	0.1194	0.0050	0.50
	SC_4	0.1194	0.0050	0.50
	SC_5	0.1194	0.0050	0.50
	SC_6	0.1194	0.0050	0.50
	SC_7	0.1194	0.0050	0.50
	SC_8	0.4032	0.0169	1.70
C_3	SC_9	0.2500	0.0230	2.30
	SC_{10}	0.2500	0.0230	2.30
	SC_{11}	0.2500	0.0230	2.30
	SC_{12}	0.25	0.0230	2.30
C_4	SC_{13}	0.1575	0.0145	1.45
	SC_{14}	0.1575	0.0145	1.45
	SC_{15}	0.0524	0.0048	0.48
	SC_{16}	0.1575	0.0145	1.45
	SC_{17}	0.0524	0.0048	0.48
	SC_{18}	0.0524	0.0048	0.48
	SC_{19}	0.0524	0.0048	0.48
	SC_{20}	0.0524	0.0048	0.48
	SC_{21}	0.0524	0.0048	0.48
	SC_{22}	0.0524	0.0048	0.48
	SC_{23}	0.1575	0.0145	1.45
C_5	SC_{24}	0.1800	0.0075	0.75
	SC_{25}	0.0680	0.0029	0.29
	SC_{26}	0.0680	0.0029	0.29
	SC_{27}	0.1800	0.0075	0.75
	SC_{28}	0.0680	0.0029	0.29
	SC_{29}	0.0300	0.0012	0.12
	SC_{30}	0.0680	0.0029	0.29
	SC_{31}	0.0680	0.0029	0.29
	SC_{32}	0.0680	0.0029	0.29
	SC_{33}	0.0680	0.0029	0.29
	SC_{34}	0.0680	0.0029	0.29
	SC_{35}	0.0680	0.0029	0.29
C_6	SC_{36}	0.3330	0.0140	1.40
	SC_{37}	0.3330	0.0140	1.40
	SC_{38}	0.3330	0.0140	1.40
C_7	SC_{39}	0.1250	0.0115	1.15
	SC_{40}	0.1250	0.0115	1.15

(Continued)

TABLE 12.6 (*Continued*)

Global Weight of Subcriteria

Criteria	Subcriteria	Local Weight	Global Weight	Global Weight (%)
C_7 (*continued*)	SC_{41}	0.1250	0.0115	1.15
	SC_{42}	0.1250	0.0115	1.15
	SC_{43}	0.1250	0.0115	1.15
	SC_{44}	0.1250	0.0115	1.15
	SC_{45}	0.1250	0.0115	1.15
	SC_{46}	0.1250	0.0115	1.15
C_8	SC_{47}	0.7100	0.0300	3.00
	SC_{48}	0.2900	0.0121	1.21
C_9	SC_{49}	0.5000	0.2500	25.00
	SC_{50}	0.5000	0.2500	25.00
C_{10}	SC_{51}	0.2500	0.1050	10.50
	SC_{52}	0.2500	0.1050	10.50
	SC_{53}	0.2500	0.1050	10.50
	SC_{54}	0.2500	0.1050	10.50
C_{11}	SC_{55}	0.3330	0.0307	3.07
	SC_{56}	0.3330	0.0307	3.07
	SC_{57}	0.3330	0.0307	3.07
C_{12}	SC_{58}	0.5000	0.2500	25.00
	SC_{59}	0.5000	0.2500	25.00
C_{13}	SC_{60}	0.7500	0.0690	6.90
	SC_{61}	0.2500	0.0230	2.30
C_{14}	SC_{62}	0.7500	0.0690	6.90
	SC_{63}	0.2500	0.0230	2.30
C_{15}	SC_{64}	0.7500	0.0690	6.90
	SC_{65}	0.2500	0.0230	2.30
C_{16}	SC_{66}	0.2000	0.0184	1.84
	SC_{67}	0.2000	0.0184	1.84
	SC_{68}	0.2000	0.0184	1.84
	SC_{69}	0.2000	0.0184	1.84
	SC_{70}	0.2000	0.0184	1.84

appointed as influencers in the management were standardization of articles (STRT), quality of articles (QUAART), recognition (RECO), teamwork (TEAW), accuracy of the information (ACURI), and visibility (VISIB), all with a participation of 9.2% each.

2. Regarding subcriteria that most influence the management of a journal, the main ones appointed by the editors were showing ISSN, e-ISSN license, creative commons attribution (APLIC)—40.32%; meeting the deadline for launching the edition (LANCED); deadline of evaluation (AVAL), and meeting the deadline when there is a need

for modification by the authors (MOD), all with 33.33% and all three belonging to the PRAZ (deadline) criteria; and the knowledge of editors to accomplish their tasks (EDIT), the knowledge of the referees to do their jobs (REFER), and the knowledge of the technical staff to perform their tasks (EQTEC), all also, with an importance of 33.33%.

3. As the electronic journal is a product of the aspect, criteria, and subcriteria that affect their management, a list indicators that aims to assist in the improvement of these journals was produced.

Therefore, this section shows the analysis of the results of the application of FDELPHI and FAHP for managing electronic scientific journals, under the approach of the concept of IC. The FDELPHI method enabled the "filtering" of the main critical factors (criteria/subcriteria) in the management of electronic scientific journals. The FAHP rated key aspects, criteria, and subcriteria that impact the management of these journals. Still, indicators were prepared based on subcriteria raised in the reality of scientific journals of electronic character to contribute to the improvement of the management of journals.

12.5 Managerial Implications

The editing process of a periodical involves activities and individuals that go from the submission of the file by the author to the final publication of the article indicated and selected by the opinion of the referees. Thus, we can say that the periodical is in a privileged environment where many scientific scholarly dialogues and social interactions occur, in addition to political negotiations between the elite members of the scientific community that publishes and decides what will be published.

Braga (2009) states that editing activities are often developed intuitively by researchers because they do not have the proper training for such procedures, even though they are experts in their areas of operation and knowledge. Consequently, this affects the management of the journal, impairing its development and growth. Thus, the results obtained in the study suggested a list of indicators (Table 12.7), which are based on the analysis of the subcriteria. It is noteworthy that the intent of these indicators is to assist editors in managing their journals.

In this chapter, we aim to contribute to the improvement of the management of electronic scientific journals and provide competitive advantages and prominence in the market. To succeed, a periodical needs to have a good editorial team led by a strong editor. A good editor should provide (Fischer, 2004; Vrana, 2012) immediate recognition of submissions, detailed feedback of how the review process works, and evaluation comments that are convenient, constructive, impartial, fair, confidential, complete, diplomatic, and ethical.

TABLE 12.7

Proposed Indicators for the Management of Electronic Journals

Categories	Subcriteria	Indicator
Organizational capital (OC)	Equipment and technological resources	Amount of equipment (computers, CPU, printer, etc.) defective in the last year.
	Physical space	Number of complaints from employees regarding the physical space.
	Usability	Number of complaints from users about the visuals of the website of the journal in the semester. Number of user complaints about the loading of the website or number of complaints from readers because of the disorganization of the navigation elements in the last 6 months.
	Accessibility	Number of requests made for the insertion of special subtitles intended for the visually impaired in the past 6 months.
	Functionality	Number of complaints or suggestions from users as to the loading of the home page of the journal in the last 6 months.
	Friendliness of the colors of the website	Number of complaints or suggestions related to the colors of the website, for the year.
	Different types of documents	Number of complaints from readers about failed loading of files in the past 6 months or number of manuscripts that had some transfer failure during the last year.
	DOI, ISSN, e-ISSN license, creative commons attribution among others	Number of times that these license types did not appear in the website of the journal; number of doubts about whether the journal has DOI in the past 6 years.
	Abstract	Number of issues that did not have an abstract in the last year; number of editions that did not meet the standards of publication of the journal in the last 2 years.
	Bibliographic caption	Number of issues in the last year where the bibliographic caption did not appear; number of editions that did not meet the standards of publication of the journal in the last 2 years.
	History of the editorial management	Number of issues in the last year where the history of management did not appear; number of editions that did not meet the standards of publication of the journal in the last 2 years.
	Mission and editorial policy	Number of issues in the last year where the mission and editorial policy did not appear; number of editions that did not meet the standards of publication of the journal in the last 2 years.

(Continued)

TABLE 12.7 (*Continued*)

Proposed Indicators for the Management of Electronic Journals

Categories	Subcriteria	Indicator
Organizational capital (OC) (*continued*)	Authorship, affiliation, and contact of the authors	Number of submissions in the last 6 months where the data of the authors appeared incomplete.
	Summary and abstract	Number of submissions in the last 6 months that did not include the summary or abstract.
	Standardization of bibliographic references	Number of submissions in the last 6 months where the bibliographic references were not standard.
	Keywords	Number of articles in the latest edition where the keywords were absent.
	Content division in the article	Number of submissions in the last 6 months out of the structure proposed by the journal.
	Electronic update of submission documents	Number of complaints from users about the progress of the submission process each semester
	Standardization of headings, subheadings, charts, figures, tables, equations	Percentage of articles per issue with problems of standardization.
	Good definition of figures and items	Number of articles sent to the authors to correct the resolution of figures.
	Size (length) of the articles proposed by the journal	Number of times where there were problems because the files exceeded the maximum capacity required
	Article dates of receipt and acceptance	Number of times per issue that the dates of receipt and acceptance of the paper was not presented.
	Correct page number order	Percentage of articles per issue with pagination errors.
	Endogeny	Number of articles with multi-institutional authors in each issue; number of articles with authors from different geographical regions in each issue.
	Information about previous editions (archive)	Number of times that users had to contact the journal to download the articles.
	Template with instructions for authors to use on their articles	Number of user requests in the month with questions on the submission and formatting process; number of articles that do not meet the formatting proposed by the journal.
	Arbitration by peer system	Number of questions of authors/readers as to the type of review used by the journal.
	Areas of publication of the journal	Number of questions of authors/readers as to the area of interest in the last year.

(Continued)

TABLE 12.7 (*Continued*)

Proposed Indicators for the Management of Electronic Journals

Categories	Subcriteria	Indicator
Organizational capital (OC) (*continued*)	Minimum of articles or editions to be published	Number of times in which the journal published under 2 editions per year over the last 2 years or number of times in which the journal published the minimum number of articles per issue per year in the last 2 years.
	CAPES criteria directed to journals	Number of editions where the CAPES standards were not met in the last 2 years.
	Copyright for authors (intellectual property)	Number of times in the last year that there were doubts or complaints about the explicitness of the copyright.
	Periodicity	Number of questions received from authors/readers on the frequency of the magazine each year.
	Editorial staff	Number of times users said the editorial staff was not explicit on the website of the journal.
	Contact info	Number of times users stated that the address or telephone number for contact was not explicit on the website of the journal
	Deadline for launching the edition	Number of editions where the release was delayed last year.
	Deadline of evaluations	Number of times that the referees delayed the return of evaluations last semester.
	Deadline for modifications by authors	Number of times that the authors delayed the return of the article with the necessary changes in the last year.
	Screening process before sending the article for a review	Number of discarded articles in the screening process for being of subjects of no interest to the editorial board in each edition; amount of articles misaligned with formatting problems and returned to authors for editing.
	Articles with an innovative character	Percentage of articles published with an innovative character in the latest edition.
	Authors with tradition in their fields	Percentage of articles published with authors with tradition in their fields in the last edition.
	Properly designate the referee	Percentage of articles submitted to referees and returned because it is not their area of evaluation.
	Current articles	Percentage of articles published on current topics, by edition.
	Articles aligned with the areas proposed by the journal	Percentage of articles published with lines and themes proposed by the journal in the last issue.

(*Continued*)

TABLE 12.7 (*Continued*)

Proposed Indicators for the Management of Electronic Journals

Categories	Subcriteria	Indicator
Organizational capital (OC) (*continued*)	Original articles	Number of submissions that have been published in part or in events submitted by the journal editor; number of input items plagiarized in the past 2 years.
	Prior review of the writing of articles	Number of articles with serious writing problems by edition.
	Financial sources	Relationship between projects submitted and laboratory the journal in the past 2 years.
	Research projects to acquire resources	Number of projects approved in the last 6 months with respect to the number of projects submitted.
Human capital (HC)	Training for editors	Number or hours of courses on editing journals taken by the editor per year.
	Training for those involved in the editing process	Number or hours of courses on editing journals taken by others involved per year (in%).
	Ability of publishers	Frequency of proactiveness in problem solving in the journal last year.
	Ability of referees	Percentage of consistent and accurate evaluations conducted by edition.
	Ability of the editorial staff	Frequency the editorial staff acted early to avoid divergences in the journal last year.
	Ability of the technical staff	Frequency at which the staff acted to solve problems in the journal in the last 6 months.
	Knowledge of editors	Frequency with which the editor cannot resolve issues of the journal.
	Knowledge of referees	Number of inconsistent or incomplete evaluations made by edition.
	Knowledge of the technical staff	Frequency of errors in the tasks performed by the staff.
	New readers	Number of new registered readers in the last 6 months.
	Qualified referees	Percentage of referees with 5 or more publications in journals equal to B2 or higher in the last 5 years that they agreed to be referees of the journal.
	Thanking for evaluations carried out	Number of motivational actions to keep the referees satisfied in the last 2 years.
	Thanking for reader's preference	Number of times that the system did not automatically trigger the gratitude for the submission.
	Synergy	Frequency with which the editor and the others involved diverge and discuss the processes of publishing; frequency with which there is extra work for only one involved.

(Continued)

TABLE 12.7 (*Continued*)

Proposed Indicators for the Management of Electronic Journals

Categories	Subcriteria	Indicator
Human capital (HC) (*continued*)	Delegate roles	Frequency with which the editor delegates functions to other stakeholders (such as scholarship students or assistants).
Relational capital	Accuracy and consistency of information	Number of user requests in a month with questions about a procedure of the journal in the past 6 months.
	Warning about the release of new editions and news	Number of pieces of news, topics and/or links outdated since the last edition.
	Disclosure of the periodical	Number of home pages of event websites that reported the journal, in the last year; number of home pages of staff and teachers who disclose the journal; number of University departments that publish the journal on their home pages, in the last year; number of home pages of research laboratories that reported the newspaper, in the last year.
	Best practices	Relationship between the number of practices used by top periodicals that have been implemented and that worked.
	Registering in databases, repositories, and national and international virtual libraries.	Percentage of approval in the databases (libraries and archives) that requested a data review in the last year.
	Databases of high recognition	Percentage of approval on the databases of high impact that requested assessment in the last year; number of potential databases for indexing.
	Impact factor	Number of actions performed towards impact factor.

12.6 Conclusions

In recent years, electronic journals have been considered a valuable means to maintain scientific communication among scholars. An electronic journal is produced, released, and distributed worldwide via the internet. In fact, electronic journals are becoming necessary to meet the demands for the dissemination of knowledge for many parts of the world. These type of journals offers more possibilities and advantages than printed ones. One, for example, is the ease of use, availability on the web, and universal acceptance of such technology, in addition to providing speedy information dissemination. Web technology allows for easier access to texts and multimedia.

Thus, the present study addressed the management of electronic journals; for this, we used the concept of IC. At the stage of the literature review, a search scheme (based on keywords) for the references was structured, that is, we started from a central theme, which in this case was IC, and we defined the object of study (electronic scientific journals) and also what tools were used to analyze them (FDELPHI, and AHP methods). Searches were made in the databases via the CAPES portal. We searched for combinations of keywords until it was proven in the literature that "What" the researcher would aim to accomplish was feasible.

Subsequently, all searches were performed and the materials were stored in reference management software, the web version of Endnote. All materials to be used were separated by folders according to the issues to be addressed. As for tools for data analysis, fuzzy logic was integrated into Delphi and AHP—FDELPHI and FAHP. This gave origin to mathematical modeling of a hybrid character. FDELPHI was used to raise the critical factors (criteria/subcriteria) present in the management of electronic scientific journals through interviews (with questionnaires) with experts, in this case, editors of electronic journals. The FAHP method was applied to calculate the relative weights of the criteria/subcriteria that affect their management the most. It is important to highlight that the fuzzy logic compensates for the vagueness and uncertainty existing in AHP, which originates in the judgment of the decision maker(s). We used the analysis method of Chang (1996) and TFNs, which have the ability to process information very well with a high degree of uncertainty and vagueness and are most suitable when it comes to management.

As to the research questionnaire, several rounds were performed for the analysis of critical points (criteria/subcriteria) with experts, until we came to a questionnaire that could reflect the management of electronic journals. Once all the stages of formulating the research instrument (questionnaire) were performed, the next step was its implementation. A sample of 14 electronic journals was used and as a criterion of choice we used the *Qualis* CAPES. Thus, all goals set in this work have been achieved. The specific aims were achieved through bibliographic research that followed a structured scheme of organization and search, and through the mathematical modeling proposed and the applied research.

Therefore, it is expected that this research will assist publishers in rethinking and leading the management of their journals; the development of science directly influences scientific communication, which may reflect on the movement of knowledge caused by this activity. The process of publishing a scientific e-journal requires ensuring the quality of published content, having as a reference the values of the different areas of science, within a communication system that is constantly changing.

References

ANPAD. (2010). *Associação Nacional de Pós-Graduação e Pesquisa em Administração. Boas práticas da publicação científica: Um manual para autores, revisores, editores e integrantes de Corpos Editoriais.* Available at: http://www.anpad.org.br/diversos/boas_praticas.pdf. Accessed September 24, 2012.

Bar-Ilan, J. and Fink, N. (2005). Preference for electronic format of scientific journals: A case study of the Science Library users at the Hebrew University. *Library and Information Science Research*, 27(3), 363–376.

Bomfá, C.R.Z. (2009). *Modelo para gestão de periódicos científicos eletrônicos com foco na promoção da visibilidade.* 238f. Tese (Engenharia de Produção), Universidade Federal de Santa Catarina Centro Tecnológico—CTC, Programa de Pós-graduação em Engenharia de Produção, Florianópolis.

Bornmann, L., Mutz, R., Hug, S.E., and Daniel, H.-D. (2011). A multilevel meta-analysis of studies reporting correlations between the h index and 37 different h index variants. *Journal of Informetrics*, 5(3), 346–359.

Braga, K.S. (2009). *A comunicação científica e a bioética brasileira: uma análise dos periódicos científicos brasileiros.* 187f. Tese (Ciência da Informação), Universidade de Brasília, Programa de Pós-graduação em Ciência da Informação.

Bravo, B.R., Díez, M.L.A., and Merino, I.O. (2015). La utilización de las revistas electrónicas en la Universidad de León (España): Hábitos de consumo y satisfacción de los investigadores. *Investigación Bibliotecológica: Archivonomía, Bibliotecología e Información*, 29(66), 17–55.

Buckley, J.J. (1985). Fuzzy hierarchical analysis. *Fuzzy Sets and Systems*, 17(3), 233–247.

Buela-Casal, G. (2004). Assessing the quality of articles and scientific journals: Proposal for weighted impact factor and a quality index. *Psychology in Spain*, 81, 60–76.

Büyüközkan, G., Feyzioğlu, O., and Nebol, E. (2008). Selection of the strategic alliance partner in logistics value chain. *International Journal of Production Economics*, 113, 148–158.

Calabrese, A., Costa, R., and Menichini, T. (2013). Using fuzzy AHP to manage intellectual capital assets: An application to the ICT service industry. *Expert Systems with Applications*, 40(9), 3747–3755. CAPES. 2012. Coordenação de Aperfeiçoamento de Pessoal de Nível Superior – Qualis Periódicos. Available at: http://www.capes.gov.br/avaliacao/qualis (Accessed 15 Oct, 2010).

Castilla-Polo, F. and Gallardo-Vázquez, D. (2016). The main topics of research on disclosures of intangible assets: A critical review. *Accounting, Auditing and Accountability Journal*, 29(2), 323–356.

Chamoli, S. (2015). Hybrid FAHP (fuzzy analytical hierarchy process)-FTOPSIS (fuzzy technique for order preference by similarity of an ideal solution) approach for performance evaluation of the V down perforated baffle roughened rectangular channel. *Energy*, 84, 432–442.

Chang, D.Y. (1996). Applications of the extent analysis method on fuzzy AHP. *European Journal of Operational Research*, 95(3), 649–655.

Chang, P.C. and Wang, Y.W. (2006). Fuzzy Delphi and back-propagation model for sales forecasting in PCB industry. *Expert Systems with Applications*, 30(4), 715–726.

Cho, J. and Lee, J. (2013). Development of a new technology product evaluation model for assessing commercialization opportunities using Delphi method and fuzzy AHP approach. *Expert Systems with Applications*, 40(13), 5314–5330.

Corera-Álvarez, E. and Molina-Molina, S. (2016). La edición universitaria de revistas científicas. *Revista Interamericana de Bibliotecologia*, 39(3), 277–288.

Dong, P., Loh, M., and Mondry, A. (2005). The "impact factor" revisited. *Biomedical Digital Libraries*, 2(1), 7.

Duarte, K.A.P.B. and Rodrigues, R.S. (2012). Periódicos em acesso aberto na área do direito. *Revista Digital de Biblioteconomia e Ciência da Informação, Campinas*, 9, 100–120.

Dubini, P., Giglia, E. 2009. Economic sustainability during transition: the case of scholarly publishing. In: *International Conference on Electronic Publishing*, 2009, Italy. Milano: ELPUB.

Fachin, G.R.B. (2002). *Modelo de avaliação para periódicos científicos on-line: Proposta de indicadores bibliográficos e telemáticos*. 210p. Dissertação (Mestrado em Engenharia de Produção)—Programa de Pós-Graduação em Engenharia de Produção, Universidade Federal de Santa Catarina, Florianópolis.

Fischer, C.C. (2004). Editor as good steward of manuscript submissions: 'Culture,' tone, and procedures. *Journal of Scholarly Publishing*, 36(1), 34–42.

Gresty, K.A. and Edwards-Jones, A. (2012). Experiencing research-informed teaching from the student perspective: Insights from developing an undergraduate e-journal. *British Journal of Educational Technology*, 43(1), 153–162.

Guanes, P.C.V.G. and Guimarães, M.C.S. (2012). Modelos de gestão de revistas científicas: Uma discussão necessária. *Perspectivas em Ciência da Informação*, 17(1), 56–73.

Homol, L. (2014). Web-based citation management tools: Comparing the accuracy of their electronic journal citations. *Journal of Academic Librarianship*, 40(6), 552–557.

Hsieh, T.Y., Lu, S.T., and Tzeng, G.H. (2004). Fuzzy MCDM approach for planning and design tenders selection in public office buildings. *International Journal of Project Management*, 22(7), 573–584.

Hsu, J.S.C., Chu, T.H., Lin, T.C., and Lo, C.F. (2014). Coping knowledge boundaries between information system and business disciplines: An intellectual capital perspective. *Information and Management*, 51(2), 283–295.

Hsu, Y., Lee, C., and Kreng, V. (2010). The application of fuzzy Delphi method and fuzzy AHP in lubricant regenerative technology selection. *Expert Systems with Applications*, 37(1), 419–425.

Huang, C.C., Chu, P.Y., and Chiang, Y.H. (2008). A fuzzy AHP application in government-sponsored R&D project selection. *Omega*, 36(6), 1038–1052.

Ingwersen, P. (1998). The calculation of web impact factors. *Journal of Documentation*, 54(2), 236–243.

Ju, Y., Wang, A., and Liu, X. (2012). Evaluating emergency response capacity by fuzzy AHP and 2-tuple fuzzy linguistic approach. *Expert Systems with Applications*, 39(8), 6972–6981.

Kahraman, C., Cebeci, U., and Ruan, D. (2004). Multi-attribute comparison of catering service companies using fuzzy AHP: The case of Turkey. *International Journal of Production Economics*, 87(2), 171–184.

Khan, A.M. and Ahmad, N. (2009). Use of e-journals by research scholars at Aligarh Muslim University and Banaras Hindu University. *Electronic Library*, 27(4), 708–717.

Kieling, C. and Gonçalves, R.R.F. (2007). Assessing the quality of a scientific journal: The case of Revista Brasileira de Psiquiatria. *Revista Brasileira de Psiquiatria*, 29, 177–181.

Krzyzanowski, R.F. and Ferreira, M.C.G. (1998). Avaliação de periódicos científicos e técnicos brasileiros. *Ciência da Informação*, 27, 165–175.

Kuo, Y.F. and Chen, P.C. (2008). Constructing performance appraisal indicators for mobility of the service industries using fuzzy Delphi method. *Expert Systems with Applications*, 35(4), 1930–1939.

Kutlu, A.C. and Ekmekçioğlu, M. (2012). Fuzzy failure modes and effects analysis by using fuzzy TOPSIS-based fuzzy AHP. *Expert Systems with Applications*, 39(1), 61–67.

Lin, M.-H., Hu, J., Tseng, M.-L., Chiu, A.S.F., and Lin, C. (2016). Sustainable development in technological and vocational higher education: Balanced scorecard measures with uncertainty. *Journal of Cleaner Production*, 120, 1–12.

Machado, M.A.A.M. and Santos, C.F. (2016). New impact factor and Editors of the Journal of Applied Oral Science. *Journal of Applied Oral Science*, 24(3), 187.

Mikaeil, R., Ozcelik, Y., Yousefi, R., Ataei, M., and Hosseini, S.M. (2013). Ranking the sawability of ornamental stone using fuzzy Delphi and multi-criteria decision-making techniques. *International Journal of Rock Mechanics and Mining Sciences*, 58, 118–126.

Moghaddam, G. and Moballeghi, M. (2008). How do we measure the use of scientific journals? A note on research methodologies. *Scientometrics*, 76(1), 125–133.

Nicholas, D., Williams, P., Rowlands, I., and Jamali, H.R. (2010). Researchers' E-journal use and information seeking behaviour. *Journal of Information Science*, 36(4), 494–516.

Pak, J.Y., Thai, V.V., and Yeo, G.T. (2015). Fuzzy MCDM approach for evaluating intangible resources affecting port service quality. *Asian Journal of Shipping and Logistics*, 31(4), 459–468.

Paksoy, T., Pehlivan, N.Y., and Kahraman, C. (2012). Organizational strategy development in distribution channel management using fuzzy AHP and hierarchical fuzzy TOPSIS. *Expert Systems with Applications*, 39(3), 2822–2841.

Parameshwaran, R., Kumar, S.P., and Saravanakumar, K. (2015). An integrated fuzzy MCDM based approach for robot selection considering objective and subjective criteria. *Applied Soft Computing*, 26, 31–41.

Raza, M.M. and Upadhyay, A.K. (2006). Usage of E-journals by researchers in Aligarh Muslim University: A study. *International Information and Library Review*, 38(3), 170–179.

Sandes-Guimarães, L.V. (2013). *Gestão de periódicos científicos: um estudo com revistas da área de Administração*. 143p. Dissertação (Mestrado em Administração de Empresas), Fundação Getúlio Vargas, São Paulo, Brasil.

Singh, S. and Dasgupta, M.S. (2016). Evaluation of research on CO_2 trans-critical work recovery expander using multi attribute decision making methods. *Renewable and Sustainable Energy Reviews*, 59, 119–129.

Stefano, N.M. (2014). *Critérios para avaliação da gestão de periódicos científicos eletrônicos sob a ótica do Capital Intelectual*. 214f. Thesis, Post-graduate Program in Production Engineering, Federal University of Santa Catarina, UFSC, Florianópolis, Santa Catarina, Brasil.

Stefano, N.M., Casarotto Filho, N., and Barrichello, R. (2015). Management of electronic journals using fuzzy AHP methodology. *IEEE Latin America Transactions (Revista IEEE America Latina)*, 13(1), 330–336.

Szymanski, B.K., De La Rosa, J.L., and Krishnamoorthy, M. (2012). An internet measure of the value of citations. *Information Sciences*, 185, 18–31.

Talley, W.K. and Ng, M.W. (2016). Port economic cost functions: A service perspective. *Transportation Research Part E: Logistics and Transportation Review*, 88, 1–10.

Taquette, S.R. and Minayo, M.C. (2016). Análise de estudos qualitativos conduzidos por médicos publicados em periódicos científicos brasileiros entre 2004 e 2013. *Physics [online]*, 26(2), 417–434.

Vasishta, S. (2013). Dissemination of electronic journals: A content analysis of the library websites of technical university libraries in North India. *The Electronic Library*, 31(3), 278–289.

Verma, T. and Dhar, S. (2016). The impact of intellectual capital on organizational effectiveness: A comparative study of public and private sectors in India. *IUP Journal of Knowledge Management*, 14(3), 7–27.

Vrana, R. (2012). Journal publishing challenges: A case of STM scientific journals in Croatia. *The International Information and Library Review*, 44(3), 147–154.

Wang, X. and Durugbo, C. (2013). Analysing network uncertainty for industrial product-service delivery: A hybrid fuzzy approach. *Expert Systems with Applications*, 40(11), 4621–4636.

Wang, Y., Yeo, G.-T., and Ng, A.K.Y. (2014). Choosing optimal bunkering ports for liner shipping companies: A hybrid Fuzzy-Delphi–TOPSIS approach. *Transport Policy*, 35, 358–365.

Yamamoto, O.H., Menandro, P.R.M., Koller, S.H., Lobianco, A.C., Hutz, C.S., Bueno, J.L.O., and Guedes, M.C. (2002). Avaliação de periódicos científicos brasileiros da área da psicologia. *Ciência da Informação*, 31, 163–177.

13

Applications of FAHP in Analysing Energy Systems

Sonal K. Thengane, Andrew Hoadley, Sankar Bhattacharya,
Sagar Mitra, and Santanu Bandyopadhyaye

CONTENTS

13.1 Introduction

Energy systems play a crucial role in the socio-economic development of human civilization. A sustainable energy system has balanced energy production and consumption with a minimal impact on the environment, giving the opportunity for a country to employ its social and economic activities. The existing fossil-based energy systems have contributed to depletion in reserves of fossil fuels in addition to increasing the risk of environmental degradation, leading to global problems such as climate change. Hence, renewable energy sources and ecologically clean technologies are promoted globally as a measure to conserve the environment. As per predictions of the International Energy Agency (IEA) in 2011, solar powered generators are expected to produce most of the world's electricity in the coming 50 years (Suganthi et al., 2015). However, the cost, lower efficiency, and difficulty in scalability for applications at commercial scale are the major concerns in the case of renewable technologies. The other alternative is to optimize the performance of existing energy systems based on fossil sources and develop the technology to minimize environmental emissions. Energy investment decisions for evaluation of energy sector plans, choice of sources and technologies, and policies are inherently multiobjective in nature. Energy modeling and energy planning are important for future economic development and environmental security, which depends on several qualitative and quantitative factors that need to be addressed and analyzed appropriately (Zhou et al., 2006). This analysis enforces the need for decision-making methods that would make it easier to select the best set of alternatives in different aspects.

Multicriteria decision-making (MCDM) methods are being widely used in energy systems selection and analysis on account of their potential to consider the multidimensionality of the sustainability goal and the complexity of socio-economic parameters. MCDM is an operational evaluation and decision support approach for addressing complex problems featuring high uncertainty, conflicting objectives, various forms of data and information, and multi-interests and perspectives, while accounting for complex and evolving biophysical and socio-economic systems. The methods based on MCDM can provide beneficial solutions to increasing energy management problems over the traditional single criterion approach normally aimed at identifying the most efficient options at a low cost. These methods, also known as multicriteria decision analysis (MCDA) methods, have found several applications in social, economic, agricultural, industrial, ecological and biological systems in addition to energy systems (Wang et al., 2009). There exists several MCDM methods out of which the analytic hierarchy process (AHP) is the most widely used method in the field of energy and the environment, specifically in areas such as energy policy analysis, electric power planning, technology choice and project appraisal, and energy utility operations and management (Zhou et al., 2006; Sheth and Huges, 2009). Pohekar and Ramachandran (2004) reviewed more than 90 MCDM studies in sustainable energy planning. Simplicity, ease of understanding, and the fact that AHP decomposes a complex problem into a simple hierarchy are some of the reasons for AHP's wide applications in the field of energy and environmental modeling (Pohekar and Ramachandran, 2004). The strength of the AHP approach is based on breaking the complex decision problem in a logical manner into subproblems in the form of levels of a hierarchy. The hierarchical structure of the AHP model permits decision makers (DMs) to compare the different prioritization criteria and alternatives more effectively (Kablan, 2004). The classical AHP, proposed by Saaty (1990), is based on crisp judgments that may not be possible in the case of many energy systems where there are numerous qualitative attributes and more than one DM. In most real-world energy sector problems, many of the decision data cannot be assessed precisely and hence the range of such parameters is assumed or calculated. The fuzzy analytic hierarchy process (FAHP) technique developed from the traditional AHP can take care of the uncertainty and the vagueness in human decisions to a greater extent. The approaches for deriving crisp weights from fuzzy pairwise comparison matrices include extent analysis (Chang, 1996) and a nonlinear method based on fuzzy preference programming (FPP) (Wang et al., 2011).

The application of AHP-based methods includes four main stages: formulation of alternatives and selection of criteria, allotting weights to criteria, evaluation, and final treatment and synthesis (Saaty and Vargas, 2000). The evaluation should cover technical, economical, or environmental problems that may not be easily identifiable and also socio-economic factors that affect various interest groups or stakeholders' needs. In view of these difficulties, the FAHP method may be useful in undertaking difficult assessment procedures in different areas involving energy systems. FAHP has proved to be an efficient tool in studies based on cost–benefit analysis; benefit, opportunity, cost and risk (BOCR) analysis; strength, weakness, opportunities and threats (SWOT) analysis; life cycle analysis (LCA); etc. MCDM methods usually find applications in areas such as technology selection, energy planning and selection, energy resource allocation, energy exploitation, energy policy, building energy management, transportation energy systems, and others. This chapter discusses the application of different types of FAHPs in some major areas concerning energy systems with examples explaining the choice of criteria and methodology of analysis.

This chapter explains the application of FAHP in the field of energy systems by categorizing the applications under four sections, namely, energy sources, renewable energy technology, energy policy and site selection, and alternative fuels. The first section explains the need to select the most sustainable energy source considering the impact of all technological, environmental, socio-political, and economic criteria. The next section discusses assessment of different renewable energy-based technologies for similar criteria. The third section talks about the role of energy policy and site selection in overall performance of a process, technology, and firm. The use of FAHP in desirable selection of these factors is also discussed. The last section explains the use of FAHP in evaluating different alternative fuels with the case study of hydrogen production by different processes. Finally, the conclusions drawn from each section and the overall chapter are presented.

13.2 Energy Sources

The importance of having efficient, abundant, eco-friendly, and economical energy sources is increasing as the global energy demand continues to rise tremendously on account of the growing population and declining reserves of fossil energy sources. Global electricity consumption is expected to cross 24,400 billion kWh by the year 2020 (Demirtas, 2013). Environmental performance is added to the list of critical parameters such as economy and efficiency for selecting better energy sources since 85% of greenhouse gas (GHG) emissions worldwide are generated by the energy sector today (Demirtas, 2013). The interaction of the energy sector with other sectors makes it inevitable to follow up the performance of major financial firms. The interaction of the energy sector with other sectors leads to uncertainty when predicting financial performance. This financial market uncertainty has increased all over the world since 2008, hence the usage of fuzzy methods can provide better results under these conditions. Eyuboglu and Celik (2016) evaluated 13 energy firms with 5 main and 15 subcriteria for the period of 2008–2013 in Turkey using FAHP and fuzzy technique for order of preference by similarity to ideal solution (TOPSIS) to find the best and poor performing firms. There are other studies in which FAHP has been applied for analyzing the energy sources for a specific application. Jaber et al. (2008) mainly focused on energy sources for space heating, with a comparison of fuzzy sets analysis and AHP. The analysis by both methods showed that heating systems based on renewable energy, that is, wind and solar energy, are most favorable, followed by traditional stoves burning petroleum products and finally the worst heating system being the electric heater. Georgilakis (2006) carried out a cost–benefit analysis of small autonomous power systems and Garcia et al. (1998) reported a feasibility study of alternative energy sources with main emphasis on greenhouse heating. Meixner and Fuzzy (2009) demonstrated the practicability and the benefits of FAHP for group decisions with the example of evaluation of energy sources. The authors evaluated photovoltaics, windmills, and coal-fired power plants for electricity production on the basis of five criteria of cost, availability, climate, dependency, and utilizability. Out of these, the cost is the only quantitative criterion, whereas the rest are qualitative that could be estimated for their importance considering different factors on which they depend. For instance, the availability of an energy source depends on climatic and geographic conditions, dependency implies relying on imported sources thus considering the possible lack of different raw materials, and utilizability indicates the fields of application of a particular energy source.

The methodologies used in the problems of thermal energy production and supply systems have been mostly of a numerical nature considering only technical and economic merits. An important problem due to the existence of so many feasible alternative sources of heat energy is how to choose the most preferable one. Problems that pertain to heat district (HD) systems are of great importance for a country like Poland. Because of the climate conditions, Polish cities need a continuous 7-month supply of energy for heating buildings. In addition, warm water is needed, especially in residential quarters, and it is produced all year long using a remote thermal energy supply. Thus, thermal energy production and delivery generate significant costs for municipal economies. Furthermore, heat production and supply systems influence the natural and social environment greatly. Therefore, it is very important to make proper short- and long-term decisions concerning the construction, maintenance, and development of such systems. Dytczak and Ginda (2006) used AHP to select the best heat energy source for a HD system for a medium sized city located in Poland on the basis of cost–benefit analysis. They treated benefits (B) and costs (C) in separate hierarchies and aggregated the partial results. For the criterion of finance, the benefits category considered operational savings, whereas in the costs category different costs are aggregated associated with investments, operations and maintenance (O&M), health protection, and environmental protection. For the social criterion, benefits included professional activation of the local population, whereas costs considered the influence on population health. Similarly, the other criteria were evaluated appropriately based on their contribution towards costs or benefits. The results obtained after applying FAHP to both these hierarchies would give the most expensive energy sources and most beneficial energy sources in increasing or decreasing order. Dytczak and Ginda (2006) aggregated the results from the benefits and costs hierarchies using the following weighted aggregation formula:

$$P_j = bP_{Bj} + (1-c)P_{Cj} \tag{13.1}$$

where P_j denotes aggregated priority and P_{Bj}, P_{Cj} are the priorities obtained from the benefits and costs analysis, respectively; b and c are respective weights expressing their relative importance. The results could also be graphically plotted to give a Pareto-optimal curve from which the most desirable option can be selected as per the DM's objective. The general trend is that the renewable sources prove to be best when social and environmental criteria are weighed higher, whereas fossil sources like natural gas prove to be a better choice when costs and high-capacity production are more important. FAHP by different approaches can be applied to the same study to get more reliable results. The weights of different criteria depend on the DM based on his objective. For the same data, different decisions can be made towards particular components of cost–benefit analysis and existing local conditions.

Most of the developing countries face challenges of rapidly increasing population and per capita energy consumption demanding an optimum usage of available energy resources. Daniel et al. (2010) considered the criteria of cost, efficiency, environmental impact, installed capacity, estimated potential, reliability, and social acceptance to evaluate renewable energy sources like solar, wind and biomass in India. Kahraman et al. (2009) compared five renewable energy sources in Turkey using fuzzy axiomatic design and the FAHP method for selecting the best and found that both approaches gave the same results, reporting wind as the best choice irrespective of the changes in pairwise comparison. In another study, Kahraman et al. (2010) applied fuzzy MCDM methodologies for evaluation of nine different energy sources available in Turkey to determine the best energy policy. Amongst the different factors considered under the domain of technological, economic, environmental, and socio-political criteria, energy proved to be a crucial factor affecting environmental and economic attributes. Based on all the above studies, Table 13.1 represents the different main and subcriteria that need to be considered for evaluation of energy sources. The application of FAHP in comparison to classical AHP

TABLE 13.1

Criteria to Be Considered for Evaluation of Energy Sources

Main Criteria	Subcriteria
A: Technological	A1: Feasibility
	A2: Reliability
	A3: Energy and raw material
	A4: Preparation phase
	A5: Implementation phase
	A6: Continuity and performance predictability
	A7: Local technical knowledge
	A8: Risk
B: Environmental	B1: Pollution
	B2: Effluent treatment
	B3: Requirement of land
	B4: Waste disposal
	B5: Plant layout
C: Socio-political	C1: National energy policy check
	C2: Political acceptance
	C3: Social acceptance
	C4: Safety
	C5: Labor impact
D: Economic	D1: Investment cost
	D2: Operational and management cost
	D3: Health and environmental protection cost
	D4: Funds availability
	D5: Economic value

for selection of the most desired energy source can be justified considering the qualitative nature of criteria such as reliability, socio-political acceptance, safety, risk, etc., and the degree of uncertainty involved in measuring them relatively.

13.3   Renewable Energy Technology

Renewable energy sources or alternative energy sources are often clean, domestic, and renewable and hence are commonly accepted as key for future life. Hence, the technologies or the energy systems based on renewables have gathered considerable importance in the current global market. The selection of the best alternative amongst the different available energy technologies is a multicriteria problem with many conflicting criteria and hence a crucial decision in energy investment. The alternatives need to be compared and analyzed by taking into account their salient features and limitations based on appropriate selection criteria. The conflicting criteria make the evaluation process tedious and the qualitative criteria make it vague. Sometimes, it is very difficult for the DM to provide exact numerical values for the criteria or attributes and hence FAHP can be efficiently applied in such cases. For the factors falling under the category of environmental, health, and social impact, it is easier and more appropriate to report them in a range or interval rather than a specific value. The fuzzy set theory is a powerful tool to treat uncertainty in the case of incomplete or vague information. There are several instances when fuzzy MCDM methodologies are suggested for selection amongst renewable energy alternatives. Suganthi et al. (2015) reviewed the applications of fuzzy logic-based models in renewable energy systems, namely solar, wind, bio-energy, micro-grid, and hybrid applications, and indicated that fuzzy-based models provide realistic estimates. Topcu and Ulengin (2004) focussed on the multi-attribute decision-making evaluation of energy resources that enabled the selection of the most suitable electricity generation alternative for Turkey. Chatzimouratidis and Pilavachi (2007) evaluated 10 types of power plants for their non-radioactive emissions using AHP and proposed to use a combination of power plants and advanced emission control techniques as no single choice was ideal. The authors also analyzed the power plants with regard to their overall impact on the living standard of local communities using AHP (Papalexandrou et al., 2008; Chatzimouratidis et al., 2008). For the objective of selecting the power plant technology leading to a maximum rise in the living standard of people, the two main criteria considered are the improvement of quality of life by reducing the drawbacks of power plant operation and the positive socio-economic aspects. Accident fatalities, non-radioactive emissions, radioactivity, and land requirements for each power plant represent the drawbacks whereas job creation, compensation rate, and social acceptance from the nearby community represent the socio-economic aspects (Chatzimouratidis et al., 2008). For the desired objective, renewables-based power plants are reported to be the first choice followed by nuclear and then fossil fuels-based ones irrespective of the weights of criteria. Promentilla et al. (2016) developed calibrated FAHP to prioritize low-carbon technology amongst biomass, solar, geothermal, hydro, and wind for electricity generation in the Philippines. They used the calibrated fuzzy scale in the fuzzy pairwise comparison matrix to address the vagueness involved in decision-making.

Electricity is the most usable and flexible form of energy with several applications in the areas of heating, transportation, instruments, etc. Technologies based on geothermal energy, solar energy, wind energy, hydropower energy, and biomass are amongst the most popular alternatives to conventional coal-, oil-, and natural gas-based electricity production. Remote islands often face severe energy challenges due to limited and expensive fossil fuels. Hence, the usage of locally available and abundant natural or renewable resources needs to be encouraged for power generation (Wimmler et al., 2015). Wimmler et al. (2015) reviewed the application of MCDM methods to renewable energy and storage technologies particularly on islands. Demirtas (2013) used classical

AHP to determine the best renewable energy technology for sustainable energy planning. It can be observed that most of the criteria are similar for both mainland and island cases, except for the environmental indicators that would be island specific and not applicable to the mainland. The weights of the selection criteria in these reported studies are determined by pairwise comparison matrices of AHP assuming the availability of crisp judgements. Hence, the application of FAHP would be more justifiable as most of the criteria are totally dependent on qualitative indicators that cannot be easily measured as crisp values. The most common renewable energy-based technologies are geothermal, solar, wind, hydro, biomass, and tidal and often the choice has to be made amongst these alternatives for investment in energy sector. Figure 13.1 shows the general main and subcriteria in the hierarchical structure for the objective of selecting the best sustainable renewable-based energy technology. The four main criteria of technical, economic, environmental, and socio-political cover almost every important aspect required to evaluate the renewable energy-based technology appropriately.

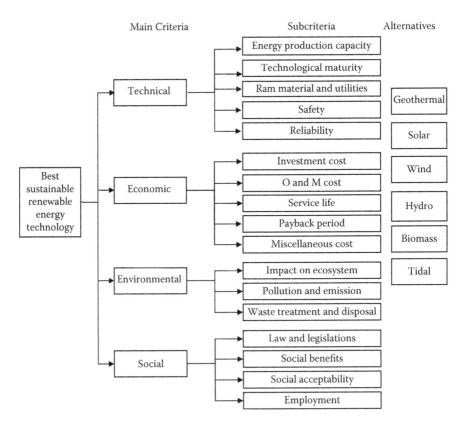

FIGURE 13.1
Hierarchical structure for evaluation of renewable energy technologies.

13.4 Energy Policy and Site Selection

Industrialization and rapid technological development increase global energy demand tremendously. However, the amounts of energy reserves and also energy policies differ from one country to another. This results in major environmental concerns, serious political conflicts, unavoidable economical dependency, and significant social consequences. The existing situation and the future estimations for energy requirements have led people to look for alternative energy resources (Demirtas, 2013). Renewable energy resources are often regarded as an alternative for conventional resources but differ in capacities for power generation, suitability of a potential site, sustainability and stability of energy resources, etc. Furthermore, renewable technologies have relatively high capital costs though lower operation and maintenance costs. These characteristics make them attractive in the long run, but less attractive in a competitive setting where the prime focus is on cost minimization (Topcu and Ulengin, 2004). The evaluation of energy systems is mainly done with the purpose of guiding the development and formulation of energy policy. Energy policy analysis covers national or regional energy systems assessment, public debate on energy and the environment, energy conservation strategies, and energy resource allocation issues in addition to assessment of climate policy, the global warming debate, and air pollution control policy (Zhou et al., 2006). Each of these attributes is non-measurable and could be understood better using fuzzy pairwise comparison matrices. Kahraman et al. (2010) employed a fuzzy MCDM methodology for the selection of best energy policies allowing evaluation scores from experts using a model that could deal with linguistic expressions and crisp and fuzzy numbers simultaneously. Elkarmi and Mustafa (1993) evaluated policy instruments such as R&D/technology adaptation, endogenous capacity building, laws/regulations, privatization, economic instruments, and incentives using AHP for increasing the utilization of solar energy technologies in Jordan. The mechanisms or policy instruments of utilizing solar energy technologies are affected by several influencing factors and constraints that can be converted into subjective non-crisp judgements using FAHP.

The primary energy sources are as follows: solid fuels, liquid fuels, gaseous fuels, hydropower, nuclear energy, solar energy, biomass energy, wind energy, ocean energy, and geothermal energy. Kahraman and Kaya (2009) considered these alternatives except ocean energy to determine the most appropriate energy policy for Turkey. Kablan (2004) illustrated the use of AHP to prioritize policy instruments for promotion of energy conservation in Jordan. Five policy instruments are compared over the criteria of satisfying energy demand, economic growth, environmental factors, and higher use of renewable energy resources. In the planning and evaluation of a project for the energy policy, several factors in addition to economics such as regional development, environmental impact, or risk are crucial in decision-making.

This decision-making process is often multidimensional based on a number of aspects at different levels such as economic, technical, environmental, political, and social. With this motivation, a multicriteria approach such as FAHP appears to be the most appropriate tool to understand all the different perspectives involved and to support those concerned with the decision-making process by creating a set of relationships between the various alternatives. The outcome of FAHP application helps to improve the quality of decisions about energy policy involving multiple criteria by making choices more explicit, rational, and efficient (Kahraman et al., 2009).

The DMs in the energy and environmental sector often face the challenge of selecting the best site or location for a new installation or expansion. This decision is often subjected to multiple conflicting criteria thereby making FAHP a potential tool to handle such problems. Both energy policy and site selection are themselves non-perceptible objectives and hence the majority of the criteria on which these would depend are bound to be intangible, which justifies the inclusion of fuzziness. Furthermore, the pairwise comparison matrices need not be crisp in FAHP. During the site selection process, the assessment of energy consumption and environmental impact needs to be carried out to find the best site with the lowest cost, lowest energy consumption, and the lowest grade in environmental impact risk assessment (Beskese et al., 2015). A geographical information system (GIS)-based MCDM approach has been used to solve conflicts and balance the trade-offs and risks related to various experts' judgments. Ibrahim et al. (2011) applied FAHP to calculate weights of the criteria considered for the selection of the most preferred site for a wastewater lift station and a used GIS to overlay and generate criteria maps and a suitability map. Asakereh et al. (2014) used FAHP and GIS to locate the most appropriate sites for solar energy farms in the Shodirwan region in Iran. Ekmekcioglu et al. (2011) integrated fuzzy TOPSIS with FAHP to develop fuzzy multicriteria SWOT analysis and applied the approach to nuclear power plant site selection by considering the internal and external parameters as criteria. Table 13.2 highlights the main criteria and indicators for a general power plant site selection. AHP has found application in the selection and location of suitable power plants in Nigeria on a regional basis made up of six geopolitical zones (Ajayi and Olamide, 2014). Each zone is considered separately, because of the differences in energy resources, weather, and landscape. The selection was made due to the technologies on ground and those with possibility to be potential power boosters. Mousavi and Rostamy (2016) proposed a framework for thermal power plant site selection using index weights based on experts, views, paired comparisons, and FAHP. The data were collected from experts using field methods and the reliable literature, which was later analyzed using FAHP-TOPSIS collectively to find preferred location. Kharat et al. (2016) applied integrated FAHP-TOPSIS methodology to find the best site for a landfill and developed the model considering socio-economic and regulatory setup for a case study of Mumbai, India. The model is supposed to act as a guiding tool for municipal

TABLE 13.2

Criteria and Indicators for Power Plant Site Selection

Criteria	Subcriteria/Indicators
Internal	Technical suitability
	Social and environmental impact
	Area for waste disposal
	Traffic conditions
	Transport facilities
	Proximity to populated area
	Distance to industry energy consumers
	Impact on nearby businesses
External	Geographical conditions
	Climate
	Proximity to water resources
	Grid conditions
	Seasonal variation
	Risk (natural calamity, terrorist attack, etc.)

solid waste (MSW) management planners. ChaoKe et al. (2013) analyzed the 10 different indicators classified into economic and social benefits having an impact on wind farm site selection using AHP. Considering the ambiguity of most of the factors and the qualitative nature of the criteria involved in these studies, it is preferable to use FAHP compared to classical AHP.

13.5 Alternative Fuels

FAHP has found applications in analyzing different alternative fuels such as hydrogen and bio-fuels as compared to the conventional ones and comparing the different processes to produce these fuels. Liquid bio-fuels are being explored as alternative fuels for internal combustion engines to substitute for a considerable amount of conventional fuels. Papalexandrou et al. (2008) assessed different bio-fuels using AHP for the criteria of bio-fuel substitution cost over conventional fuels, potential of substitution, total cycle GHG emissions, and total cycle energy consumed. Hydrogen is the forerunner in the list of new generation fuels being considered as a possible solution to problems with the traditional fossil-based energy sources. The process to produce hydrogen is the determining factor for its cost and environmental performance. Buyukozkan and Guleryuz (2016) used FAHP to determine the evaluation criteria weights and fuzzy TOPSIS to select the most suitable energy technology alternative for the case of Turkey. Amongst different sustainable energy resources available, nuclear energy proved to be the best alternative for Turkey. Heo et al. (2012) evaluated six hydrogen-producing methods in Korea using FAHP under benefits, opportunities, costs, and risk concepts for 12 different criteria. Steam methane reforming (SMR) proved to be the most preferable method because more importance was attributed to equipment investment cost and market size compared to human resource development and environmental contribution. Pilavachi et al. (2009) compared seven different hydrogen-producing technologies for the criteria of CO_2 emissions, O&M costs, capital cost, feedstock cost, and hydrogen production cost using AHP. The renewable-based electrolysis process was the most preferred technology when CO_2 emissions were of major concern. Thengane et al. (2014) carried out cost–benefit analysis for eight different hydrogen-producing technologies using AHP and FAHP to find a cost-effective environmentally benign hydrogen production technology. Figure 13.2 presents the hierarchy tree structure for the evaluation of eight hydrogen production processes with different criteria considered. The results are calculated on the basis of certain selected criteria for an assumed plant capacity and the projected costs using mathematical calculations. Wind-based electrolysis (W-EL), hydro-based electrolysis (H-EL), and water splitting by chemical looping (WS-CL) proved to be the most beneficial technologies with the higher final scores. In this study, three of the five criteria, that is, GHG emissions, raw material and utilities consumption, and waste disposal and atmospheric emissions, were in favor of renewable approaches over fossil-based technologies. Figure 13.3 reports the cost–benefit analysis results using Chang's extent analysis method and logarithmic FPP method. For the same pairwise comparison matrices, different FAHP approaches give slightly different results, which might change the preference. Hence, it is advisable to compare the results using different AHP-based methods and then consult with the experts to arrive at a final decision. The rankings given by FAHP-based approaches are solely dependent on the DM's objectives and the importance of different criteria.

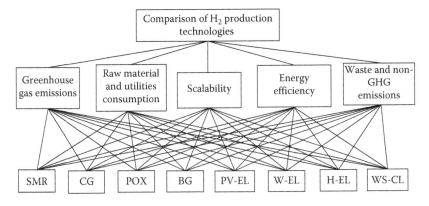

FIGURE 13.2
Hierarchy tree structure for the evaluation of eight hydrogen production processes. (From Thengane SK et al., *Int J Hydrogen Energy*, 39, 15293–15306, 2014.)

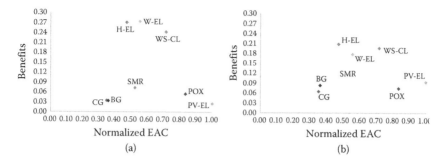

FIGURE 13.3

Cost–benefit analysis results for different fuzzy analytic hierarchy processes. (a) Extent analysis method and (b) LFPP method. (From Thengane SK et al., *Int J Hydrogen Energy*, 39, 15293–15306, 2014.)

In addition to the range of energy systems reviewed, there is further potential for new studies such as in energy conservation, energy buildings, transportation energy systems, etc., where FAHP can be potentially applied to address the complex decision-making processes. The inherent multi-objective nature of energy investment decisions and the dependence on qualitative attributes of socio-economic and environmental aspects give FAHP an edge over the classical AHP. The criteria need to be appropriately chosen for the desired objective and the weights would depend on the requirements of the DM. The different fuzzy-based AHP approaches may give different overall scores for the alternatives. Hence, it is advisable to check the results by applying at least two different FAHP-based approaches in addition to classical AHP. The weights of the criteria influence directly the decision-making results of energy projects' alternatives. Hence, a case for equal criteria weights should be included in addition to the other cases of varied criteria importance depending on the requirements. The sensitivity analysis study accounting for variation of importance given to different criteria helps to analyze and understand the system/process satisfactorily.

13.6 Conclusions

Though the renewable energy-based technologies and alternative fuels are promoted globally because of their lower impact on the environment, there are major concerns relating to economy, efficiency, and scale-up potential that need attention. Not all the alternatives are equally good in all tangible and non-tangible aspects and hence there is a need to evaluate critically the different criteria using appropriate weights. FAHP is a promising approach in MCDM considering its ability to characterize the inherent uncertainties to a greater extent for both qualitative and quantitative criteria. In the case of energy sources and renewable technologies, the general trend is that the renewables prove to be best when social and environmental criteria are weighed higher, whereas fossil fuel-based alternatives prove to be a better choice when costs and high-capacity production are more important. Both AHP and fuzzy set theory have evolved over the years and hence the application of FAHP by different approaches to the same study would result in more reliable results. The integration of FAHP with other methods, such as BOCR, SWOT analysis, LCA, TOPSIS, etc., beneficially contributes to the analysis of energy systems. However, FAHP could alone give satisfactory results provided the appropriate criteria and subcritera are chosen and the rational weighting factors are allotted based on the DM's objective.

References

Ajayi K, Olamide O. Investigation of power plant selection and location using eigenvector method of analytic hierarchy process. In: *8th International Conference on Engineering and Technology Research, Dubai*; 2014.

Asakereh A, Omid M, Alimardani R, Sarmadian F. Developing a GIS-based fuzzy AHP model for selecting solar energy sites in Shodirwan Region in Iran. *Int J Adv Sci Technol*. 2014;68:37–48.

Beskese A, Demir H, Ozcan K, Okten E. Landfill site selection using fuzzy AHP and fuzzy TOPSIS: A case study for Istanbul. *Environ Earth Sci*. 2015;73(7):3513–3521. doi:10.1007/s12665-014-3635-5.

Buyukozkan G, Guleryuz S. Fuzzy multi criteria decision making approach for evaluating sustainable energy technology alternatives. *Int J Renew Energ Sour*. 2016;1:1–6.

Chang D. Applications of the extent analysis method on fuzzy AHP. *Eur J Oper Res*. 1996;2217(95):649–655.

ChaoKe, Hanchao Y, Dang F. Application of AHP in selection of wind farm's location. *Int J Energy Sci*. 2013;3(1):18–22.

Chatzimouratidis AI, Pilavachi P. Multicriteria evaluation of power plants impact on the living standard using the analytic hierarchy process. *Energy Policy*. 2008;36(3):1074–1089. doi:10.1016/j.enpol.2007.11.028.

Chatzimouratidis AI, Pilavachi PA. Objective and subjective evaluation of power plants and their non-radioactive emissions using the analytic hierarchy process. *Energy Policy*. 2007;35:4027–4038. doi:10.1016/j.enpol.2007.02.003.

Daniel J, Nandigama V, Vishal V, Albert B, Selvarsan I. Evaluation of the significant renewable energy resources in India using AHP. In: *Multiple Criteria Decision Making for Sustainable Energy and Transportation Systems*, Vol. 634; 2010, pp. 269–276. doi:10.1007/978-3-642-04045-0.

Demirtas O. Evaluating the best renewable energy technology for sustainable energy planning. *Int J Energy Econ Policy*. 2013;3:23–33.

Dytczak M, Ginda G. Benefits and costs in selecting fuel for municipality heating systems with the analytic hierarchy process. *J Syst Sci Syst Eng.* 2006;15(2):165–177. doi:10.1007/s11518-006-5005-7.

Ekmekcioglu M, Kutlu AC, Kahraman C. A fuzzy multi-criteria SWOT analysis: An application to nuclear power plant site selection. *Int J Comput Intell Syst.* 2011;4(4):583. doi:10.2991/ijcis.2011.4.4.15.

Elkarmi F, Mustafa I. Increasing the utilization of solar energy technologies (SET) in Jordan. *Energy Policy.* 1993;21(9):978–984.

Eyuboglu K, Celik P. Financial performance evaluation of turkish energy companies with fuzzy AHP and TOPSIS methods. *Bus Econ Res J.* 2016;7(3):21–37.

Garcia JL, De la Plaza S, Navas LM, Benavente RM, Luna L. Evaluation of the feasibility of alternative energy sources for greenhouse heating. *J Agric Eng Res.* 1998;69:107–114. doi:10.1006/jaer.1997.0228.

Georgilakis PS. State-of-the-art of decision support systems for the choice of renewable energy sources for energy supply in isolated regions. *Int J Distrib Energy Resour.* 2006;2(2):129–150.

Heo E, Kim J, Cho S. Selecting hydrogen production methods using fuzzy analytic hierarchy process with opportunities, costs, and risks. *Int J Hydrogen Energy.* 2012;37(23):17655–17662. doi:10.1016/j.ijhydene.2012.09.055.

Ibrahim E, Mohamed S, Atwan A. Combining fuzzy analytic hierarchy process and GIS to select the best location for a wastewater lift station in El-Mahalla El-Kubra, North Egypt. *Int J Eng Sci Technol.* 2011;11(5):44–50.

Jaber J, Jaber Q, Sawalha S, Mohsen M. Evaluation of conventional and renewable energy sources for space heating in the household sector. *Renew Sustain Energy Rev.* 2008;12(1):278–289. doi:10.1016/j.rser.2006.05.004.

Kablan MM. Decision support for energy conservation promotion: An analytic hierarchy process approach. *Energy Policy.* 2004;32(10):1151–1158. doi:10.1016/S0301-4215(03)00078-8.

Kahraman C, Kaya İ, Cebi S. A comparative analysis for multiattribute selection among renewable energy alternatives using fuzzy axiomatic design and fuzzy analytic hierarchy process. *Energy.* 2009;34(10):1603–1616. doi:10.1016/j.energy.2009.07.008.

Kahraman C, Kaya I. A fuzzy multicriteria methodology for selection among energy alternatives. *Expert Syst Appl.* 2010;37(9):6270–6281. doi:10.1016/j.eswa.2010.02.095.

Kharat MG, Kamble SJ, Raut RD, Kamble SS, Dhume SM. Modeling landfill site selection using an integrated fuzzy MCDM approach. *Model Earth Syst Environ.* 2016;2:52–68. doi:10.1007/s40808-016-0106-x.

Meixner O. Fuzzy AHP group decision analysis and its application for the evaluation of energy sources. In: *ISAHP, Vienna;* 2009.

Mousavi H, Rostamy AAA. A framework for assessment and selection of thermal power plant location based on MADM methods. *Comput Res Prog Appl Sci Eng.* 2016;2(1):35–39.

Papalexandrou MA, Pilavachi PA, Chatzimouratidis AA. Evaluation of liquid bio-fuels using the analytic hierarchy process. *Process Saf Environ Prot.* 2008;86(5):360–374. doi:10.1016/j.psep.2008.03.003.

Pilavachi PA, Chatzipanagi AI, Spyropoulou AI. Evaluation of hydrogen production methods using the analytic hierarchy process. *Int J Hydrogen Energy.* 2009;34(13):5294–5303. doi:10.1016/j.ijhydene.2009.04.026.

Pohekar SD, Ramachandran M. Application of multi-criteria decision making to sustainable energy planning—A review. *Renew Sustain Energy Rev.* 2004;8(4): 365–381. doi:10.1016/j.rser.2003.12.007.

Promentilla MAB, Antonio MRSE, Chuaunsu II RMM, de Serra AJ. A calibrated fuzzy AHP approach to derive priorities in a decision model for low carbon technologies. In: *DLSU Research Congress.* Vol. 4. Manila, Philippines; 2016:7–12.

Saaty T, Vargas L. *Models, Methods, Concepts and Applications of the Analytic Hierarchy Process.* Boston, MA: Kluwer Academic Publishers; 2000.

Saaty TL. How to make a decision: The analytic hierarchy process. *Eur J Oper Res.* 1990;48:9–26.

Sheth N, Hughes L. Quantifying energy security: An analytic hierarchy process approach. In: *5th Dubrovnik Conference on Sustainable Development of Energy, Water, and Environment Systems in Dubrovnik, Croatia*; 2009.

Suganthi L, Iniyan S, Samuel A. Applications of fuzzy logic in renewable energy systems—A review. *Renew Sustain Energy Rev.* 2015;48:585–607. doi:10.1016/j. rser.2015.04.037.

Thengane SK, Hoadley A, Bhattacharya S, Mitra S, Bandyopadhyay S. Cost-benefit analysis of different hydrogen production technologies using AHP and fuzzy AHP. *Int J Hydrogen Energy.* 2014;39(28):15293–15306. doi:10.1016/ j.ijhydene.2014.07.107.

Topcu YI, Ulengin F. Energy for the future: An integrated decision aid for the case of Turkey. *Energy.* 2004;29(1):137–154. doi:10.1016/S0360-5442(03)00160-9.

Wang J, Jing Y, Zhang C, Zhao J. Review on multi-criteria decision analysis aid in sustainable energy. *Renew Sustain Energy Rev.* 2009;13:2263–2278. doi:10.1016/ j.rser.2009.06.021.

Wang Y, Chin K. Fuzzy analytic hierarchy process: A logarithmic fuzzy preference programming methodology. *Int J Approx Reason.* 2011;52(4):541–553. doi:10.1016/j.ijar.2010.12.004.

Wimmler C, Hejazi G, Fernandes E, Moreira C, Connors S. Multi-criteria decision support methods for renewable energy systems on islands. *J Clean Energy Technol.* 2015;3(3):185–195. doi:10.7763/JOCET.2015.V3.193.

Zhou P, Ang B, Poh K. Decision analysis in energy and environmental modeling: An update. *Energy.* 2006;31(14):2604–2621. doi:10.1016/j.energy.2005.10.023.

14

FAHP-Based Decision Making Framework for Construction Projects

Long D. Nguyen and Dai Q. Tran

CONTENTS

14.1 Introduction

Decision-making in construction management often takes place in complex environments and is fraught with risk and uncertainty. The construction industry by nature confronts more risky events and uncertain situations than others (Dey et al., 2002; Tran and Molenaar, 2015). For example, construction workers are subject to exposure to serious injuries due to weather conditions (e.g., cold, heat, rain, snow), the surrounding environment (e.g., traffic on a highway), or heavy equipment. Research has shown that approximately 80% of projects involve a high level of uncertainty at the beginning of the construction phase (Mulholland and Christian, 1999). Characteristics such as dynamic work environments, industry fragmentation, multilingual work crews, customized projects, and exposure to weather create a number of challenges for decision makers. The political and economic environments also play vital roles in the construction process (Tran et al., 2016). Factors such as fluctuating prices of materials, inflation, and interest rates have a substantial impact on decision-making in construction projects (Öztaş and Ökmen, 2005).

As construction management involves many multicriteria decision-making (MCDM) problems, making a good decision in construction management is a challenging task. Decision makers have to consider and evaluate a wide range of factors (e.g., project conditions, input variables, alternatives, and various criteria). Consequently, a number of MCDM methods have been employed in construction management. Among these MCDM methods, the analytic hierarchy process (AHP) and specifically fuzzy AHP (FAHP) have received substantial attention in construction management research and practice. Jato-Espino et al. (2014) conducted a comprehensive literature review of MCDM issues in the construction industry. Through analyzing 88 publications over the last two decades, this study found that AHP and FAHP are the two most popular multicriteria selection methods in the construction industry.

AHP, introduced by Thomas Saaty in 1980, is one of the most effective MCDM methods for selecting best alternatives (Saaty, 1980). However, the traditional AHP cannot reflect the uncertainty in human thinking (Lam et al., 2008; Chou et al., 2013). As a result, FAHP was developed to solve hierarchical uncertain problems (van Laarhoven and Pedrycz, 1983). This chapter is presented in five sections. After the introduction, Section 14.2 presents an overview of the FAHP-based decision-making approach to construction management. It is followed by sections that will illustrate an application of this approach in evaluating and measuring project complexity. Specifically, Section 14.3 summarizes the evolution of measuring project complexity as an application domain. Section 14.4 demonstrates the step-by-step guide to the use of FAHP in measuring project complexity. Section 14.5 concludes with a discussion on key findings and practical recommendations.

14.2 Application of FAHP in Construction Management

FAHP is a systematic approach to MCDM problems by using the concepts of fuzzy set theory and hierarchical structure analysis. The applications of FAHP have been well documented in construction management literature. These applications can be classified into the following domains: project site selection, contractor selection/bid evaluation, selection of construction means and methods, risk assessment and management, and other emerging areas (Table 14.1). The following subsections briefly discuss the application of FAHP in these domains.

14.2.1 Project Site Selection

Kuo et al. (1999) developed a decision support system for selecting the location of a convenience store through integration of FAHP and an artificial neural network. The system consists of four components: FAHP, weights

TABLE 14.1

Fuzzy Analytic Hierarchy Process (FAHP) Applications in Construction

Application Domains	Selected Studies	Remark
Project site selection	Kuo et al. (1999)	FAHP outperformed regression models
	Vahidnia et al. (2009)	Integrated GIS with FAHP
	Yu and Liu (2012)	Integrated a nonlinear model with FAHP
Contractor selection/ bid evaluation	Jaskowski et al. (2010)	Facilitating group decision-making
	Nassar and Hosny (2013)	AHP with fuzzy-C means to classify contractors
	Chou et al. (2013)	Integrated regression-based Monte Carlo simulation with FAHP for bidding strategies
	Nasab and Ghamsarian (2015)	FAHP model for contractor qualification
Selection of construction means and methods	Pan (2008)	FAHP model for choosing a suitable bridge construction method
	Shi et al. (2014)	Integrated FAHP with tunnel seismic prediction to predict the surrounding rock classification
	Ebrahimian et al. (2015)	Integrated FAHP with compromise programming for construction methods of urban storm water system
Risk assessment and management	Zeng et al. (2007)	Integrated FAHP with risk likelihood, risk severity, and risk factor index
	Zhang and Zou (2007)	FAHP as the optimal technique for evaluating risks associated with joint venture projects
	Li and Zou (2011)	FAHP-based risk assessment approaches for public–private partnership (PPP) projects
	Hsueh et al. (2013)	Delphi FAHP model to assess risk of abandoned public buildings
	Andrić and Lu (2016)	Using FAHP approaches to conduct risk assessment of bridges under multiple hazards
Other areas	Zheng et al. (2012)	FAHP application for the evaluation of worker's health and safety
	Akadiri et al. (2013)	FAHP application for the selection of sustainable building materials
	Nguyen et al. (2015)	FAHP application for measurement of project complexity

determination, data collection, and decision-making. The hierarchical structure of FAHP was established by reviewing literature and interviewing the experts. This study concluded that the decision support system was able to provide more accurate results than regression models. Vahidnia et al. (2009) combined Geographical Information System (GIS) analysis with the FAHP method to choose the location of a hospital site subject with multiple conflicting criteria. By using three methods to estimate the total weights and priorities of the project location candidates, the study highlighted that FAHP was a suitable tool for the project site selection problem. Yu and Liu (2012) proposed a FAHP model to select the most suitable projects among multiple sites for safety improvement. The authors integrated a nonlinear model with FAHP to improve the consistency in pairwise comparison and weight estimation for MCDM problems. The study suggested that the proposed FAHP provided an effective tool for decision makers to prioritize safety improvement investments.

14.2.2 Contractor Selection and Bid Evaluation

Jaskowski et al. (2010) developed an FAHP method to facilitate the process of group decision-making for select contractors. This study found that FAHP allowed the user to analyze a contractor selection problem in detail to enhance precision of estimation and assessment consistency. Nassar and Hosny (2013) integrated an AHP method with the fuzzy-C means algorithm to evaluate performance of 14 contractors. This study showed that the method could be used to classify contractors into different performance groups in a rational and unbiased way. Chou et al. (2013) integrated FAHP and a regression-based model to develop a new bidding strategy to support the decision-making process. The authors employed FAHP to analyze questionnaire data and obtain the weighted values of various influencing factors required for the bid decision-making process. This study demonstrated that the proposed model could be used as a strategic tool for quantifying project risks and calculating bids for construction projects. Recently, Nasab and Ghamsarian (2015) developed an FAHP model for contractor prequalification. The model considered six major criteria and 22 subcriteria to select a suitable contractor. This study indicated that FAHP provided a more accurate, effective, and systematic evaluation process to select contractors for construction projects.

14.2.3 Selection of Construction Means and Methods

Pan (2008) developed an FAHP model for choosing a suitable bridge construction method. The model used triangular and trapezoidal fuzzy numbers and the α-cut concept to capture the uncertainty and subjective judgment in the decision-making process. The study concluded that the FAHP model provided a structured and systematic approach to effectively determine the most appropriate bridge construction technique. Shi et al. (2014) integrated the FAHP method with the tunnel seismic prediction (TSP) tool to develop an optimized classification method for predicting the surrounding rock classification. Trapezoidal fuzzy numbers were used in the FAHP. This study found that the FAHP–TSP approach had two main advantages, namely the easy accessibility of the data of the main factor and the applicability of the method to improve the accuracy of the evaluation results. Recently, Ebrahimian et al. (2015) combined FAHP and compromise programming (CP) to select a suitable construction method for an urban storm water system. This combination improved the accuracy and reduced the time required for completing the pairwise comparison process of the traditional AHP. This study suggested that the combined FAHP and CP approach can be an effective tool for the selection of construction means and methods in other fields of urban construction.

14.2.4 Construction Risk Analysis and Assessment

A number of researchers and practitioners have employed FAHP methods to analyze and assess construction risk. Zeng et al. (2007) proposed an FAHP-based decision-making methodology to construction project risk assessment. This study introduced a new risk parameter, called the factor index, to structure and evaluate diverse risk factors in construction projects. The factor index, risk likelihood, and risk severity were integrated with FAHP to provide a systematic tool to deal with qualitative and quantitative data in the construction process. Through an illustrative example, this study argued that FAHP could be an effective and efficient risk assessment tool to enable decision makers to perform reliable risk analysis and management. Zhang and Zou (2007) developed an FAHP model for the appraisal of environmental risk pertaining to construction joint venture projects to support the rational decision-making process. FAHP was considered the optimal technique for evaluating risks associated with joint venture projects as it provided a judgmental level of information for assessing uncertain conditions. The study suggested that the proposed FAHP method can be applicable to the research and analysis of risks associated with any type of construction projects. Li and Zou (2011) developed an FAHP-based risk assessment methodology for public–private partnership (PPP) projects. This study compared the results from FAHP with traditional AHP and found that FAHP improved risk assessment accuracy and reduced the expert's subjectivity involved in the decision-making process. FAHP was found to be suitable technique to assess and rank the risk factors for PPP projects. Hsueh et al. (2013) proposed the DELPHI FAHP model to redevelop derelict public buildings. The model, which combined the DELPHI method, fuzzy theory, and AHP, included four main steps: identifying decision criteria, establishing the hierarchical framework, choosing an appropriate membership function, and developing decision rules. This study affirmed that the DELPHI FAHP model was beneficial in reducing the risk of the public sector investing in the reuse of abandoned public buildings. Recently, Andrić and Lu (2016) employed FAHP approaches to perform risk assessment of bridges under multiple hazards. FAHP was used to evaluate risk weights to rank risk factors using triangular fuzzy numbers. The study indicated that the FAHP approach was a systematic, accurate, and effective technique to perform multihazard risk analysis and assessment of bridges. The study additionally pointed out that FAHP can be applied for risk assessment in other industry or engineering areas.

14.2.5 Other Emerging Areas

In addition to the four main areas discussed previously, FAHP can be employed in other areas in construction management. For example, Zheng et al. (2012) used a trapezoidal FAHP method to evaluate the health and safety of workers in hot and humid environments. Akadiri et al. (2013) proposed an FAHP multicriteria evaluation model to select sustainable materials for building construction projects. Nguyen et al. (2015) employed an FAHP-based model to measure the complexity of transportation projects.

In summary, Figure 14.1 graphically illustrates a general FAHP-based decision-making framework in construction management. FAHP is one of the most useful approaches to decision support systems that include multiple criteria and/or factors as input, especially when decisions are collectively made as a group of decision makers. FAHP is an effective technique to deal with uncertain, subjective, and linguistic data (e.g., judgments), which are frequently observed in construction management areas. Depending on

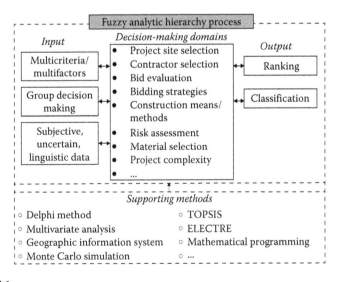

FIGURE 14.1
Fuzzy analytic hierarchy process (FAHP)-based decision-making framework in construction management.

decision-making domains, FAHP can provide decision makers with direct and reliable outputs such as ranking and classification of alternatives.

FAHP can be effectively integrated with other methods to improve the quality of its decision supports (Figure 14.1). For input, the Delphi method can be used to structure data collection from experts. GIS can be used when dealing with spatial data (e.g., in project site selection). Multivariate analysis (e.g. factor analysis, principal component analysis) can be employed to structure various criteria in decision-making problems. For output, previous studies have employed other MCDM techniques such as Technique for Order of Preference by Similarity to Ideal Solution (TOPSIS) and ELimination Et Choix Traduisant la REalité (ELECTRE) to assist in ranking alternatives. In those circumstances, FAHP has been used to determine the weights of decision-making criteria. Monte Carlo simulation (MCS) can be employed to account for uncertainty in the output. Mathematical programming (e.g., multiobjective programming) can be also used with FAHP to optimize the decision-making problems. The following sections demonstrate the use of FAHP in measuring the complexity of transportation construction projects.

14.3 Evolution of Measuring Project Complexity

Project complexity is among the most important and controversial topics in management (Bakhshi et al., 2016). Several models have been proposed to evaluate and measure project complexity. A coefficient of network complexity (CNC) was proposed to calculate the degree of complexity of a critical path network (Kaimann, 1974). A cyclomatic number, which provided the number of independent cycles in a graph, was developed as a measurement for project complexity (Temperley, 1982). To avoid counting redundant arcs as in CNC, a measure of assessing project schedules' complexity was later proposed based on connectivity of activities (Nassar and Hegab, 2006). Using a systematic thinking approach, a project complexity model was developed where complexity could be visualized based on a spider chart (Hass, 2009). Vidal et al. (2011) employed AHP to measure project complexity. Recognizing uncertainty and subjectivity in judgment from the users and decision makers, recent studies employed FAHP (Nguyen et al., 2015) or the fuzzy analytic network process (FANP) (He et al., 2015) to measure project complexity. AHP, fuzzy set theory, and hybrid methods have been extensively employed in modeling and measuring project complexity (Qazi et al., 2016). Figure 14.2 graphically presents the timeline of these studies. The earlier studies focused on measuring the complexity of a model (e.g., scheduling network) of the project system, whereas the later studies tended to measure the complexity of the project system itself (Vidal et al., 2011).

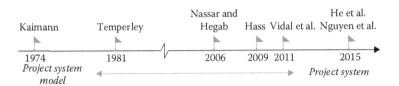

FIGURE 14.2
Timeline of studies in project complexity measurement.

14.4 FAHP for Measuring Project Complexity

In line with the recent studies, this section presents the use of FAHP in measuring complexity in transportation construction projects. The process includes 10 steps as follows:

- Step 1—Develop a hierarchical structure of project complexity
- Step 2—Obtain experts' judgment with pairwise comparisons
- Step 3—Convert experts' judgment into fuzzy numbers
- Step 4—Aggregate experts' judgment into fuzzy judgment matrices
- Step 5—Check for consistency
- Step 6—Defuzzify judgment matrices
- Step 7—Determine local weights for criteria and subcriteria
- Step 8—Determine global weights for subcriteria
- Step 9—Rate complexity scores for each project
- Step 10—Measure complexity levels (CLs) for each project

The following subsections discuss these 10 steps in detail. The data from Nguyen et al. (2015) were used for illustrative purposes.

Step 1—Develop a Hierarchical Structure of Project Complexity

There are various factors that contribute to project complexity. Using relevant statistical analysis including factor analysis, Nguyen et al. (2015) reduced 50 factors to 18 factors (subcriteria) and proposed a hierarchical structure of project complexity as shown in Figure 14.3. These subcriteria are grouped under the six major criteria: sociopolitical complexity ($C1$), environmental complexity ($C2$), organizational complexity ($C3$), infrastructural complexity ($C4$), technological complexity ($C5$), and scope complexity ($C6$). Subcriteria are denoted as Cij. For example, in infrastructural complexity ($C4$), three complexity subcriteria include site compensation and clearance ($C41$), transpor-

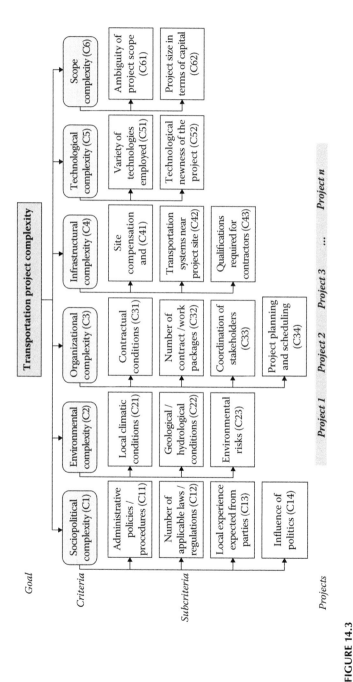

FIGURE 14.3
Hierarchical structure of project complexity.

Criteria and Subcriteria	Extremely more complex	Much more complex	More complex	Slightly more complex	Equally complex	Slightly more complex	More complex	Much more complex	Extremely more complex	
…										…
Infrastructural complexity (C4)										
Site compensation and clearance	☐	☒	☐	☐	☐	☐	☐	☐	☐	Transportation systems near project site
Site compensation and clearance	☐	☐	☐	☐	☒	☐	☐	☐	☐	Qualifications required for contractors
Transportation systems near project site	☐	☐	☐	☐	☐	☒	☐	☐	☐	Qualifications required for contractors
Technological complexity (C5)										
…										…

FIGURE 14.4
Judgments of infrastructural complexity (C4) from Expert 1.

TABLE 14.2

Expert's Judgment and Fuzzy Judgment Numbers for Infrastructural Complexity

Expert	Step 2: Expert's Judgment			Step 3: Triangular Fuzzy Judgment Numbers[a]								
	C41 vs. C42	C41 vs. C43	C42 vs. C43	C41 vs. C42			C41 vs. C43			C42 vs. C43		
				l	m	u	l	m	u	l	m	u
1	$\bar{7}$	$\bar{1}$	$1/\bar{3}$	6	7	8	1	1	1	1/4	1/3	1/2
2	$\bar{1}$	$\bar{1}$	$\bar{1}$	1	1	1	1	1	1	1	1	1
3	$\bar{3}$	$\bar{3}$	$\bar{2}$	2	3	4	2	3	4	1	2	3
4	$\bar{1}$	$\bar{1}$	$\bar{1}$	1	1	1	1	1	1	1	1	1
5	$\bar{8}$	$\bar{8}$	$1/\bar{2}$	7	8	9	7	8	9	1/3	1/2	1
...	...	...	...	...	...	...	...	...	...	...	...	...
20	$\bar{9}$	$\bar{5}$	$\bar{1}$	8	9	9	4	5	6	1	1	1
21	$\bar{7}$	$\bar{1}$	$1/\bar{5}$	6	7	8	1	1	1	1/6	1/5	1/4
Min. (l), max. (u), or geometric mean (m):				1	5.61	9	0.33	3.17	9	0.17	0.62	4

[a] With the fuzzification factor Δ = 1.

tation near project site (C42), and qualifications required for contractors (C43). After establishing this hierarchical structure with 6 criteria and 18 subcriteria, FAHP can be used to determine their relative importance to support the measurement of project complexity.

Step 2—Obtain Experts' Judgments with Pairwise Comparisons

Experts' were asked to rate pairwise comparisons between criteria and between subcriteria within a criterion (e.g., C4). Figure 14.4 illustrates the partial judgments of Expert 1 for subcriteria in criterion C4 with a "fuzzy" nine-point scale. The judgments for the other criteria and subcriteria by Expert 1 and the other 20 experts were conducted in a similar manner. For simplicity, only C4's subcriteria are illustrated from this step onwards.

Figure 14.4 shows that Expert 1 rated the site compensation and clearance subcriteria (C41) as "much more complex" (score of $\bar{7}$) in comparison with the transportation systems near project site subcriteria (C42). The result from Expert 1's judgment for "C41 vs. C42" was then entered in Table 14.2 as $\bar{7}$ (Table 14.2). Similarly, Expert 1 rated the qualification requirements for contractors subcriteria (C43) as "slightly more complex" (score of $\bar{3}$) in comparison with the transportation systems near project site subcriteria (C42). Table 14.2 presents that Expert 1's judgment for "C42 vs. C43" was 1/$\bar{3}$, the reciprocal value of $\bar{3}$.

Step 3—Convert Experts' Judgments into Fuzzy Numbers

As experts' are linguistic, subjective, and uncertain, their inputs were converted from the fuzzy nine-point scale into triangular fuzzy numbers (l, m, u) with a fuzzification factor Δ as follows:

$$
\begin{cases}
(1,1,1) \text{ if relative complexity was judged as } \bar{1} \\
(x - \Delta, x, x + \Delta) \text{ if relative complexity was judged as } \bar{x}\,(\bar{x} = \bar{2}, \bar{3}, \dots \bar{8}) \\
(8,9,9) \text{ if relative complexity was judged as } \bar{9} \\
(1/(x+\Delta), x, 1/(x-\Delta)) \text{ if relative complexity was judged as } 1/\bar{x} \\
(1/9, 1/9, 1/8) \text{ if relative complexity was judged as } 1/\bar{9}
\end{cases}
\tag{14.1}
$$

Table 14.2 shows the judgments from 21 experts and the triangular fuzzy numbers of judgments based on Equation 14.1 for the subcriteria of infrastructural complexity (C4). Only the first five and the last two experts' judgments are actually shown in Table 14.2. The last row in Table 14.2 either shows the minimum of lower value (l), geometric mean of modal value (m), or maximum of upper value (u) among the 21 experts' judgments. They are used for aggregating experts' judgments in Step 4.

Step 4—Aggregate Experts' Judgments into Fuzzy Judgment Matrices

The following equations were used to combine experts' judgments (Büyüközkan and Feyzioğlu, 2004; Chang et al., 2009):

$$
\bar{J}_{ij} = (l_{ij}, m_{ij}, u_{ij}) \text{ such that } l_{ij} \leq m_{ij} \leq u_{ij} \text{ and } l_{ij}, m_{ij}, u_{ij} \in [1/9, 9]
\tag{14.2}
$$

$$
l_{ij} = \min\left(l_{ijk}\right)
\tag{14.3}
$$

$$
m_{ij} = \sqrt[K]{\prod_{k=1}^{K} m_{ijk}}
\tag{14.4}
$$

$$
u_{ij} = \max\left(u_{ijk}\right)
\tag{14.5}
$$

TABLE 14.3

Fuzzy Judgment Matrix for Infrastructural Complexity

	C41			C42			C43		
	l	m	u	l	m	u	l	m	u
C41	1.00	1.00	1.00	1.00	5.61	9.00	0.33	3.17	9.00
C42	0.11	0.18	1.00	1.00	1.00	1.00	0.17	0.62	4.00
C43	0.11	0.32	3.00	0.25	1.62	6.00	1.00	1.00	1.00

where $(l_{ijk}, m_{ijk}, u_{ijk})$ is the pairwise comparison between criteria (or subcriteria) i and j evaluated by the kth expert from step 3 and K is the number of experts. Alternatively, geometric means can also be used to determine l_{ij} and u_{ij} as in Nguyen et al. (2015).

Table 14.3 presents the combined fuzzy judgment matrix for C4. The relative complexity of C41 and C42 of (1, 5.61, 9), for example, is from the last row in Table 14.2. It should be noted that $(l_{ji}, m_{ji}, u_{ji}) = (1/u_{ij}, 1/m_{ij}, 1/l_{ij})$.

Step 5—Check for Consistency

An important measure of consistency for pairwise comparisons of the experts' judgments is the consistency ratio (CR). CR is determined by Equation 14.6 (Saaty, 1980):

$$CR = \frac{CI}{RI} = \frac{\lambda_{max} - n}{RI(n-1)} \tag{14.6}$$

where CI = consistency index, n = size of matrixes $< \lambda_{max}$ = maximum or principal eigenvalue of the judgment matrix, and RI = random index based on the size of matrixes n, for example, RI = 0.52 for $n = 3$ (Saaty, 1980). CR can be assessed for the modal value m (Tesfamariam and Sadiq, 2006) or a crisp number of (l, m, u), that is, $(l + 4m + u)/6$ (Akadiri et al., 2013). If the modal value m is used, from the above fuzzy judgment matrix (Table 14.3) we have

$$\bar{M}_{ij} = \begin{vmatrix} 1 & 5.61 & 3.17 \\ 0.18 & 1 & 0.62 \\ 0.32 & 1.62 & 1 \end{vmatrix}$$

The principal eigenvalue of $\bar{M}_{ij}$ is $\lambda_{max} = 3.009$ and CR for the 3×3 matrix is $(3.009-3)/[0.52(3-1)] = 0.0008$ or 0.08%. Since CR is less than the threshold of 10%, the judgment matrix is acceptable.

Step 6—Defuzzify Judgment Matrices

The defuzzification process is to convert the fuzzy numbers in pairwise comparison matrixes into crisp numbers. The degree of confidence (α-cut) and attitude toward risk (λ) of the decision maker are used. Both α-cut and λ are from 0 to 1. A greater α-cut or λ shows more confidence or a more optimistic view of the decision maker, respectively. In this study, the defuzzification process is carried out based on the following equations (Liou and Wang, 1992):

$$z_{ijl}^{\alpha} = (m_{ij} - l_{ij})\alpha + l_{ij} \tag{14.7}$$

$$z_{ijr}^{\alpha} = u_{ij} - (u_{ij} - m_{ij})\alpha \tag{14.8}$$

TABLE 14.4

Defuzzification and Local Weights of the Subcriteria

	Step 6: Defuzzification[a]			Step 7: Local Weights			
	C41	C42	C43	C41	C42	C43	Local Weight[b]
C41	1.00	5.30	3.92	0.69	0.75	0.63	0.690
C42	0.19	1.00	1.35	0.13	0.14	0.22	0.163
C43	0.26	0.74	1.00	0.18	0.11	0.16	0.147
Column sum:	1.44	7.04	6.27	1	1	1	1

[a] Defuzzification example with α-cut = 0.5 and λ = 0.5.
[b] Averaging across rows.

$$z_{ij,\alpha}^{\lambda} = \lambda z_{ijr}^{\alpha} + (1-\lambda)z_{ijl}^{\alpha} \tag{14.9}$$

Table 14.4 demonstrates a case of the defuzzification with α-cut of 0.5 (moderate confidence) and λ of 0.5 (moderate risk attitude) for C4. For example,

$$z_{12l}^{\alpha=0.5} = (5.61-1)\times 0.5 + 1 = 3.305$$

$$z_{12r}^{\alpha=0.5} = 9 - (9-5.61)\times 0.5 = 7.305$$

$$z_{12,\alpha=0.5}^{\lambda=0.5} = 0.5 \times 7.305 + (1-0.5)\times 3.305 = 5.30$$

Step 7—Determine Local Weights for Criteria and Subcriteria

Various methods are used to obtain local weights (w_i) of criteria and subcriteria. A simple method is as follows:

$$w_i = \frac{1}{n}\sum_{j=1}^{n}\frac{z_{ij}}{\sum_{k=1}^{n}z_{kj}} \tag{14.10}$$

where z_{ij} and z_{kj} are elements in the defuzzifed judgment matrix. Table 14.4 also demonstrates the use of Equation 14.10 by normalizing the column vectors and then averaging across the rows. For example,

$$c_{41} = \frac{1}{3}\left(\frac{1}{1+.19+.26} + \frac{5.3}{5.3+1+.74} + \frac{3.92}{3.92+1.35+1}\right) = 0.690$$

Similarly, Steps 2–7 help determine local weights for complexity criteria (c_i) and subcriteria (c_{ij}) as shown in Table 14.5:

Step 8—Determine Global Weights for Subcriteria

Using Equation 14.11, global weights (w_{ij}) of the subcriteria with α-cut = 0.5 and λ = 0.5 are shown in Table 14.5:

$$w_{ij} = c_i \times c_{ij} \tag{14.11}$$

where c_i is the local weight of criterion i and c_{ij} is the local weight of subcriterion j within criterion i.

Sensitivity analysis is conducted to determine how local and global weights of each criterion or subcriterion vary when α-cut and λ vary in [0, 1]. Figure 14.5 displays the contours of global weights of subcriteria C41, C42, and C43. C41 is generally not sensitive to the risk attitude (λ) of decision makers while it is rather sensitive to the confidence of the judgment (α-cut). The global weight of C41 increases when α-cut increases. C42 and C43 have similar patterns and are sensitive to both λ and α-cut. Their global weights tend to be more sensitive to λ when α-cut is small and more sensitive to α-cut when λ is high.

TABLE 14.5

Weights of Complexity Criteria and Complexity Scores of Projects

Step 8: Calculating Weights of Complexity Criteria[a]				Step 9: Complexity Rating		
		Local Weight		Global Weight (w_{ij})	Complexity Scores (k_{ij})		
Criteria	Subcriteria	c_i	c_{ij}		Project 1	Project 2	Project 3
C1		0.375					
	C11		0.418	0.157	9	8	2
	C12		0.270	0.101	9	5	2
	C13		0.191	0.072	5	9	3
	C14		0.122	0.046	6	4	3
C2		0.170					
	C21		0.448	0.076	4	4	8
	C22		0.388	0.066	8	4	7
	C23		0.164	0.028	4	4	6
C3		0.156					
	C31		0.437	0.068	9	8	2
	C32		0.218	0.034	3	3	2
	C33		0.215	0.034	7	3	3
	C34		0.130	0.020	9	10	7
C4		0.148					
	C41		0.690	0.102	9	10	2
	C42		0.163	0.024	7	2	2
	C43		0.147	0.022	9	2	8
C5		0.081					
	C51		0.742	0.060	7	3	7
	C52		0.259	0.021	7	3	8
C6		0.070					
	C61		0.793	0.055	8	3	2
	C62		0.207	0.014	10	5	3

[a] Weights with α-cut = 0.5 and λ = 0.5.

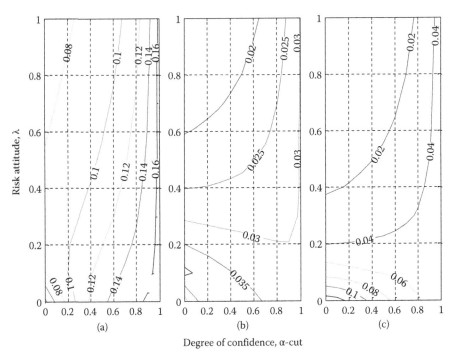

FIGURE 14.5
Contours of global weights for infrastructural complexity. (a) Global weight for C41. (b) Global weight for C42. (c) Global weight for C43.

Step 9—Rate Complexity Scores for Each Project

To measure the complexity of a project in project portfolio, a decision maker will evaluate the degree of complexity for each subcriterion for such a project. With the use of a distinctive scale of 11 points (from "0 = extremely low" to "10 = extremely high", where "5 = neutral") (Nguyen et al., 2015), the complexity scores (k_{ij}) of the 18 subcriteria can be rated for a given project. Table 14.5 presents the ratings by a senior project director of a construction firm for each of their projects: project 1 (highway), project 2 (highway), and project 3 (bridge and highway).

Step 10—Measure CLs for Each Project

The overall complexity of a project can be measured by the CL, ranging from 0 to 10, where the higher value of CL shows higher project complexity. CL is determined as in Equation 14.12:

$$CL = \sum_{i=1}^{6}\sum_{j=1}^{J}\left(w_{ij} \times k_{ij}\right)$$

(14.12)

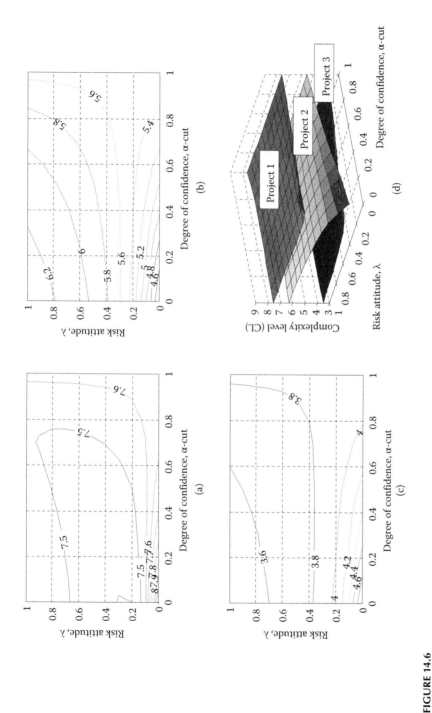

FIGURE 14.6
Sensitive analysis of the project complexity level. (a) Complexity level contours for Project 1. (b) Complexity level contours for Project 2. (c) Complexity level contours for Project 3. (d) Project complexity level.

where J is the number of the subcriteria within a criterion. With α-cut = 0.5 and λ = 0.5, the CLs for Projects 1, 2, and 3 are 7.47, 5.81, and 3.72, respectively, on a scale 0–10.

Figure 14.6 presents the results of sensitivity analysis of complexity of Projects 1, 2, and 3 with regard to α-cut and λ. The CLs of Projects 1, 2, and 3 range from 7.40 to 8.31, 4.23 to 6.31, and 3.52 to 5.01, respectively. Project 1 is more complex than Project 2 and Project 2 tends to be more complex than Project 3 although there is some overlap in the CLs of Projects 2 and 3 when α-cut and λ are small (Figure 14.6d). This implies that construction of Project 1 would be more problematic. As such, CLs should be an important measure for decision makers in construction firms to properly allocate their scarce resources among projects in their current project portfolio.

14.5 Conclusions

A wide range of FAHP applications have taken place in construction engineering and management in both research and practice for decades. These applications include project site selection, contractor selection and bid evaluation, selection of construction means and methods, risk assessment and management, and many other areas. Previous studies show that FAHP can also be used with other methods to develop a number of decision support systems that are useful for MCDM problems in construction management. As illustrated in this chapter, FAHP was employed to quantify the complexity of transportation construction projects. The illustrative example shows that FAHP is an effective technique to deal with uncertain, subjective, and linguistic data in construction management. The CL, as a single indicator of project complexity, provides top management personnel and decision makers in construction firms a general picture with regard to the degree of complexity of different projects in their portfolio. This result in turn provides leaders and managers with effective strategies to best allocate resources in their current and future projects.

References

Akadiri, P. O., Olomolaiye, P. O., and Chinyio, E. A. (2013). Multi-criteria evaluation model for the selection of sustainable materials for building projects. *Automation in Construction*, 30, 113–125.

Andrić, J. M. and Lu, D. G. (2016). Risk assessment of bridges under multiple hazards in operation period. *Safety Science*, 83, 80–92.

Bakhshi, J., Ireland, V., and Gorod, A. (2016). Clarifying the project complexity construct: Past, present and future. *International Journal of Project Management*, 34(7), 1199–1213.

Büyüközkan, G. and Feyzioğlu, O. (2004). A fuzzy-logic-based decision-making approach for new product development. *International Journal of Production Economics*, 90(1), 27–45.

Chang, C.-W., Wu, C.-R., and Lin, H.-L. (2009). Applying fuzzy hierarchy multiple attributes to construct an expert decision making process. *Expert Systems with Applications*, 36(4), 7363–7368.

Chou, J. S., Pham, A. D., and Wang, H. (2013). Bidding strategy to support decision-making by integrating fuzzy AHP and regression-based simulation. *Automation in Construction*, 35, 517–527.

Dey, P. K., Ogunlana, S. O., and Takehiko, N. (2002). Risk management in build-operate-transfer projects. *International Journal of Risk Assessment and Management*, 3(2–4), 269–291.

Ebrahimian, A., Ardeshir, A., Rad, I. Z., and Ghodsypour, S. H. (2015). Urban stormwater construction method selection using a hybrid multi-criteria approach. *Automation in Construction*, 58, 118–128.

Hass, K. B. (2009). *Managing Complex Projects: A New Model*. Vienna, VA: Management Concepts.

He, Q., Luo, L., Hu, Y., and Chan, A. P. C. (2015). Measuring the complexity of mega construction projects in China—A fuzzy analytic network process analysis. *International Journal of Project Management*, 33(3), 549–563.

Hsueh, S. L., Lee, J. R., and Chen, Y. L. (2013). DFAHP multicriteria risk assessment model for redeveloping derelict public buildings. *International Journal of Strategic Property Management*, 17(4), 333–346.

Jaskowski, P., Biruk, S., and Bucon, R. (2010). Assessing contractor selection criteria weights with fuzzy AHP method application in group decision environment. *Automation in construction*, 19(2), 120–126.

Jato-Espino, D., Castillo-Lopez, E., Rodriguez-Hernandez, J., and Canteras-Jordana, J. C. (2014). A review of application of multi-criteria decision making methods in construction. *Automation in Construction*, 45, 151–162.

Kaimann, R. A. (1974). Coefficient of network complexity. *Management Science*, 21(2), 172–177.

Kuo, R. J., Chi, S. C., and Kao, S. S. (1999). A decision support system for locating convenience store through fuzzy AHP. *Computers and Industrial Engineering*, 37(1), 323–326.

Lam, K. C., Lam, M. C. K., and Wang, D. (2008). MBNQA-oriented self-assessment quality management system for contractors: Fuzzy AHP approach. *Construction Management and Economics*, 26(5), 447–461.

Li, J. and Zou, P. X. (2011). Fuzzy AHP-based risk assessment methodology for PPP projects. *Journal of Construction Engineering and Management*, 137(12), 1205–1209.

Liou, T. S. and Wang, M. J. J. (1992). Ranking fuzzy numbers with integral value. *Fuzzy Sets and Systems*, 50(3), 247–255.

Mulholland, B. and Christian, J. (1999). Risk assessment in construction schedules. *Journal of Construction Engineering and Management*, 125(1), 8–15.

Nasab, H. H. and Ghamsarian, M. M. (2015). A fuzzy multiple-criteria decision-making model for contractor prequalification. *Journal of Decision Systems*, 24(4), 433–448.

Nassar, K. M. and Hegab, M. Y. (2006). Developing a complexity measure for schedules. *Journal of Construction Engineering and Management*, 132(6), 554–561.

Nassar, K. and Hosny, O. (2013). Fuzzy clustering validity for contractor performance evaluation: Application to UAE contractors. *Automation in Construction*, 31, 158–168.

Nguyen, A. T., Nguyen, L. D., Le-Hoai, L., and Dang, C. N. (2015). Quantifying the complexity of transportation projects using the fuzzy analytic hierarchy process. *International Journal of Project Management*, 33(6), 1364–1376.

Öztaş, A. and Ökmen, Ö. (2005). Judgmental risk analysis process development in construction projects. *Building and Environment*, 40(9), 1244–1254.

Pan, N. F. (2008). Fuzzy AHP approach for selecting the suitable bridge construction method. *Automation in Construction*, 17(8), 958–965.

Qazi, A., Quigley, J., Dickson, A., and Kirytopoulos, K. (2016). Project complexity and risk management (ProCRiM): Toward modelling project complexity driven risk paths in construction projects. *International Journal of Project Management*, 34(7), 1183–1198.

Saaty, T. L., 1980. *The Analytic Hierarchy Process*. New York, NY: McGraw-Hill.

Shi, S. S., Li, S. C., Li, L. P., Zhou, Z. Q., and Wang, J. (2014). Advance optimized classification and application of surrounding rock based on fuzzy analytic hierarchy process and tunnel seismic prediction. *Automation in Construction*, 37, 217–222.

Temperley, H. N. V. (1982). *Graph Theory and Applications*. Somerset, NJ: John Wiley & Sons.

Tesfamariam, S. and Sadiq, R. (2006). Risk-based environmental decision-making using fuzzy analytic hierarchy process (F-AHP). *Stochastic Environmental Research and Risk Assessment*, 21(1), 35–50.

Tran, D. and Molenaar, K. R. (2015). Risk-based project delivery selection model for highway design and construction. *Journal of Construction Engineering and Management*, 141(12), 04015041.

Tran, D., Molenaar, K. R., and Alarcon, L. F. (2016). A hybrid cross-impact approach to predicting cost and variance of project delivery decisions for highways. *Journal of Infrastructure Systems*, 22(1), 04015017.

Vahidnia, M. H., Alesheikh, A. A., and Alimohammadi, A. (2009). Hospital site selection using fuzzy AHP and its derivatives. *Journal of Environmental Management*, 90(10), 3048–3056.

van Laarhoven, P. J. M. and Pedrycz, W. (1983). A fuzzy extension of Saaty's priority theory. *Fuzzy Sets and Systems*, 11(1), 199–227.

Vidal, L. A., Marle, F., and Bocquet, J. C. (2011). Using a Delphi process and the analytic hierarchy process (AHP) to evaluate the complexity of projects. *Expert Systems with Applications*, 38(5), 5388–5405.

Yu, J. and Liu, Y. (2012). Prioritizing highway safety improvement projects: A multicriteria model and case study with safety analyst. *Safety Science*, 50(4), 1085–1092.

Zeng, J., An, M., and Smith, N. J. (2007). Application of a fuzzy based decision making methodology to construction project risk assessment. *International Journal of Project Management*, 25(6), 589–600.

Zhang, G. and Zou, P. X. (2007). Fuzzy analytical hierarchy process risk assessment approach for joint venture construction projects in China. *Journal of Construction Engineering and Management*, 133(10), 771–779.

Zheng, G., Zhu, N., Tian, Z., Chen, Y., and Sun, B. (2012). Application of a trapezoidal fuzzy AHP method for work safety evaluation and early warning rating of hot and humid environments. *Safety Science*, 50(2), 228–239.

15

Plantation Land Segmentation for an Orchard Establishment Using FAHP

Theagarajan Padma, S. P. Shantharajah, and Shabir Ahmad Mir

CONTENTS

15.1 Introduction

The decision-making process aims to understand conflicts that occur due to various vital factors such as diverse opinions, fluctuating environmental conditions, subjective assessments, etc., that arise in many of the important fields such as health care, agriculture, textiles, finance, defense, business, and many more. In real life, decisions have often been made under various alternatives with their associated criteria. An inappropriate final decision may cause an unpleasant outcome with undesired results such as abuse of resources, manpower, and money, as well as valuable time. Hence, it is important to achieve an optimal decision in real-world problems that involve multiple alternatives, using major and subcriteria in qualitative and quantitative domains.

Orchard establishment is a long-term investment and deserves very critical planning. The selection of a proper location and site, the planting system and planting distance, choosing the varieties, and the nursery plants have to be considered carefully to ensure maximum production (Kumar, 1997). Proper planning includes evaluating soil type, selection of elite rootstock used for fruit trees, irrigation facility, fertilization, practices that influence the water and nutrient supply to the plant, the agro-ecological situation, maintaining soil pH and salinity, integrated production practices, insect and disease management, evaluations of business goals, management style, site characteristics, market potential, and more (Marcela et al., 2015; Ahumada et al., 2012; Català et al., 2013; Zhang and Wilhelm, 2011). Farm segmentation for fruit growing is a multidisciplinary process. Due to the involvement of heterogeneous farm management practices that are exceedingly uncertain and complex, and involve diverse processes of combined experience, comprised of multiple disciplines and persons involving multiple levels of abstraction, a farm segmenting model can be described as a spatial, dynamic, and nonlinear programming model. This farm allocation model can be used to assist farmers to assess various farm cropping scenarios in terms of economic return and ecological effects; determine environmentally optimal irrigation methods; follow efficient utilization of the farm regardless of various constraints; guide in producing commodities with appropriate methods and suitable seasonal time periods; and manage fluctuations in product price and planting cost, unexpected seasonal changes, calamities, pests, plant diseases, etc., so that the produce yield can be maximized and the farmers gain economic benefits (Kanellopoulos et al., 2014; Sulaiman et al., 2014; Wu et al., 2014; Lopez-Luquea et al., 2015; Gregorio et al., 2013). Thus, orchard establishment is a very imprecise and vague situation comprising various criteria having different levels of abstraction. Therefore, the situation cannot be dealt with using the common quantitative methods. Therefore, this research applies the fuzzy analytic hierarchy process (FAHP) method, which is one of the multicriteria decision-making (MCDM) techniques to guide the evaluation and assessment of various factors deemed vital for orchard establishment.

Numerous MCDM techniques have been developed to date. The basic MCDM technique is the analytic hierarchy process (AHP) developed by Prof. Thomas L. Saaty, who defines AHP as a decision-making method that decomposes a complex multicriteria decision problem into a hierarchical structure consisting of criteria, subcriteria, and alternatives cascading from the decision objective and employs a weighted linear average algorithm to prioritize the alternatives in the presence of multiple decision criteria using informed judgments (Saaty, 1990). The preference of the decision maker in the criteria for each pairwise comparison is rated on a crisp point scale of 1/9–9. For problems with uncertain data, AHP will not provide a solution. It is often criticized for its lack of ability to adequately handle the intrinsic uncertainty and imprecision associated with relating the decision maker's perception to exact numbers (Ayağ, 2005). Decision makers usually feel more confident to give interval judgments rather than expressing their judgments in the form of single numeric values. Such uncertainties are addressed by fuzzy set theory (FST), which was proposed and used by Zadeh (1988). FST emphasized the thoughts, inference, and cognitions of surroundings of humans and can be used as a modeling tool for uncertain and complex systems that are difficult to accurately define. In FST, the concept of a membership function is used to describe the solutions to uncertain and vague problems. Thus, FST is introduced into the pairwise comparison to deal with the deficiency in the traditional AHP where crisp judgments are transformed into fuzzy judgments. This is referred to as FAHP.

This research makes an attempt to formulate an orchard segmentation model based on FAHP that assists farmers in decision-making while allocating land in an orchard establishment for different kinds of fruits, considering investment and economic feasibility, locational suitability, market availability, and availability of technical know-how.

The remainder of this chapter is organized as follows. Section 15.2 presents a review of literature for the application of FAHP in an agricultural domain. Section 15.3 presents the fuzzy extent analysis method used in the study. The suitability of using FAHP in orchard establishment has been discussed in Section 15.4. Section 15.5 presents discussion and results of an orchard segmentation model. Section 15.6 presents the conclusion and direction for future results.

15.2 Review of Literature

Numerous authors have recently presented different ranking methods to rank alternatives in different areas of agriculture using fuzzy methods. Shabir et al. prioritized opportunities and constraints for cultivation of rice under temperate climatic conditions and demonstrated how different alternatives can be evaluated and ranked under uncertain and vague conditions (Mir and Padma, 2016). Sánchez-Moreno et al. used FAHP for comparison between three land suitability assessments for upland rice and rubber (Sánchez-Moreno, 2014). Srdjevican and Medeiros applied FAHP for assessment of water management plans (Srdjevic and Medeiros). Alavi used a FAHP method for plant species selection in mine reclamation plans and presented a case study of a Sungun copper mine (Alavi, 2014). A FAHP approach for agricultural production planning that was applied in the Atrak watershed, Iran, has also successfully attained feasible solutions amidst vague situations. (Mohaddes and Mohayidin, 2008). Morteza et al. applied FAHP to assess objectives and attributes of the location indicators of Agricultural Service Centers (Morteza et al., 2015). A FAHP method was used to rank alternatives to find the most reasonable and efficient method of agricultural reservoir water resources assessment by Choi et al. (2013). Soliman et al. determine the optimal policy for regional sustainability development under climate change in the agriculture sector (Soliman et al., 2011). Edward et al. developed a new prioritization methodology for climate change variation in proposing various adaptation approaches in developing countries (Edward et al., 2014). Abdul et al. determined prioritization of a watershed within a large drainage basin that is required for proper planning and management of natural resources for sustainable development (Abdul et al., 2015). Weerawat et al. applied FAHP to analyze the suitable area for rubber plantation enlargement in Thailand by eco-efficiency assessment (Weerawat et al., 2012). Aliasghar et al. found out the precise performance assessment of irrigation projects by applying FAHP (Aliasghar et al., 2013). Goran et al. investigated a greenhouse microclimatic environment controlled by a mobile measuring station with the aim of improving performance by using wireless sensor networks technology; here the control strategy selection system is supported by FAHP (Goran and Janos, 2014). Makram et al. describe a methodology to rank suitable sites for irrigation with treated waste water using FAHP based on the geographical information system (GIS) where the Nabeul–Hammamet aquifer catchment (Tunisia) is selected as the target area (Makram et al., 2012). The study of various aspects of river basins to find the most efficient use of a water system using FAHP was proposed by Alias et al. (2009). In addition, FAHP has also been used for priority setting in agricultural land-use planning (Akpinar et al, 2009), evaluation of important risk factors in agriculture (Toledo et al., 2011), location indicator development for an agricultural service center (Zangeneh et al., 2015), comprehensive evaluation of drought vulnerability assessment (Jing and Jian-ping, 2010), forage selection (Juan et al., 2004), crop area planning (Gupta et al., 2000), and comparable global change vulnerability assessments (Colin et al., 2007). The studies were carried out to assess and prioritize complex and vague factors in order to equip decision makers with more precise and important decision information. On the other hand, measuring intellectual capital using FAHP was studied by Chen (2009). This literature reveals a significant finding that triangular fuzzy numbers (TFNs) are used to represent vague data or linguistic information by all of the researchers.

15.3 Multicriteria Comparison of Land Segmentation in Orchard Founding Using FAHP

A fuzzy set is a class of objects with a continuum of grades of membership. Such a set is characterized by a membership (characteristic) function that operates over the range of real numbers [0, 1]. The main characteristic of fuzziness is the grouping of individuals into classes that do not have sharply defined boundaries. The uncertain comparison judgment can be represented by the fuzzy number. The TFN used as the membership function is illustrated in Figure 15.1. A TFN is the special class of fuzzy number whose membership function is defined by the triplet (l, m, u) defined as in Equation 15.1. TFNs help the decision maker to make easier decisions.

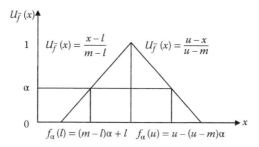

FIGURE 15.1
Left and right representation of triangular fuzzy number.

$$U(x) = \begin{cases} (x-l)/(m-l) & l \le x \le m \\ (u-x)/u-m) & m \le x \le u \\ 0 & \text{otherwise} \quad u < x > l \end{cases} \tag{15.1}$$

The calculation of fuzzy numbers can be done according to the extension principle of TFNs.

A fuzzy number can always be given by its corresponding left and right representation of each degree of membership:

$$\tilde{M} = (f_\alpha(l), f_\alpha(u)) = ((m-l)\alpha + l, u - (u-m)\alpha), y \in [0,1] \tag{15.2}$$

where $f_\alpha(l)$ and $f_\alpha(u)$ denote the left-side representation and the right-side representation of a fuzzy number, respectively.

Introduced originally by Chang (1996), the extent analysis method for FAHP is used in this study to understand the priority weights of different major and subcriterion of an orchard land partitioning problem. The notations used in this study are given hereunder.

Let $X = \{x_1, x_2, ..., x_n\}$ be the set of objectives and $U = \{u_1, u_2, \cdots, u_m\}$ be a set of goals. According to the method of Chang's extent analysis, each object is taken separately and extent analysis for each goal g_i is performed one by one, respectively. Therefore, m extent analysis values for each object can be obtained, with the following signs:

$$M_{g_i}^1, M_{g_i}^2, ..., M_{g_i}^m, \quad i = 1, 2, ..., n \tag{15.3}$$

where all the $M_{g_i}^j (j = 1, 2, ..., m)$ are TFNs.

The steps of Chang's extent analysis can be given as follows:

Step 1: The value of the fuzzy synthetic extent with respect to the *i*th object is defined as

$$S_i = \sum_{j=1}^{m} M_{g_i}^j \otimes \left[\sum_{i=1}^{n} \sum_{j=1}^{m} M_{g_i}^j \right]^{-1} \tag{15.4}$$

To obtain $\sum_{j=i}^{m} M_{g_i}^j$, perform the fuzzy addition operation of m extent analysis values for a particular matrix such that

$$\sum_{j=i}^{m} M_{g_i}^j = \left(\sum_{j=1}^{m} l_j, \sum_{j=1}^{m} m_j, \sum_{j=1}^{m} u_j \right) \tag{15.5}$$

To obtain $\left[\sum_{i=1}^{n} \sum_{j=1}^{m} M_{g_i}^j \right]^{-1}$ perform the fuzzy addition operation of $M_{g_i}^j$ $(j = 1, 2, \ldots, m)$ values such that

$$\sum_{i=1}^{n} \sum_{j=1}^{m} M_{g_i}^j = \left(\sum_{i=1}^{n} l_i, \sum_{i=1}^{n} m_i, \sum_{i=1}^{n} u_i \right) \tag{15.6}$$

and then compute the inverse of the vector in Equation 15.6 such that

$$\left[\sum_{i=1}^{n} \sum_{j=1}^{m} M_{g_i}^j \right]^{-1} = \left(\frac{1}{\sum_{i=1}^{n} u_i}, \frac{1}{\sum_{i=1}^{n} m_i}, \frac{1}{\sum_{i=1}^{n} l_i} \right). \tag{15.7}$$

Step 2: The degree of possibility of $M_2 = (l_2, m_2, u_2) \geq M_1 = (l_1, m_1, u_1)$ is defined as

$$V(M_2 \geq M_1) = \sup_{y \geq x} \left\lfloor \min(\mu_{M_1}(x), \mu_{M_2}(y)) \right\rfloor \tag{15.8}$$

and can be equivalently expressed as follows:

$$V(M_2 \geq M_1) = \text{hgt}(M_1 \cap M_2) = \mu_{M_2}(d)$$

$$= \begin{cases} 1, & \text{if } m_2 \geq m_1 \\ 0, & \text{if } l_1 \geq u_2 \\ \dfrac{l_1 - u_2}{(m_2 - u_2) - (m_1 - l_1)}, & \text{otherwise} \end{cases} \tag{15.9}$$

where d is the ordinate of the highest intersection point D between μ_{M_1} and μ_{M_2} (Figure 15.2). To compare M_1 and M_2, both the values of $V(M_1 \geq M_2)$ and $V(M_2 \geq M_1)$ are needed.

Step 3: The degree of possibility that a convex fuzzy number is greater than k convex fuzzy numbers $M_i(i=1,2,\ldots,k)$ can be defined by

$$V(M \geq M_1, M_2, \ldots, M_k)$$

$$= V[(M \geq M_1) \text{ and } (M \geq M_2) \text{ and } L \text{ and } V(M \geq M_k)]$$

$$= \min V(M \geq M_i), \quad i=1,2,3,\ldots,k. \tag{15.10}$$

Assume that

$$d'(A_i) = \min(S_i \geq S_k). \tag{15.11}$$

for $k=1,2,\ldots,n; k \neq i$. Then the weight vector is given by

$$W' = (d'(A_1), d'(A_2), \ldots, d'(A_n))^T \tag{15.12}$$

Here $A_i(i=1,2,\ldots,n)$ are n elements.

Step 4: Via normalization, the normalized weight vectors are

$$W = (d(A_1), d(A_2), \ldots, d(A_n))^T \tag{15.13}$$

where W is a nonfuzzy number.

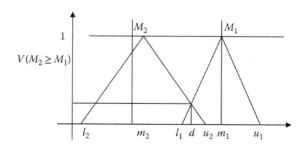

FIGURE 15.2
The intersection between M_1 and M_2.

15.4 Suitability of FAHP in Orchard Establishment

The agricultural domain presents an interdisciplinary approach involving various domains such as climate, soil, meteorology, land suitability, etc. Mostly the agricultural domain includes uncertainty, ambiguity, and incomplete information, which is vague in nature. The orchard land allocation problem considers many factors like investment, economic criteria, location suitability, market availability, technical components, etc. All of these factors can be considered as separate groups and can have various characteristics, and these characteristics, are not always equally important; the contribution of each criterion differs significantly toward the goal of land allocation. The relative level of involvement of various criteria can be highly addressed when they are grouped in hierarchies of various stages. The orchard land allocation problem entails the most important decisions at various stages for deciding about the allocation of a land. FAHP deals with problems that are multidimensional in nature and is also capable of managing and representing uncertainty (Chan and Kumar, 2005; Padma and Balasubramanie, 2011; Ghodsypour and O'Brien, 1998), ensuring that the subjective and conceptual information is valued. It assists in providing solutions to issues that are crucial in agriculture and hence have great applicability in the agricultural domain, helping farmers to make the right decisions for their farming process. The ability of FAHP in combining different types of input data and the uncertainty method of pairwise comparisons were used to simultaneously compare multiple parameters for the purposes of classifying land allocation for establishing an orchard for various alternatives considered in this study. The use of FAHP makes it possible to improve the analysis and prioritization toward achieving the goal of having a defined hierarchical structure of the problem domain.

15.5 Multiattribute Comparisons for Plantation of Several Alternative Fruits

An orchard, a method of fruit growing, is one of the most important and profitable branches of agriculture. It has been practiced in most parts of the world since times. The level of affluence of people of countries is evaluated by the production and per capita conservation of fruits. The economics of orchard establishment include a number of issues that need to be decided before the trees are even ordered including cultivar, rootstock, training system, trees per acre, and method of orchard founding, for example, soil fumigation, rotation crops, etc., which economically influence factors such as prices, yields, and costs. Following are the points that show the fiscal importance of an orchard. (1) Per unit yields are high: from a well-established and maintained orchard, more yield and income is realized than any of the field crops. (2) High net profits: though the initial cost of establishment of an orchard is high, it is compensated by higher productivity and value. (3) Source of raw material for the agro-based industries: fruit farming provides raw material for various agro-based industries like canning and fresh fruit preservation. (4) Efficient utilization of resources: unlike seasonal crops, fruit growing being perennial in nature enables the grower to remain engaged throughout the year in farm operations and utilizes the resources and assets entirely. (5) Utilization of waste and barren lands for production: although most of the fruit crops require perennial and good soils for production, there are many fruit that grow in lands that are hardy in nature with proper upkeep. (6) Ability of earning foreign exchange: fresh fruits, processed products, and spices earn a good amount of foreign exchange through export to various countries.

Orchard establishment also provides social benefits and well-being of human life in a variety of ways. Planting of trees maintains ecological balance and increases precipitation within the locality, while reducing soil erosion, silting tanks, and air pollution. Fruit tree faming being a highly intensive and skillful enterprise generates employment. Trees give much-needed oxygen and seize carbon dioxide, increase biodiversity, fix nitrates into soil making, and make it more fertile to grow other plants, like vegetables, provide nutritious fruit to eat, and improve an area's water quality.

Though orchard founding raises the well-being of a country in several aspects, creating a new orchard needs proper planning of several critical characteristics. Variety, rootstock, training system, and planting spacing all need to be decided well ahead of time (Burcu et al., 2012; Wearing et al., 2012; Adel et al., 2015). Factors such as weather conditions, soil type, soil pests, water quality and its availability, irrigation system, and pest pressures can all influence the decision (Ayars et al., 2015; Michele and Zhenli, 2014; Patrik et al., 2012). Even management factors including labor availability, labor skill level, packing options, and marketing outlets should be considered (Dvoralai et al., 2012; Deborah, 2012; Xiangping et al., 2012; Mireille et al., 2014; Florian and Michael, 2012). In some cases, corrective measures such as ripping, soil pH adjustment, installing tile drainage, land leveling, upgrading the irrigation system, or soil fumigation may be needed before planting (Beng et al., 2012; John et al., 2013; Brian, 2012; Weiwen et al., 2014; Charles et al., 2014). All these factors might interact in subtle ways.

When a farmer plans to establish an orchard on a farm and wishes to decide about the segment of the farm that should be allocated for different fruits, a range of criteria considered for planting are decided through a knowledge acquisition process performed through literature analysis, as well as traditional and concept mapping interviews with domain experts comprising of horticulturists, agricultural economists, and entomologists to identify major and critical characteristics. Hybrid knowledge acquisition methodologies ensure thorough coverage of the knowledge necessary to identify the deciding critical characteristics in land segmentation. The major criteria identified include investment, economic criteria, location suitability, market availability, and technical criteria. These criteria are very likely not independent. Also the array of critical characters recognized within each major criterion as given in Figure 15.3 may have tremendous variability from fruit to fruit; a viable means for evaluating and finding the relative degree of significance of the major criterion and the critical characters in planting a fruit is necessary. Thus, it is important to carry out a methodical evaluation. The FAHP model has been used to determine relative measures of significance and priority weights for different critical characters while planting fruits in an orchard.

The first step in FAHP is to develop representation of the problem in terms of overall goal, criteria, subcriteria, and decision alternatives. Figure 15.3 is the hierarchical representation for the problem of farm segmentation to various fruits in establishing an orchard.

The top level of the hierarchy represents the overall goal that determines the portion of the farm to be segmented for a different variety of fruits. The second level shows five major criteria. The third level explains subcriteria that are critical characteristics in contributing the achievement of the overall goal. The five decision alternatives that are to be planted in the orchard are apple, pear, cherry, plum, and peach as shown at the bottom level.

Since land segmentation is basically determined by subjective perceptions and thoughts toward each evaluated criterion, the fuzzy MCDM approach can be more suitable to elucidate how farmers make decisions in best distributing a land for an orchard. While crisp data are not enough to model the real-life situations in MCDM, this work applies linguistic variables to specifically describe the degrees of a criterion in order to facilitate subjective assessment by the decision makers using fuzzy numbers. A linguistic variable is a variable that uses natural language to describe its degree of value and the kind of expression used to compare each criteria is shown in Table 15.1.

The fuzzy comparison judgments with respect to the overall goal are shown in Table 15.2. Following that, the decision maker compares the critical characteristics with respect to major criteria. Table 15.3 gives the fuzzy comparison data of the critical characteristics of investment criterion. The other matrices of pairwise comparisons of economic criteria, location suitability, market availability, and technical criteria and the weight vector of each matrix are given in Tables 15.4 through 15.7. The three agricultural field experts compared the fruit varieties with respect to the critical characteristics and the results are given in Tables 15.8 through 15.12.

The weight vector from Table 15.2 is calculated and the results of the main criteria reveal that the investment is the most important criteria followed by economic criteria and locational suitability. Market availability ranked fourth and technical criteria are regarded as the least important among the criteria factors.

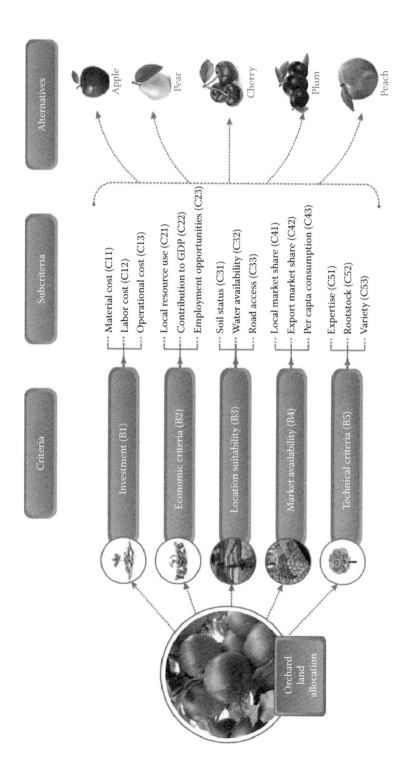

FIGURE 15.3
Hierarchy of farm segmentation in orchard foundation.

TABLE 15.1

Linguistic Variables for Importance of the Criteria

Linguistic Variable	Fuzzy Numbers	Domain	Triangular Fuzzy Number Scale (l, m, u)
Equally important	$\tilde{1}$	$1 \leq x \leq 3$	(1, 1, 3)
Weakly important	$\tilde{3}$	$1 \leq x \leq 5$	(1, 3, 5)
Strongly important	$\tilde{5}$	$3 \leq x \leq 7$	(3, 5, 7)
Very strongly important	$\tilde{7}$	$5 \leq x \leq 9$	(5, 7, 9)
Extremely important	$\tilde{9}$	$7 \leq x \leq 9$	(7, 9, 9)

TABLE 15.2

The Fuzzy Evaluation Matrix with Respect to the Main Criteria

Goal	B1	B2	B3	B4	B5	Weight
B1	1,1,1	4.20,5.25,6.47	3.70,4.20,5.25	0.13,0.14,0.17	3.80,4.20,4.80	0.32
B2	0.15,0.19,0.24	1,1,1	4.55,5.54,6.22	3.43,4.23,5.12	0.10,0.14,0.20	0.19
B3	0.19,0.24,0.27	0.16,0.18,0.22	1,1,1	4.12,5.37,6.22	3.20,4.20,5.20	0.19
B4	5.88,7.14,7.69	0.20,0.24,0.29	0.16,0.19,0.24	1,1,1	1.20,2.30,2.70	0.16
B5	0.21,0.24,0.26	5.00,7.14,10.0	0.19,0.24,0.31	0.37,0.43,0.83	1,1,1	0.14

The weight vector of investment criteria reflects that material cost and operational cost are most important from an investment point of view with a weight of 0.37 each: $W_{B1} = (0.37, 0.26, 0.37)^T$.

A cursory look at the economic subcriteria shown in Table 15.4 envisages that local resource use and employment opportunities are more important than that of contribution to the gross domestic product (GDP).

It is evident from the locational suitability subcriteria that soil status is of paramount importance followed by water availability. Access to the road is of least priority.

Priority weights assigned to evaluated subcriteria for market suitability reflect that all types of markets are necessary and are preferred equally. Local market and export market and per capita consumption are equally important due to the fact that based on the quality and the requirements of fruits, they are exported or sent to the local market.

Results of the technical subcriteria highlight the fact that expertise and variety are of significant importance while rootstock is the least important issue.

In order to determine the priority weights for the farm segmentation of fruits, the combination of relative precedence for major criteria, critical characteristics, and alternative variety of fruits is presented in Table 15.13. The results of the FAHP analysis envisage that about 38% of the farm should be allocated for apple fruit plantation, 21% for pear fruit plantation, about 19% for cherry fruit plantation, 10% for plum fruit plantation, and 12% for peach fruit plantation.

TABLE 15.3

Evaluation of Critical Characteristics with Respect to Investment Criteria

B1	C11	C12	C13	Weight
C11	1,1,1	4.90,5.50,6.30	0.14,0.13,0.11	0.37
C12	0.16,0.18,0.20	1,1,1	3.45,4.75,5.85	0.26
C13	9.00,8.00,7.00	0.17,0.21,0.29	1,1,1	0.37

TABLE 15.4

Evaluation of Critical Characteristics with Respect to Economic Criteria

B2	C21	C22	C23	Weight
C21	1,1,1	5.88,7.41,8.54	1.23,3.48,5.80	0.39
C22	0.12,0.13,0.17	1,1,1	3.56,5.28,7.94	0.22
C23	0.17,0.29,0.81	0.13,0.19,0.28	1,1,1	0.39

TABLE 15.5

Evaluation of Critical Characteristics with Respect to Location Suitability

B3	C31	C32	C33	Weight
C31	1,1,1	6.78,7.32,8.45	0.17,0.14,0.13	0.39
C32	0.12,0.14,0.15	1,1,1	6.21,7.18,8.20	0.37
C33	8.00,7.00,6.00	0.12,0.14,0.16	1,1,1	0.25

TABLE 15.6

Evaluation of Critical Characteristics with Respect to Market Suitability

B4	C41	C42	C43	Weight
C41	1,1,1	0.17,0.14,0.13	1.80,2.50,3.70	0.33
C42	8.00,7.00,6.00	1,1,1	4.12,5.70,6.60	0.33
C43	0.27,0.40,0.56	0.15,0.18,0.24	1,1,1	0.33

TABLE 15.7

Evaluation of Critical Characteristics with Respect to Technical Subcriteria

B5	C51	C52	C53	Weight
C51	1,1,1	4.90,5.50,6.30	0.14,0.13,0.11	0.37
C52	0.16,0.18,0.20	1,1,1	3.45,4.75,5.85	0.26
C53	9.00,8.00,7.00	0.17,0.21,0.29	1,1,1	0.37

TABLE 15.8

Evaluation of Different Alternatives with Respect to Subcriteria of B1

Subcriteria	Alternatives	Apple	Pear	Cherry	Plum	Peach	Weight
C11	Apple	1,1,1	2.00,3.00,4.00	4.00,5.00,6.00	6.00,7.00,8.00	8.00,9.00,9.00	0.39
	Pear	0.20,0.33,1.00	1,1,1	1.00,3.00,5.00	5.00,7.00,9.00	3.00,5.00,7.00	0.24
	Cherry	0.14,0.20,0.33	0.20,0.33,1.00	1,1,1	7.00,8.00,9.00	5.00,7.00,9.00	0.21
	Plum	0.13,0.14,0.17	0.11,0.14,0.20	0.11,0.13,0.14	1,1,1	1.00,3.00,5.00	0.01
	Peach	0.11,0.11,0.13	0.14,0.20,0.33	0.11,0.14,0.20	0.20,0.33,1.00	1,1,1	0.15
C12	Apple	1,1,1	2.20,3.40,4.00	4.50,4.00,6.00	6.00,7.60,8.40	8.20,9.00,9.00	0.44
	Pear	0.25,0.29,0.45	1,1,1	1.40,3.00,4.70	3.35,5.50,7.65	3.65,5.20,7.45	0.21
	Cherry	0.17,0.19,0.25	0.21,0.33,0.71	1,1,1	6.7,7.7,8.7	5.70,6.70,8.70	0.19
	Plum	0.12,0.13,0.17	0.13,0.18,0.30	0.11,0.13,0.15	1,1,1	1.20,3.20,5.20	0.04
	Peach	0.11,0.11,0.12	0.13,0.19,0.27	0.11,0.15,0.18	0.19,0.31,0.83	1,1,1	0.12
C13	Apple	1,1,1	2.35,3.55,4.55	4.15,5.55,6.15	6.15,7.85,8.55	8.15,8.55,9.00	0.46
	Pear	0.22,0.28,0.43	1,1,1	1.55,3.15,4.85	3.45,5.65,7.80	3.80,5.45,7.70	0.22
	Cherry	0.16,0.18,0.24	0.21,0.32,0.65	1,1,1	6.85,7.85,8.85	5.70,6.70,8.70	0.19
	Plum	0.12,0.13,0.16	0.13,0.18,0.29	0.11,0.13,0.15	1,1,1	1.25,3.25,5.20	0.03
	Peach	0.11,0.12,0.12	0.13,0.18,0.26	0.11,0.15,0.18	0.19,0.31,0.80	1,1,1	0.11

TABLE 15.9

Evaluation of Different Alternatives with Respect to Subcriteria of B2

Subcriteria	Alternatives	Apple	Pear	Cherry	Plum	Peach	Weight
	Apple	1,1,1	1.05,1.24,1.96	1.21,1.56,1.98	1.11,1.33,1.99	1.09,1.40,2.00	0.10
	Pear	0.51,0.81,0.95	1,1,1	1.00,1.30,1.75	1.20,1.40,1.90	2.20,3.40,4.75	0.15
C21	Cherry	0.51,0.64,0.83	0.57,0.77,1.00	1,1,1	4.75,5.50,6.80	3.40,5.60,7.90	0.29
	Plum	0.50,0.75,0.90	0.53,0.71,0.83	0.15,0.18,0.21	1,1,1	4.25,6.45,8.90	0.18
	Peach	0.50,0.71,0.92	0.21,0.29,0.45	0.13,0.18,0.29	0.11,0.16,0.24	1,1,1	0.29
	Apple	1,1,1	2.10,3.30,3.80	3.80,5.20,5.80	5.80,7.40,8.20	8.10,8.80,9.00	0.42
	Pear	0.26,0.30,0.48	1,1,1	1.20,2.80,4.50	3.15,5.30,7.45	3.45,5.00,7.25	0.20
C22	Cherry	0.17,0.19,0.26	0.22,0.36,0.83	1,1,1	6.50,7.50,8.50	5.50,6.50,8.50	0.19
	Plum	0.12,0.14,0.17	0.13,0.19,0.32	0.12,0.13,0.15	1,1,1	1.00,3.00,5.00	0.05
	Peach	0.11,0.11,0.12	0.14,0.20,0.29	0.12,0.15,0.18	0.20,0.33,1.00	1,1,1	0.15
	Apple	1,1,1	2.22,3.47,4.8	4.22,5.57,6.47	6.25,7.5,8.75	8.15,8.5,9	0.46
	Pear	0.21,0.29,0.45	1,1,1	1.5,3.15,4.5	3.5,5.5,7.6	3.6,5.5,7.7	0.22
C23	Cherry	0.15,0.18,0.24	0.22,0.32,0.67	1,1,1	6.5,7.8,8.6	5.7,6.5,8.7	0.19
	Plum	0.11,0.13,0.16	0.13,0.18,0.29	0.12,0.13,0.15	1,1,1	1.5,3.5,5.5	0.05
	Peach	0.11,0.12,0.12	0.13,0.18,0.28	0.11,0.15,0.18	0.18,0.29,0.67	1,1,1	0.07

TABLE 15.10

Evaluation of Different Alternatives with Respect to Subcriteria of B3

Subcriteria	Alternatives	Apple	Pear	Cherry	Plum	Peach	Weight
	Apple	1,1,1	5.50,6.25,7.00	7.25,8.00,8.75	7.25,8.00,8.75	7.25,8.00,8.75	0.20
	Pear	0.14,0.16,0.18	1,1,1	5.50,6.25,7.00	8.00,8.75,7.25	8.00,8.75,7.25	0.20
C31	Cherry	0.11,0.13,0.14	0.14,0.16,0.18	1,1,1	3.25,4.00,4.75	3.25,4.00,4.75	0.20
	Plum	0.11,0.13,0.14	0.14,0.11,0.13	0.21,0.25,0.31	1,1,1	3.25,4.00,4.75	0.20
	Peach	0.11,0.13,0.14	0.14,0.11,0.13	0.21,0.25,0.31	0.21,0.25,0.31	1,1,1	0.20
	Apple	1,1,1	6.25,7.25,8.25	5.30,7.20,8.60	7.50,8.25,8.90	6.50,7.50,8.50	0.40
	Pear	0.12,0.14,0.16	1,1,1	3.50,5.50,7.50	3.75,5.60,7.75	2.80,3.90,4.90	0.12
C32	Cherry	0.12,0.14,0.19	0.13,0.18,0.29	1,1,1	1.60,3.30,4.50	3.15,5.50,7.75	0.24
	Plum	0.11,0.12,0.13	0.13,0.18,0.27	0.22,0.30,0.63	1,1,1	1.75,3.25,4.55	0.17
	Peach	0.12,0.13,0.15	0.20,0.26,0.36	0.13,0.18,0.32	0.22,0.31,0.57	1,1,1	0.07
	Apple	1,1,1	1.25,3.50,4.45	3.12,4.15,5.17	5.10,6.20,7.40	2.38,3.50,4.50	0.32
	Pear	0.22,0.29,0.80	1,1,1	1.60,3.30,4.50	3.15,5.50,7.75	3.50,5.25,7.50	0.28
C33	Cherry	0.19,0.24,0.32	0.22,0.30,0.63	1,1,1	6.25,7.25,8.25	5.30,7.20,8.60	0.28
	Plum	0.14,0.16,0.20	0.13,0.18,0.32	0.12,0.14,0.16	1,1,1	1.75,3.25,4.55	0.04
	Peach	0.22,0.29,0.42	0.13,0.19,0.29	0.12,0.14,0.19	0.22,0.31,0.57	1,1,1	0.09

TABLE 15.11

Evaluation of Different Alternatives with Respect to Subcriteria of B4

Subcriteria	Alternatives	Apple	Pear	Cherry	Plum	Peach	Weight
	Apple	1,1,1	3.33,4.55,5.60	5.20,6.90,7.80	7.30,8.50,9.00	8.10,8.45,9.00	0.29
	Pear	0.18,0.22,0.30	1,1,1	2.45,4.25,5.75	4.25,5.75,6.90	4.60,5.75,8.10	0.09
C41	Cherry	0.13,0.14,0.19	0.17,0.24,0.41	1,1,1	7.25,8.12,8.95	6.75,7.45,8.75	0.04
	Plum	0.11,0.12,0.14	0.14,0.17,0.24	0.11,0.12,0.14	1,1,1	2.20,3.45,4.75	0.21
	Peach	0.11,0.12,0.12	0.12,0.17,0.22	0.11,0.13,0.15	0.21,0.29,0.45	1,1,1	0.22
	Apple	1,1,1	6.50,7.75,8.50	1.25,1.75,2.25	7.25,8.00,8.75	7.25,8.00,8.75	0.49
	Pear	0.12,0.13,0.15	1,1,1	0.17,0.14,0.13	7.50,8.00,8.50	7.50,8.00,8.50	0.23
C42	Cherry	0.44,0.57,0.80	8.00,7.00,6.00	1,1,1	7.50,8.00,8.50	7.25,8.00,8.75	0.14
	Plum	0.11,0.13,0.14	0.12,0.13,0.13	0.12,0.13,0.13	1,1,1	2.50,3.00,3.50	0.02
	Peach	0.11,0.13,0.14	0.12,0.13,0.13	0.11,0.13,0.14	0.29,0.33,0.40	1,1,1	0.11
	Apple	1,1,1	2.30,3.50,4.00	4.40,5.50,6.80	5.50,7.50,8.25	8.15,8.75,9.00	0.50
	Pear	0.25,0.29,0.43	1,1,1	1.50,2.50,4.20	3.15,5.50,7.75	3.50,5.25,7.50	0.23
C43	Cherry	0.15,0.18,0.23	0.24,0.40,0.67	1,1,1	6.25,7.25,8.25	5.30,7.20,8.60	0.20
	Plum	0.12,0.13,0.18	0.13,0.18,0.32	0.12,0.14,0.16	1,1,1	1.75,3.25,4.55	0.02
	Peach	0.11,0.11,0.12	0.13,0.19,0.29	0.12,0.14,0.19	0.22,0.31,0.57	1,1,1	0.05

TABLE 15.12

Evaluation of Different Alternatives with Respect to Subcriteria of B5

Subcriteria	Alternatives	Apple	Pear	Cherry	Plum	Peach	Weight
	Apple	1,1,1	4.50,5.00,5.50	7.25,8.00,8.75	7.25,8.00,8.75	7.25,8.00,8.75	0.32
	Pear	0.18,0.20,0.22	1,1,1	0.50,1.00,1.50	0.75,1.25,1.75	1.50,1.75,2.50	0.17
C51	Cherry	0.11,0.13,0.14	0.67,1.00,2.00	1,1,1	0.17,0.14,0.13	0.14,0.13,0.11	0.15
	Plum	0.11,0.13,0.14	0.57,0.80,1.33	8.00,7.00,6.00	1,1,1	0.50,1.00,1.50	0.32
	Peach	0.11,0.13,0.14	0.40,0.57,0.67	9.00,8.00,7.00	0.67,1.00,2.00	1,1,1	0.04
	Apple	1,1,1	1.25,3.50,4.45	3.12,4.15,5.17	5.10,6.20,7.40	2.38,3.50,4.50	0.29
	Pear	0.22,0.29,0.80	1,1,1	6.25,7.25,8.25	5.30,7.20,8.60	7.50,8.25,8.90	0.41
C52	Cherry	0.19,0.24,0.32	0.12,0.14,0.16	1,1,1	1.60,3.30,4.50	3.15,5.50,7.75	0.05
	Plum	0.14,0.16,0.20	0.12,0.14,0.19	0.22,0.30,0.63	1,1,1	1.75,3.25,4.55	0.17
	Peach	0.22,0.29,0.42	0.11,0.12,0.13	0.13,0.18,0.32	0.22,0.31,0.57	1,1,1	0.07
	Apple	1,1,1	4.25,5.00,5.75	6.10,7.00,8.20	8.25,8.60,9.00	8.25,8.60,9.00	0.51
	Pear	0.17,0.20,0.24	1,1,1	6.00,7.00,8.00	8.00,8.50,9.00	8.00,8.50,9.00	0.29
C53	Cherry	0.12,0.14,0.16	0.13,0.14,0.17	1,1,1	0.50,1.00,1.50	0.50,1.00,1.50	0.09
	Plum	0.11,0.12,0.12	0.11,0.12,0.13	0.67,1.00,2.00	1,1,1	0.50,1.00,1.50	0.02
	Peach	0.11,0.12,0.12	0.11,0.12,0.13	0.67,1.00,2.00	0.67,1.00,2.00	1,1,1	0.09

TABLE 15.13

Summary Combination of Priority Weights

Subcriteria	Criteria Weight	Apple	Pear	Cherry	Plum	Peach
C11	0.1203	0.39	0.24	0.21	0.01	0.15
C12	0.0831	0.44	0.21	0.19	0.04	0.12
C13	0.1203	0.46	0.22	0.19	0.03	0.11
C21	0.0729	0.10	0.15	0.29	0.18	0.29
C22	0.0402	0.42	0.20	0.19	0.05	0.15
C23	0.0729	0.46	0.22	0.19	0.05	0.07
C31	0.0724	0.20	0.20	0.20	0.20	0.07
C32	0.0681	0.40	0.12	0.24	0.17	0.07
C33	0.0458	0.32	0.28	0.28	0.04	0.09
C41	0.0538	0.29	0.09	0.04	0.29	0.29
C42	0.0538	0.49	0.23	0.14	0.02	0.11
C43	0.0538	0.50	0.23	0.20	0.02	0.05
C51	0.0532	0.32	0.17	0.15	0.32	0.04
C52	0.0367	0.29	0.41	0.05	0.17	0.07
C53	0.0532	0.51	0.29	0.09	0.02	0.09
Final fruit weight		**0.38**	**0.21**	**0.19**	**0.10**	**0.12**
FAHP rank		**I**	**II**	**III**	**V**	**IV**

15.6 Conclusions

This research develops an evaluation criterion to determine the sustainability of the five varieties of fruit trees in the considered criteria and their critical characteristics. Decision-making is a cumbersome process in any increasingly complex environment. Orchard founding is one such problem where the knowledge of domain experts in various agricultural, economic, and policy making fields is necessary and a stringent system network is required for making effective decisions. This research formulates the orchard founding problem as a nonlinear multiple criteria decision-making problem under uncertainty and the FAHP-based analysis results in detailed alternative priority weights of the major and subcritical characteristics with respect to the goal of land segmentation. The results show that the proposed model is capable and flexible for application to different types of fruits to prioritize their significance level in orchard founding and it is a constructive supporting tool to farmers when establishing an orchard.

References

Abdul Rahaman S., Abdul Ajeez S., Aruchamy S., Jegankumar R. (2015) Prioritization of sub watershed based on morphometric characteristics using fuzzy analytical hierarchy process and geographical information system—A study of Kallar Watershed, Tamil Nadu. *Aquat Procedia* 4: 1322–1330.

Adel D. A.-Q., Mohamed A. A., Saleh M. I. (2015) Growth, yield, fruit quality and nutrient uptake of tissue culture-regenerated 'Barhee' date palms grown in a newly established orchard as affected by NPK fertigation. *Scientia Hort* 184: 114–122.

Ahumada O., Villalobos R., Mason N. (2012) Tactical planning of the production and distribution of fresh agricultural products under uncertainty. *Agr Syst* 112: 17–26.

Akpinar N., Talay I., Gun S. (2005) Priority setting in agricultural land-use types for sustainable development. *Renew Agric Food Syst* 20(3): 136–147.

Alavi I. (2014) Fuzzy AHP method for plant species selection in mine reclamation plans: Case study Sungun copper mine. *Iran J Fuzzy Syst* 11(5): 23–38.

Alias M. A., Hashim S. Z. M., Samsudin, S. (2009) Using fuzzy analytic hierarchy process for Southern Johor River Ranking. *Int J Adv Soft Comput Appl* 1(1): 62–76.

Aliasghar M., Omid N. G., Richard L. S. (2013) A fuzzy analytical hierarchy methodology for the performance assessment of irrigation projects. *Agric Water Manage* 121: 113–123.

Ayağ Z. (2005) A fuzzy AHP-based simulation approach to concept evaluation in a NPD environment. *IIE Trans* 37: 827–842.

Ayars J. E., Fulton A., Taylorc B. (2015) Subsurface drip irrigation in California—Here to stay? *Agric Water Manage* 157: 39–47.

Beng P. U., Danielle P. O., Sean F., David J. C., John L. H., Rai S. K., Bertram O. (2012) The effect of terrain and management on the spatial variability of soil properties in an apple orchard. CATENA 93: 38–48.

Brian B. (2012) Citrus best management practices. In: Anoop Kumar Srivastava, ed. *Advances in Citrus Nutrition*. Netherlands: Springer 391–413..

Burcu Ç., Yasemin T., Gaye K., Ercan V., Murat A., Zeki K. (2012) Magnitude and efficiency of genetic diversity captured from seed stands of *Pinus nigra* (Arnold) subsp. *pallasiana* in established seed orchards and plantations. *New Forests* 43(3): 303–317.

Català L. P., Durand G. A., Blanco A. M., Bandoni J. A. (2013) Mathematical model for strategic planning optimization in the Pome fruit industry. *Agr Syst* 115: 63–71.

Chan F., Kumar N. (2005) Global supplier development considering risk factors using fuzzy extended AHP-based approach. *Int J Manage Sci* 35: 1–15.

Chang D. Y. (1996) Applications of the extent analysis method on fuzzy AHP. *Eur J Oper Res* 95: 649–655.

Charles B., Ratna Reddy V., Conor L., Murli D., Sumit R., Rebecca M. (2014) Do water-saving technologies improve environmental flows? *J Hydrol* 518: 140–149.

Chen H. H. (2009) Measuring intellectual capital using fuzzy analytic hierarchy process. *Int J Innov Learn* 6(1): 51–61.

Choi E. H., Bae S. S., Jee H. K. (2013) Prioritization for water storage increase of agricultural reservoir using FAHP method. *J Korea Water Resour Assoc* 46(2): 171–182.

Colin P., Rob N., Brent Y. (2007) Building comparable global change vulnerability assessments: The vulnerability scoping diagram. *Glob Environ Chang* 17: 472–485.

Deborah D. (2012) Sunshine coast food production, An investigation into production, processing, marketing and distribution. Queensland Government, South East Regional Services.

Dvoralai W., Felipe A. Z., Camilla P. T., Inés Z. L., Marta G.-F. (2012) Multilevel systematic sampling to estimate total fruit number for yield forecasts. *Prec Agric* 13(2): 256–275.

Edward S. S., Allen H. H., Chia-Wei H., Momodou N. (2014) Prioritization of climate change adaptation approaches in the Gambia. *Mitig Adapt Strat Glob Chang* 19(8): 1163–1178.

Florian S., Michael B. (2012) Farming and marketing system affects carbon and water footprint—A case study using Hokaido pumpkin. *J Cleaner Prod* 28: 113–119.

Ghodsypour S. H., O'Brien C. (1998) A decision support system for supplier selection using an integrated analytic hierarchy process and linear programming. *Int J Prod Econ* 56(57): 199–212.

Goran M., Janos S. (2014) Greenhouse microclimatic environment controlled by a mobile measuring station. *NJAS Wagen J Life Sci* 70–71(6): 61–70.

Gregorio E., Pedro A. N., Rafael D., Alain B., Alejandro P.-P., María M. G.-R. (2013) Almond agronomic response to long-term deficit irrigation applied since orchard establishment. *Irrigation Sci* 31(3): 445–454.

Gupta A., Harboe R., Tabucanon M. T. (2000) Fuzzy multiple-criteria decision making for crop area planning in Narmada river basin. *Agric Syst* 36(1): 1–18.

Jing C., Jian-ping T. (2010) Fuzzy comprehensive evaluation of drought vulnerability based on the analytic hierarchy process. An empirical study from Xiaogan City in Hubei Province. *Agric Agric Sci Procedia* 1: 126–135.

John W., Johan F., Marlise E. J. (2013) Effects of soil surface management practices on soil and tree parameters in a 'Cripps Pink'/M7 apple orchard 1. Mineral nutrition. *S Afr J Plant Soil* 30(3): 163–170.

Juan S., Quangong C., Ruijun L., Wenlan J. (2004) An application of the analytic hierarchy process and fuzzy logic inference in a decision support system for forage selection. *New Zeal J Agric Res* 47(3): 327–331.

Kanellopoulos A., Reidsma P., Wolf J., Van Ittersum M. K. (2014) Assessing climate change and associated socio-economic scenarios for arable farming in the Netherlands: An application of benchmarking and bio-economic farm modeling. *Eur J Agron* 52: 69–80.

Kumar N. (1997) *Introduction to Horticulture*. Nagercoil: Rajalakshmi Publications 15.47-15.50. http://www.agritech.tnau.ac.in/horticulture/horti_orchard%20management.html

Lopez-Luquea R., Recab J., Martínezb J. (2015) Optimal design of a standalone direct pumping photovoltaic system for deficit irrigation of olive orchards. *Appl Energy* 149(1): 13–23.

Makram A., Lamia B., Atef L., Salah J. (2012) Ranking suitable sites for irrigation with reclaimed water in the Nabeul-Hammamet region (Tunisia) using GIS and AHP-multi criteria decision analysis. *Resour Conserv Recy* 65: 36–46.

Marcela C., González-Araya M. C., Wladimir E. S.-S., Luis G. A. E. (2015) Harvest planning in apple orchards using an optimization model. In: Lluis M.

Plà-Aragonés, ed. *Handbook of Operations Research in Agriculture and the Agri-Food Industry, International Series in Operations Research & Management Science* New York: Springer-Verlag 224: 79–105.

Mir S. A., Padma, T. (2016) Evaluation and prioritization of rice production practices and constraints under temperate climatic conditions using fuzzy analytical hierarchy process (FAHP). Span J Agric Res 14(4): e0909.

Mireille N., Lucie D., Claire L. (2014) Crop management, labour organization, and marketing: Three key issues for improving sustainability in organic vegetable farming. *Int J Agric Sustain* 13(3): 257–274.

Mohaddes A., Mohayidin G. (2008) Application of the fuzzy approach for agricultural production planning in a watershed, a case study of the Atrak watershed, Iran. *Am Eurasian J Agric Environ Sci* 3(4): 636–648.

Morteza Z., Asadolah A., Peter N., Alireza K. (2015) Developing location indicators for Agricultural Service Center: A Delphi–TOPSIS–FAHP approach. *Prod Manuf Res* 3(1): 124–148.

Omar S. S., Aboul E. H., Neveen I. G., Nashwa E.-B., Ruhul A. S. (2011) A model-based decision support tool using fuzzy optimization for climate change. *Rough Sets Knowl Technol Lecture Notes Comput Sci* 6954: 388–393.

Padma T., Balasubramanie P. (2011) A fuzzy analytic hierarchy processing decision support system to analyze occupational menace forecasting the spawning of shoulder and neck pain. *Expert Syst Appl* 38(12): 15303–15309.

Patrik M., Bart H., Andreas N., Jörn S., Frank H., Jesus A., Aude A., Heinrich H., José H., Gabriele M., Gérard G., Joan S., Benoit S., Andrea P., Jörg S., Esther B., Claire L., Marko B., Burkhard G., Christian S., Ursula A., Franz B. (2012) Sustainability assessment of crop protection systems: Sustain OS methodology and its application for apple orchards. *Agric Syst* 113: 1–15.

Saaty T. L. (1990) *Decision Making for Leaders*. Pittsburgh, PA: RWS Publications.

Sánchez-Moreno J. F., Farshad A., Petter P. P. (2014) Farmer or expert–A comparison between three land suitability assessments for upland rice and rubber in Phonexay District, Lao Pdr. *Ecopersia* 1(3): 235–260.

Srdjevic B., Medeiros Y. D. P. (2008) Fuzzy AHP assessment of water management plans. *Water Resour Manage* 22: 877–894.

Sulaiman H., Malec K., Maitah M. (2014) Appropriate tools of Marketing Information System for Citrus Crop in the Lattakia Region, R. A. SYRIA. *Agris On-line Papers Econom Inform* 4(3): 69–78.

Toledo R., Engler A., Ahumada V. (2011) Evaluation of risk factors in agriculture: An application of the analytical hierarchical process (AHP) methodology. *Chil J Agric Res* 71(1): 114–121.

Wearing C. H., Marshall R. R., Attfield B. A., Colhoun K. (2012) Ecology and management of the leafroller (Tortricidae) complex over ten years during establishment of an organic pipfruit orchard in Central Otago, New Zealand. *Crop Prot* 33: 82–93.

Weerawat O., Thunwadee T. S., Kitikorn C. (2012) Selection of the sustainable area for rubber plantation of Thailand by eco-efficiency. *Procedia Social Behav Sci* 40: 58–64.

Weiwen Z., Wen W., Xuewen L., Fangzhi Y. (2014) Economic development and farmland protection: An assessment of rewarded land conversion quotas trading in Zhejiang, China. *Land Use Policy* 38: 467–476.

Wu Y. Q., Weng Y. H., Hennigar C., Fullarton M. S., Lantz V. (2014) Benefit–cost analysis of a white spruce clonal seed orchard in New Brunswick, Canada. *New Forests* 46(1): 141–156.

Xiangping J. I. A., Jikun H. U. A. N. G., Zhigang X. U. (2012) Marketing of farmer professional cooperatives in the wave of transformed agro food market in China. China Econ Rev 23(3): 665–674.

Zadeh L. A. (1988) Fuzzy logic. *Computer* 21(4): 83.

Zangeneh M., Akram A., Nielsen P., Keyhani A. (2015) Developing location indicators for Agricultural Service Center: A Delphi-TOPSIS-FAHP approach. *Prod Manufact Res* 3(1): 124–148.

Zhang W., Wilhelm W. E. (2011) OR/MS decision support models for the specialty crops industry: A literature review. *Ann Oper Res* 190(1): 131–148.

16

Developing a Decision on the Type of Prostate Cancer Using a FAHP

Lazim Abdullah

CONTENTS

16.1 Introduction

One of the most prevalent deadly diseases that has impacted most of the people in the world today is cancer. There is no scarier disease than cancer in most people's minds. The disease is often thought of as an untreatable, unbearably painful disease with no cure, and undoubtedly a serious and potentially life-threatening illness. There are many patients who have been diagnosed with multiple types of cancer worldwide. In Malaysia, the five most frequent types of cancers among males, according to the statistics released in 2007, were lung, colorectal, nasopharynx, prostate, and lymphoma, while the five most common cancers among females were breast, colorectal, cervix, ovary, and lung (Omar et al., 2007). Among Malaysian males, prostate cancer is ranked the fourth most prevalent cancer and this is expected to increase in the future. The incidence of prostate cancer increases after the age of 45 years. The finding also shows that environmental and genetic factors may play an important role in the likelihood of prostate cancer (National Cancer Registry, 2002) and may need some public health intervention. Other than genetic factors, several other factors have shown to be associated with prostate cancer, such as sociodemographic factors, lifestyle, diet, occupational exposure, and medical and health status (Subahir et al., 2009; Ross et al., 2016).

There are at least three types of prostate cancer that are known thus far. The American Society of Clinical Oncology noted that adenocarcinoma is the most common type of prostate cancer, accounting for more than 95% of prostate cancer cases. Adenocarcinoma is a cancer that begins in the glandular tissue of the prostate cancer. Glands are structures that secrete fluids; in the case of the prostate gland, the gland cells secrete prostate fluid that combines with seminal fluid during ejaculation. The American Cancer Society explains that most adenocarcinomas grow slowly. In fact, many autopsies of men who died from causes other than prostate cancer actually had prostatic adenocarcinoma without their knowledge (Oncology, 2009). Adenocarcinomas of the prostate are characterized by the development of dense blastic metastases, whereas small-cell carcinomas of the prostate typically produce lytic bone metastases (Sozen et al., 2000). Sarcoma is another type of prostate cancer. An international health-care center in the United States reports that prostatic sarcoma is a rare form of prostate cancer, making up less than 0.1% of all prostate cancers. This type of prostate cancer primarily affects men who are in the age range of 35–60 years. The tumor in prostatic sarcoma often grows very large, causing obstruction in the flow of urine from the bladder out through the urethra. It may also increase the urge to urinate at night, also called nocturia. This type of cancer arises from cells that have the potential of developing into muscles, lymphatic vessels, blood, and connective tissue (Galil Medical, 2007).

Small-cell carcinoma of the prostate is the third type of prostate cancer. It is a pathologic subtype of prostate cancer with unique clinical features which account for about 1%–2% of malignancies of the prostate gland. Small-cell carcinoma of the prostate is a particularly aggressive form of prostate cancer. Brownback et al. (2009) wrote about small-cell carcinoma of the prostate. They report that small-cell carcinoma of the prostate is very rare and has been documented in only 150 cases since 1997. The prognosis of this type of prostate cancer is quite poor. The authors indicate that the median survival time of patients diagnosed with small-cell carcinoma of the prostate is between 6 and 17 months after the diagnosis. The poor prognosis is generally attributed to the tendency for this form of prostate cancer to metastasize early in the disease, spreading to other sites within the body (Brownback et al., 2009).

All types of prostate cancer typically share almost common symptoms. The most common symptom of prostate cancer is difficulty in urinating. The prostate is located underneath the bladder and surrounds a portion of the urethra, a small tube that carries urine from the bladder out of your body. When the prostate becomes enlarged because of cancerous cell growth, the urethra can be pinched, which can lead to urinary problems. It can be difficult to begin urinating, even if the bladder is full. Very often patients with prostate cancer experience a weak flow of urine when urinating. The weak flow of urine may prevent patients from fully emptying the bladder. Certain men with prostate cancer experience sensations of burning or pain while urinating that can be accompanied by blood in the urine (hematuria).

They also experience difficulty in starting the urine stream (hesitancy) and urinate excessively at night (nocturia).

The identification of symptoms for the specific type of prostate cancer is a very tricky process. There is no clear indication to precisely identify the type of prostate cancer just by observing the most common symptoms. In other words, the observed symptoms and types of prostate cancer are the two layers of attributes that need to be concurrently considered. Therefore, the multiple symptoms of prostate cancer and its common multiple symptoms can be assumed as a multicriteria decision-making (MCDM) problem. It is very difficult to determine the type of cancer just based on the observed symptom due to no clear one-to-one mapping between symptom and type of prostate cancer. However, experts' opinions are definitely very critical in suggesting the type of cancer based on the prevailing symptoms. Many medical decisions depend on collaborative efforts among experts. Consensus opinions really help in identifying the right type of cancer.

A major task of medical experts of prostate cancer, such as urologists, is to treat the disease. However, the treatment process is not a direct and simple task because the information available to medical experts about their patients and about medical relationships in general is inherently uncertain (Adlassnig, 1986). Many approaches and theories have been proposed to offer solutions to the problems of the multiple symptoms and types of prostate cancer. Abdullah and Sulaiman (2016), for example, identified the most likely type of prostate cancer using the analytic hierarchy process (AHP). However, decision-making always inherits some extent of vagueness and uncertainty. One of the popular theories that deal with uncertainty, and has been widely used in medical research, is the fuzzy set theory (Cuit et al., 1990; Zadeh, 1965). The fuzzy set theory combined with the knowledge of decision-making has prompted a lot of new research into diagnosis thanks to the uncertain and vague information pertaining to the disease and also the symptoms. One of the MCDM methods that successfully combined with fuzzy sets is the fuzzy analytic hierarchy process (FAHP). The FAHP was introduced to handle uncertainty in linguistic judgment. Initial research on FAHP was conducted by Van Laarhoven and Pedrycz (1983), who discussed the use of triangular fuzzy numbers and Lootsma's logarithmic least squares method to derive fuzzy weights and fuzzy performance scales. The fuzzy number was introduced to express linguistic variables. The FAHP has been successfully applied in diverse areas. Cebeci (2009), for example, applied FAHP to compare enterprise resource planning systems in a textile manufacturing company. Celik et al. (2009) investigated shipping registry selection using FAHP. Recently, Lupo (2013) proposed the use of FAHP within the higher education sector. However, the applications of FAHP in medical and health sciences are very limited. To the best of the author's knowledge, there is no application of FAHP that is specifically tailored for identifying types of prostate cancer based on its common symptoms. Therefore, this chapter aims to propose relative weights for types of cancer based on the selected typical symptoms of prostate cancer using FAHP. The rest of this chapter is organized as follows. In Section 16.2, we review some of the research related to medical sciences, AHP, and fuzzy sets. The FAHP procedure is presented in Section 16.3. In Section 16.4, implementation of the case of prostate cancer using FAHP is presented. Finally, Section 16.5 concludes the chapter.

16.2 Related Research

Fuzzy sets have made a huge contribution to MCDM including in solving medical decision problems. An early application of fuzzy sets to the medical science fields was proposed by Zadeh (1969) and Sanchez (1979) who invented a fully developed relationship modeling theory of symptoms and disease using fuzzy sets. One of the popular MCDM methods that combines with fuzzy sets is FAHP. Direct applications of FAHP to prostate cancer are very limited. However, there have been several research studies conducted to apply FAHP and its derivative methods such as intuitionistic FAHP, interval-valued intuitionistic FAHP, and the interval type-2 FAHP in the medical and health sciences. Recently, Li et al. (2016) applied intuitionistic FAHP to get the weight of the index for the evaluation of basic health service equalization. Nguyen and Nahavandi (2016) proposed a modification to FAHP to select the most informative genes that serve as inputs to an interval type-2 fuzzy logic system for cancer classification where the modified FAHP allows individual gene selection. Tsai and Lin (2012), for example, employed FAHP to establish an evaluation model of optimal region selection for joint-venture hospitals or clinics in China. On the basis of the combined perspectives of 30 experts, they adopted 6 criteria and 19 subcriteria for selection. Büyüközkan et al. (2011) used FAHP to evaluate the service quality of the health-care sector. Hong and Wu (2013), for example, proposed a method for syndrome differentiation in the Traditional Chinese Medical clinical analysis based on interval-valued intuitionistic FAHP. Intuitionistic fuzzy cognitive was used to match between syndromes. Pathinathan et al. (2014) proposed a technique to diagnose symptoms of disease using interval-valued intuitionistic FAHP with logical operators. Ahn et al. (2011) proposed a new approach for medical diagnosis using the distance between interval-valued intuitionistic FAHP. For that purpose, they developed an interview chart with interval fuzzy degrees based on the relation between symptoms and diseases (three types of headache) and utilized the interval-valued intuitionistic fuzzy weighted arithmetic average operator to aggregate fuzzy information from the symptoms. In addition, they proposed a measure based on the distance between interval-valued intuitionistic fuzzy sets for medical diagnosis. It seems that the applications of FAHP specifically for selecting the type of prostate cancer have been given no attention.

16.3 FAHP Procedure

The FAHP method is known as a procedure where the eigenvector correspond-ing to the largest eigenvalue of the pairwise comparison matrix provides the relative priorities of the factors and preserves ordinal preferences among the alternatives. This means that if one alternative is preferred to another possibil-ity, then its eigenvector component is larger than that of the other possibility. A vector of weights obtained from the pairwise comparison matrix reflects the relative performance of the various factors. In FAHP, triangular fuzzy numbers are utilized to improve the scaling scheme in the judgment matrices. Interval arithmetic is used to solve the fuzzy eigenvector (Cheng and Mon, 1994). Table 16.1 shows the triangular fuzzy number for every linguistic variable.

The seven-step procedure of this approach is summarized as follows.

Step 1: Construct the fuzzy comparison matrix.

Triangular fuzzy numbers are used to indicate the relative strength of each pair of elements in the same hierarchy. Triangular fuzzy num-bers via pairwise comparison are used to construct the fuzzy judg-ment matrix $\widetilde{A}(a_{ij})$.

$$(A) = \begin{bmatrix} 1 & a_{12} & \cdots & \cdots & a_{1n} \\ a_{21} & 1 & \cdots & \cdots & a_{2n} \\ \cdots & \cdots & \cdots & \cdots & \cdots \\ \cdots & \cdots & \cdots & \cdots & \cdots \\ a_{n1} & a_{n2} & \cdots & \cdots & 1 \end{bmatrix}$$

where $\tilde{a}_{ij}^{\alpha} = 1$, if i is equal to j, and $\tilde{a}_{ij}^{\alpha} = \tilde{1}, \tilde{3}, \tilde{5}, \tilde{7}, \tilde{9}$ or $\tilde{1}^{-1}, \tilde{3}^{-1}, \tilde{5}^{-1}, \tilde{7}^{-1}, \tilde{9}^{-1}$, if i is not equal to j.

Step 2: Construct the priority of weight for each criterion, $(\widetilde{W}_{cl}, \widetilde{W}_{cm}, \widetilde{W}_{cu}^k)$.

TABLE 16.1

Triangular Fuzzy Number of Linguistic Variables

Linguistic Variables	Triangular Fuzzy Numbers	Reciprocal Triangular Fuzzy Numbers
Extremely strong	(9,9,9)	(1/9,1/9,1/9)
Very strong	(6,7,8)	(1/8,1/7,1/6)
Strong	(4,5,6)	(1/6,1/5,1/4)
Moderately strong	(2,3,4)	(1/4,1/3,1/2)
Equally strong	(1,1,1)	(1,1,1)
Intermediate	(7,8,9), (5,6,7), (3,4,5), (1,2,3)	(1/9,1/8,1/7), (1/7,1/6,1/5), (1/5,1/4,1/3), (1/3,1/2,1)

Source: Chang, D.-Y. *European Journal of Operational Research,* 55, 649–655, 1996.

Step 3: Compute the weight priority for the hierarchy of alternatives, $\widetilde{W}_{Al}, \widetilde{W}_{Am}, \widetilde{W}_{Au}^{k}$, with respect to each criterion.

Step 4: Compute the weighted performance ($\widetilde{WP}^k$) for each alternative with respect to each subcriterion.

Step 5: Compute the global priority (priority of overall weights in the entire hierarchy).

In this step, α-cuts are used to represent a level of confidence. The α-cut is known to incorporate the expert's or decision maker's confidence in his or her preference or the judgments.

Let us assume α-cut = 0.95, then $\alpha_{Left} = [\alpha(A_m - A_l) + A_l]$ and $\alpha_{Right} = A_r - [\alpha(A_r - A_m)]$ where (A_l, A_m, A_r) are triangular fuzzy numbers.

Step 6: Compute the global priority (priority of overall weights in the entire hierarchy).

Find the composite priority for each alternative using the lambda function (λ).

The value of λ is from 0 to 1.

$$A_t = \lambda(\alpha_{left}) + \lambda(\alpha_{right})$$

The value of λ is used as a measure to check the consistency of experts in giving evaluations or opinions.

Step 7: Rank the alternatives.

Alternatives can be ranked in descending order where the highest global weight is chosen as the best alternative.

16.4 Implementation

A questionnaire was designed to assist in linguistic data collection and subsequently used to construct pairwise comparison among the symptoms associated with prostate cancer. The questionnaire was used as a guideline in a personal interview with experts or in this case medical officers. Three medical officers (MO1, MO2, MO3) were invited to provide linguistic judgment data based on the FAHP framework. All three officers were attached to a government-funded hospital in Malaysia. The officers need to judge the relative measurement between the criterion and the alternatives using the pairwise comparison proposed by Chang (1996). The experts were asked to specify rating the alternatives with the linguistic expression (e.g., trouble urinating versus hematuria—blood in urine) and to indicate whether they felt that one factor was "strongly more important" or "extremely more important" compared to another factor on a six-point degree of association scale. The scale and the relative importance are presented in Table 16.1.

To find the contributing symptoms associated with prostate cancer, criteria and alternatives in the framework of FAHP are defined. The alternatives of prostate cancer are "adenocarcinoma" (A1), "sarcoma" (A2), and "small cell sarcoma" (A3). The selected criteria are "trouble urinating" (C1), "hesitancy— difficulty starting urine stream" (C2), "hematuria—blood in urine" (C3), and "nocturia—excessive urination at night" (C4). The hierarchy structure of goal, criteria, and alternatives is shown in Figure 16.1.

Computations for linguistic data provided by medical officer 1 (MO1) are presented as follows.

Step 1: Construct the pairwise comparison matrix.

The judgment matrix provided by MO1 is presented in Table 16.2.

Pairwise comparison of alternatives with respect to every criterion is given in Table 16.3.

Step 2: Compute the priority weight for each criterion.

The priority weight for criterion C1 is computed as

$$
C1 = \left[\left(\frac{\left(\frac{1}{31/4} + \frac{2}{15} + \frac{1/6}{2} + \frac{4}{53/4} \right)}{4} \right) \left(\frac{\left(\frac{1}{31/2} + \frac{3}{12} + \frac{1/5}{7/4} + \frac{5}{56/5} \right)}{4} \right) \left(\frac{\left(\frac{1}{38/7} + \frac{4}{9} + \frac{1/5}{11/7} + \frac{6}{55/6} \right)}{4} \right) \right]
$$

$$
= (0.1619, 0.2412, 0.3604).
$$

Similarly, priority weights for criteria C2, C3, and C4 are given as C2 = (0.0591, 0.0861, 0.1366), C3 = (0.3628, 0.5097, 0.7096), and C4 = (0.1117, 0.1630, 0.2450), respectively.

Step 3: Compute the priority weight for the hierarchy of alternatives with respect to each criterion.

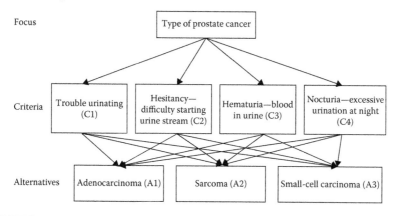

FIGURE 16.1
Hierarchy structure of focus, criteria, and alternatives.

TABLE 16.2

Pairwise Comparison of Criteria

Criteria	Trouble Urinating (C1)	Hesitancy (C2)	Hematuria (C3)	Nocturia (C4)
Trouble urinating (C1)	(1,1,1)	(2,3,4)	(1/6,1/5,1/4)	(4,5,6)
Hesitancy (C2)	(1/4,1/3,1/2)	(1,1,1)	(1/4,1/3,1/2)	(1/6,1/5,1/4)
Hematuria (C3)	(4,5,6)	(2,3,4)	(1,1,1)	(4,5,6)
Nocturia (C4)	(1/6,1/5,1/4)	(4,5,6)	(1/6,1/5,1/4)	(1,1,1)
Hold	(38/7,13/2,31/4)	(9,12,15)	(11/7,7/4,2)	(55/6,56/5,53/4)

TABLE 16.3

Pairwise Comparison of Alternatives with Respect to Each Criterion

C1	A1	A2	A3
A1	(1,1,1)	(1/6,1/5,1/4)	(6,7,8)
A2	(4,5,6)	(1,1,1)	(6,7,8)
A3	(1/8,1/7,1/6)	(1/8,1/7,1/6)	(1,1,1)
Sum of column	(41/8,43/7,43/6)	(9/7,4/3,10/7)	(13,15,17)
C2	A1	A2	A3
A1	(1,1,1)	(6,7,8)	(4,5,6)
A2	(1/8,1/7,1/6)	(1,1,1)	(4,5,6)
A3	(1/6,1/5,1/4)	(1/6,1/5,1/4)	(1,1,1)
Sum of column	(9/7,4/3,10/7)	(43/6,41/5,37/4)	(9,11,13)
C3	A1	A2	A3
A1	(1,1,1)	(1/6,1/5,1/4)	(1/4,1/3,1/2)
A2	(4,5,6)	(1,1,1)	(4,5,6)
A3	(2,3,4)	(1/6,1/5,1/4)	(1,1,1)
Sum of column	(7,9,11)	(4/3,7/5,3/2)	(21/4,19/3,15/2)
C4	A1	A2	A3
A1	(1,1,1)	(2,3,4)	(1/6,1/5,1/4)
A2	(1/4,1/3,1/2)	(1,1,1)	(4,5,6)
A3	(4,5,6)	(1/6,1/5,1/4)	(1,1,1)
Sum of column	(21/4,19/3,15/2)	(19/6,21/5,21/4)	(31/6,31/5,29/4)

The priority weight for alternative A1 with respect to criteria C1 is computed as

$$
A1 = \frac{\left(\dfrac{a_{11l}}{b_{1u}} + \dfrac{a_{12l}}{b_{1u}} + \dfrac{a_{13l}}{b_{1u}}\right)\left(\dfrac{a_{11m}}{b_{1m}} + \dfrac{a_{12m}}{b_{1m}} + \dfrac{a_{13m}}{b_{1m}}\right)\left(\dfrac{a_{13u}}{b_{1l}} + \dfrac{a_{13u}}{b_{1l}} + \dfrac{a_{13u}}{b_{1l}}\right)}{3}
$$

$$
= \left[\left(\frac{\left(\dfrac{1}{43/6} + \dfrac{1/6}{10/7} + \dfrac{6}{17}\right)}{3}\right)\left(\frac{\left(\dfrac{1}{43/7} + \dfrac{1/5}{4/3} + \dfrac{7}{15}\right)}{3}\right)\left(\frac{\left(\dfrac{1}{41/8} + \dfrac{1/4}{9/7} + \dfrac{8}{13}\right)}{3}\right)\right]
$$

$$
= (0.2034, 0.2595, 0.3347)
$$

The priority weights for other alternatives with respect to C1, C2, C3, and C4 are computed similarly. This is summarized in Table 16.4.

Step 4: Compute the weighted performance for each alternative with respect to each criterion.

The weighted performance for alternative A1 with respect to C1 is computed as

$$A1 = \left[0.2034(0.1619), \ 0.2595(0.2412), \ 0.3347(0.3604) \right]$$

$$= (0.0329, \ 0.0626, \ 0.1206)$$

The weighted performances for alternatives A2 and A3 with respect to C1are given as A2 = (0.0873, 0.1628, 0.3076) and A3 = (0.0089, 0.0158, 0.0286), respectively.

The weighted performances for alternatives A1, A2, and A3 with respect to C2 are given as A1 = (0.0328, 0.0589, 0.1165), A2 = (0.0099, 0.0196, 0.0426), and A3 = (0.0042, 0.0076, 0.0155), respectively.

The weighted performances for alternatives A1, A2, and A3 with respect to C3 are given as A1 = (0.0285, 0.0521, 0.1007), A2 = (0.1891, 0.3499, 0.6504), and A3 = (0.0516, 0.1077, 0.2246), respectively.

TABLE 16.4

Priority Weights of Alternatives with Respect to Criteria

Priority Weights of Alternatives with Respect to C1	
A1	(0.2034, 0.2595, 0.3347)
A2	(0.5390, 0.6751, 0.8534)
A3	(0.0548, 0.0654, 0.0795)
Priority Weights of Alternatives with Respect to C2	
A1	(0.5541, 0.6843, 0.8524)
A2	(0.1680, 0.2276, 0.3117)
A3	(0.0548, 0.0654, 0.0795)
Priority Weights of Alternatives with Respect to C3	
A1	(0.0785, 0.1022, 0.1419)
A2	(0.5212, 0.6864, 0.9167)
A3	(0.1421, 0.2114, 0.3165)
Priority Weights of Alternatives with Respect to C4	
A1	(0.1791, 0.3015, 0.5007)
A2	(0.2585, 0.3657, 0.5241)
A3	(0.2343, 0.3328, 0.4718)

The weighted performances for alternatives A1, A2, and A3 with respect to C4 are given as A1 = (0.0200, 0.0491, 0.1226), A2 = (0.0289, 0.0596, 0.1284), and A3 = (0.0262, 0.0542, 0.1156), respectively.

Hence, the overall weighted performance for alternative A1 is computed as A1 = (0.0329 + 0.0328 + 0.0285 + 0.0200), (0.0626 + 0.0589 + 0.0521 + 0.0491), (0.1206 + 0.1165 + 0.1007 + 0.1226) = (0.1142, 0.2228, 0.4604).

Similarly, the overall weighted performances for alternatives A2 and A3 are (0.3152, 0.5919, 1.1290) and (0.0908, 0.1853, 0.3842), respectively.

Step 5: Compute the global priority (priority of overall weights in the entire hierarchy).

With α-cut = 0.95, the confidence level representation with respect to A1 is computed as

$$\alpha_{Left} = \left[0.95(0.2228 - 0.1142) + 0.1142 \right] = 0.2173$$

$$\alpha_{Right} = 0.4604 - [0.95(0.4604 - 0.2228)] = 0.2346$$

The confidence level representation with respect to A2 is

$$\alpha_{Left} = 0.5781$$

$$\alpha_{Right} = 0.6188$$

The confidence level representation with respect to A3 is

$$\alpha_{Left} = 0.1806$$

$$\alpha_{Right} = 0.1953$$

Step 6: Compute the global priority (priority of overall weights in the entire hierarchy).

Find the composite priority for each alternative using the lambda function, $\lambda = 0.5$.

The composite priorities for alternatives A1, A2, and A3 are given as A1 = 0.2260, A2 = 0.5984, and A3 = 0.1879.

The total composite priority is A1 + A2 + A3 = 0.2260 + 0.5984 + 0.1879 = 1.0

Thus, the proportions of composite priority (global priority) of A1, A2, and A3 are A1 = 0.2232, A2 = 0.5911, and A3 = 0.1857, respectively.

Step 7: Rank the alternatives.

The ranking of the alternatives is summarized in Table 16.5.

The rankings of the alternatives obtained from medical officer 2 (MO2) and medical officer 3 (MO3) are computed using similar procedures. The global priorities obtained from the judgments of MO2 and MO3 are given in Tables 16.6 and 16.7, respectively.

The average global priority weight provided by the three experts is summarized in Table 16.8.

TABLE 16.5

Ranking of Alternatives and Global Weights Provided by MO1

Alternatives	Global Priority	Ranking
A1	0.2232	2
A2	0.5911	1
A3	0.1857	3

TABLE 16.6

Ranking of Alternatives and Global Weights Provided by MO2

Alternatives	Global Priority	Ranking
A1	0.6127	1
A2	0.2698	2
A3	0.1175	3

TABLE 16.7

Ranking of Alternatives and Global Weights Provided by MO3

Alternatives	Global Priority	Ranking
A1	0.4499	1
A2	0.4332	2
A3	0.1170	3

TABLE 16.8

Average Score of Global Priority and Rank of Alternatives

Alternatives	Global Priority	Ranking
A1	0.4286	2
A2	0.4314	1
A3	0.1400	3

The three medical officers prioritized A2 as the best alternative. It can be concluded that the sarcoma type of prostate cancer (A2) has the highest likelihood based on the four symptoms.

16.5 Conclusions

Very often medical experts have experienced problems in identifying a specific disease due to multiple common symptoms that could be shared. This chapter has developed a fuzzy MCDM model using FAHP that has been characterized with pairwise comparison. The most common symptoms that lead to several types of prostate cancer were defined prior to computing linguistic data with FAHP. Triangular fuzzy numbers were used to define linguistics variables. This chapter mainly focused on establishing the relative weight of types of prostate cancer using FAHP. The decision model was utilized to overcome medical experts' predicaments for identifying the type of prostate cancer. The FAHP method used a seven-step computation with the ultimate aim to establish the relative weights of alternatives. The model successfully identified "sarcoma" as the most likely type of prostate cancer based on the highest relative weights (global weights) among alternatives. However, the result is subject to further investigation especially in dealing with the validity and veracity of the model. Further research could be undertaken to further ascertain the results.

References

Abdullah, L. and Sulaiman, F. (2016). Identifying type of prostate cancer using analytic hierarchy process. *Journal of Emerging Trends in Computing and Information Sciences*, 7(2), 27–32.

Adlassnig, K.P. (1986). Fuzzy set theory in medical diagnosis. *IEEE Transactions on Systems, Man, and Cybernetics*, 16, 260–265.

Ahn, J.Y., Han K.S., Oh S.Y., and Lee C.D. (2011). An application of interval-valued intutionistic fuzzy sets for medical diagnosis of headache. *International Journal of Innovative Computing, Information and Control*, 7(5(B)), 2755–2762.

Brownback, K.R., Renzulli J., DeLellis R., and Myers J.R. (2009). Small-cell prostate carcinoma: A retrospective analysis of five newly reported cases. *Indian Journal of Urology*, 25(2), 259–263.

Büyüközkan, G., Çifçi, G., and Güleryüz, S. (2011). Strategic analysis of healthcare service quality using fuzzy AHP methodology. *Expert Systems with Applications*, 38(8), August 2011, 9407–9424.

Cebeci, U. (2009). Fuzzy AHP-based decision support system for selecting ERP systems in textile industry by using balanced scorecard. *Expert Systems with Applications*, 36(5), 8900–8909.

Celik, M., Er, I.D., and Ozok, A.F. (2009). Application of fuzzy extended AHP methodology on shipping registry selection: The case of Turkish maritime industry, *Expert Systems with Applications*, 36, 190–198.

Chang, D.-Y. (1996). Applications of the extent analysis method on fuzzy AHP. *European Journal of Oprational Research*, 55, 649–655.

Cheng, C.H. and Mon, D.L. (1994). Evaluating weapon system by analytic hierarchy process based on fuzzy scale. *Fuzzy Sets and Systems*, 63, 1–10.

Cuit W. and Blockley D.I. (1990). Interval probability theory for evidential support. *International Journal of Intelligent Systems*, 5, 183–192.

Galil Medical. (2007). *Galil Medical*. Retrieved February 11, 2013, from Galil Medical website: http://www.galil-medical.com/prostate-cancer-cryotherapy/prostate-cancer-prostatic-sarcoma.html.

Hong, Z. and Wu, M. (2013). An approach to TCM syndrome differentiation based on in interval-valued intuitionistic fuzzy sets. *Advances in Intelligent Systems and Computing*, 191, 77–81.

Li, W.X., Huang, Z.K., Feng, Z.M., and Zhang, C.Y. (2016). Weight of basic health service equalization index based on the intuitionistic fuzzy analytic hierarchy process. *International Conference on Oriental Thinking and Fuzzy Logic*, 243–250. doi:10.1007/978-3-319-30874-6_24.

Lupo, T. (2013). A fuzzy ServQual based method for reliable measurements of education quality in Italian higher education area. *Expert Systems with Applications*, 40(17), 7096–7110.

National Cancer Registry Malaysia. (2002). Second Report of Cancer Incidence in Malaysia.

Nguyen, T. and Nahavandi, S. (2016). Modified AHP for gene selection and cancer classification using type-2 fuzzy logic. *IEEE Transactions on Fuzzy Systems*, 24(2), 273–287, doi:10.1109/TFUZZ.2015.2453153.

Omar, Z.A. and Ibrahim Tamim, N.S. (2007). National Cancer Registry Report, Malaysia Cancer Statistics-Data and Figure. Putrajaya: Ministry of Health.

Oncology, A.S. (2009). American Society of Clinical Oncology. Retrieved February 11, 2013, from American Society of Clinical Oncology website: http://www.cancer.net/cancer-types/prostate-cancer.

Pathinathan, T., Arokiaraj, J., and Ilavarasi, P. (2014) Application of interval valued intuitionistic fuzzy sets in medical diagnosis using logical operators, *International Journal of Computing Algorithm*, 3, 495–498.

Ross, L.E., Howard, D.L., Bowie, J.V., Thorpe, R.J., Kinlock, B.L., Burt, C., and LaVeist, T.A. (2016). Factors associated with men's assessment of prostate cancer treatment choice. *Journal of Cancer Education*, 31(2), 301–307. doi:10.1007/s13187-015-0837-9.

Sanchez, E. (1979). Medical diagnosis and composite fuzzy relations. *Advances in Fuzzy Set Theory and Applications*, Gupta M.M., Ragade R.R., and Yager R.R. (eds.), Elsevier Science, Amsterdam, The Netherlands.

Sozen, S., Uner, A., and Alkibany, T. (2000). Small cell carcinoma of the prostate: Report of two cases. *Turkish Journal of Cancer*, 30(3),131–134.

Subahir, M.N., Shah, S.A., and Zainuddin, Z.M. (2009). Risk factors for prostate cancer in Universiti Kebangsaan Malaysia Medical Centre. *Asian Pacific Journal of Cancer Prevention*, 10, 1015–1020.

Tsai, M.C. and Lin, C.T. (2012). Selecting an optimal region by fuzzy group deci-
    sion making: Empirical evidence from medical investors. *Group Decision and
    Negotiation*, 21(3), 399–416, doi:10.1007/s10726-010-9214-6.
Van Laarhoven, P.J.M. and Pedrycz, W. (1983). A fuzzy extension of Saaty's priority
    theory. *Fuzzy Sets and Systems*, 11, 229–241.
Zadeh, L.A. (1965). Fuzzy sets. *Information and Control*, 8:338–353.
Zadeh, L.A. (1969). Biological applications of the theory of fuzzy sets and systems.
    *Proceedings of an International Symposium on Biocybernetices of the Central Nervous
    System*. Little, Brown and Company, Boston, MA, 199–206.

17

Assessing Environmental Actions in the Natura 2000 Network Areas Using FAHP

J. M. Sánchez-Lozano and J. A. Bernal-Conesa

CONTENTS

17.1 Introduction

The main objective of the Habitats Directive 92/43/CE was to establish a European ecological network to preserve the natural habitats and wild fauna and flora (Commission of the European Communities, 2003). In order to do that, the Natura 2000 network was created in 1992 (EU Council Directive, 1992). That network, composed of sites hosting the natural habitat types listed in Annex I and habitats of the species listed in Annex II, shall enable the natural habitat types and the species' habitats concerned to be maintained or, where appropriate, restored to a favorable conservation status in their natural range (Article 3.1 of the Habitats Directive).

Due to Directive 2009/147/CE (Commission of the European Communities, 2009), which is an amended version of a directive on the conservation of wild birds (Commission of the European Communities, 1979), Natura 2000 also recognizes that habitat loss and degradation are the most serious threats to the conservation of wild birds. Therefore, it places great emphasis on the protection of habitats for endangered as well as migratory species, especially through the establishment of a coherent network of special protection areas (SPAs) comprising the most suitable territories for these species. Although the Natura 2000 network sites are not rigidly protected reserves devoid of human activity (Russo et al., 2011), the Habitats Directive guarantees, in its article 2, the protection of nature yet taking into account "economic, social and cultural needs as well as regional and local particularities."

When it is necessary to prioritize environmental actions to preserve the habitats of these areas, the criteria that influence the decision making are not always numerical values but can also include qualitative criteria in the form of labels or linguistic variables that can be represented through fuzzy membership. This is exactly the reason why it is very advisable to combine techniques such as fuzzy logic, which allow us to deal with vague, imprecise, and uncertain problems, with multicriteria decision analysis (Malczewski and Rinner, 2015) through multicriteria decision-making (MCDM) methods.

17.2   Literature Review

A MCDM problem consists of a set of alternatives to be evaluated with respect to a list of criteria. All that information is contained in the so-called decision matrix. The main goal is to find out the best option among all the alternatives once they have been assessed by all the decision criteria. Sometimes, it is not possible to find a solution satisfying all the criteria simultaneously (Keeney and Raiffa, 1976). In these cases, the so-called multiple criteria decision analysis (MCDA) plays a key role since it allows the decision makers to properly tackle the decision-making process. It is worth mentioning that MCDA involves a wide range of techniques to deal with each decision problem in an orderly manner. They also facilitate a consensus regarding the final decision as well as the treatment of the large amount of information, which is usually expressed by different magnitudes of measurement and meanings. Several MCDM techniques have been developed in scientific literature over the years. Each MCDM approach uses distinct paradigms and conceptions. Among them, we can quote ELimination Et Choix Traduisant la Realité (ELECTRE) (Roy, 1968), the Preference Ranking Organization Method for Enrichment Evaluation (PROMETHEE) (Brans et al., 1984), VIseKriterijumska Optimizacija I Kompromisno Resenje (VIKOR) (Opricovic and Tzeng, 2004), the analytic hierarchy process (AHP) (Saaty, 1980), and the Technique for Order of Preference by Similarity to Ideal Solution (TOPSIS) (Hwang and Yoon, 1981). The AHP methodology has been accepted by the international scientific community as a robust and flexible MCDM tool for dealing with complex decision problems. On the other hand, fuzzy logic and its combination with MCDM methods has been adopted in energy modeling since the beginning of this century (Basim and Alsyouf, 2003).

Recent studies have allowed evaluation of large protected areas in Natura 2000 network sites like Mediterranean habitat (Barredo et al., 2016; Foresta et al., 2016), or even specific countries such as the Italian coastal area (Drius et al., 2016; Riccioli et al., 2016), Austria (Geitzenauer et al., 2016), Greece (Votsi et al., 2016), or Poland (Żmihorski et al., 2016), just to quote some of them. However, there is no consensus to define what types of criteria should be taken into account to prioritize environmental actions in such a network with the aim of preserving its habitats. Furthermore, it is quite common for the information provided by the criteria to have different natures, with qualitative criteria coexisting with quantitative criteria, and therefore, linguistic labels and numerical values should be employed to model, by means of triangular fuzzy numbers (TFNs), the coefficients of importance of each criteria. For these reasons, in this chapter, the combination of fuzzy logic with a MCDM method such as AHP is proposed. This combination allows us to deal with vague, imprecise, and uncertain criteria such as qualitative criteria. Regarding this, we apply the FAHP methodology to obtain the weights of the criteria through an expert group with the aim of carrying out environmental managements in the Natura 2000 network areas.

The remaining parts of this chapter are divided into four sections: Section 17.3 defines the criteria that influence the decision-making, while in Section 17.4, the fuzzy sets and the multicriteria methodology are described. In Section 17.5, we explain how to obtain the weights of the criteria and comment on the results, and finally, Section 17.6 presents the conclusions of this study.

17.3 Decision Criteria for Carrying Out Environmental Actions in the Natura 2000 Network Areas

Some research and studies, directly or indirectly, have allowed us to define the characteristics that these criteria must contain (Aretano et al., 2013; Mücher et al., 2009; Lung et al., 2013; Semeraro et al., 2016; Ladisa et al., 2012; Santini et al., 2010; Kosmas et al., 1999; IUCN, 2012) (Figure 17.1):

Each of the specific criteria is briefly described:

Criterion C_1—Distance to game preserves (m): Space or interval between the nearest game preserve and the alternatives (military heritage areas). Carrying out environmental actions in areas in which the distance to game preserves is the minimum possible will be the priority.

Criterion C_2—Distance to towns or villages (m): Space or interval between the population centers (cities or towns) and the alternatives. It also seems logical that the areas nearest to population centers will have a higher risk than others.

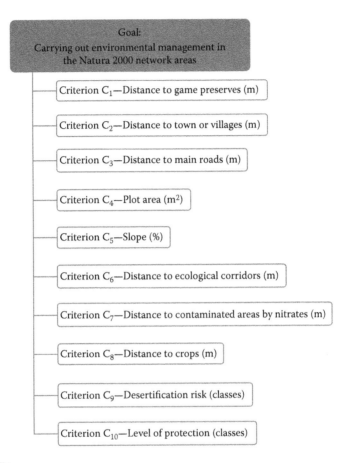

FIGURE 17.1
Criterion tree resulting from the problem structuring phase.

Criterion C_3—Distance to main roads (m): Space or interval between the nearest road and the alternatives. Due to the pollution and the easy access to protected areas, the areas where the distance to roads is short should be prioritized.

Criterion C_4—Plot area (m²): The area contained within a perimeter of land that can carry out environmental actions. A larger area has a greater propensity to be invaded by cyclists, hill walkers, etc.; therefore, it has a higher tourism pressure.

Criterion C_5—Slope (%): The higher the percentage of land slope a territory has, the more difficult public access will be. Therefore, those areas will have a lesser risk for being invaded than others..

Criterion C_6—Distance to ecological corridors (m): Space or interval between the nearest ecological corridor and the alternatives. An ecological corridor could be defined as a territory that eases the dispersion of living beings through its habitat. Its objective is to facilitate the genetic flow between populations, increasing the probability of survival in the long term. Therefore, the areas that are located far away from any ecological corridor will have a higher risk and so the highest priority.

Criterion C_7—Distance to areas contaminated by nitrates (m): Space or interval between the nearest area contaminated by nitrates and the alternatives. Zones located close to these types of contaminated areas will have priority.

Criterion C_8—Distance to crops (m): Space or interval between the nearest crop and the alternatives. Although the use of fertilizers and pesticides improves the growing of crops, it could cause damage to the surrounding areas since fertilizers represent the main cause of ground water pollution. Therefore, areas located close to crops have more propensity to be damaged than others.

Criterion C_9—Desertification risk (classes): Desertification is generated by the consequence of a set of processes that occur in arid and semiarid areas, where water is the main limiting factor. Carrying out environmental actions will not be urgent if a territory presents low desertification risk (class 1), whereas it will be necessary if the desertification risk is high (class 4).

Criterion C_{10}—Level of protection (classes): Any territory can be classified by categories according to the risk of extinction of the species located inside (critically endangered or class 9, endangered or class 7, vulnerable or class 5, near threatened or class 3, least concern or class 1).

17.4 Methodology

17.4.1 Fuzzy Sets

The fuzzy set theory, introduced by Zadeh (1965), to deal with vague, imprecise, and uncertain problems has been used as a modeling tool for complex systems that can be controlled by humans but are hard to define precisely. For instance, the class of objects characterized by such commonly used adjectives as significant, serious, substantial, simple, etc., are fuzzy sets since there is no sharp transition from membership to nonmembership. The main point is that in a real world, there are no crisp or real boundaries that separate those objects that belong to the classes in question from those that do not (Bellman and Zadeh, 1970). A collection of objects (universe of discourse) X has a fuzzy set A described by a membership function f_A with values in the interval [0,1] (García-Cascales and Lamata, 2011).

In this chapter, we only make reference to the operations on a triangular membership function through the fuzzy number sets that will be used in the study case. The basic theory regarding triangular fuzzy numbers is described in detail in Klir and Yuan (1995).

17.4.2 Analytic Hierarchy Process

The AHP methodology, proposed by Saaty (1980), has been accepted by the international scientific community as a robust and flexible MCDM tool to deal with complex decision problems. Basically, AHP has three underlying concepts: structuring the complex decision as a hierarchy of goal, criteria, and alternatives, pairwise comparison of elements at each level of the hierarchy with respect to each criterion on the preceding level, and finally, vertically synthesizing the judgments over the different levels of the hierarchy. AHP attempts to estimate the impact of each one of the alternatives on the overall objective of the hierarchy. In this case, we will only apply the method in order to obtain the criteria weights.

We assume that the quantified judgments provided by the decision maker on pairs of criteria (C_i, C_j) are contained in an $n \times n$ matrix as follows:

$$
C = \begin{array}{c c} & \begin{array}{cccc} c_1 & c_2 & \cdots & c_n \end{array} \\ \begin{array}{c} C_1 \\ C_2 \\ \cdot \\ \cdot \\ C_n \end{array} & \left[\begin{array}{cccc} c_{11} & c_{12} & \cdots & c_{1n} \\ c_{21} & c_{22} & \cdots & c_{2n} \\ \cdot & \cdot & \cdots & \cdot \\ \cdot & \cdot & \cdots & \cdot \\ c_{n1} & c_{n2} & \cdots & c_{n3} \end{array} \right] \end{array}
$$

For instance, the c_{12} value represents an approximation of the relative importance of C_1–C_2, that is, $c_{12} \approx (w_1/w_2)$. This can be generalized and the statements below can be concluded:

1. $c_{ij} \approx (w_i / w_j)$, $i, j = 1, 2, \ldots, n$
2. $c_{ii} = 1, i = 1, 2, \ldots, n$
3. If $c_{ij} = \alpha, \alpha \neq 0$, then $a_{ji} = 1 / \alpha, i = 1, 2, \ldots, n$
4. If C_i is more important than C_j, then $c_{ij} \cong (w_i / w_j) > 1$

This implies that the matrix C should be a positive and reciprocal matrix with ones on the main diagonal. Hence, the decision maker only needs to provide value judgments in the upper triangle of the matrix. The values assigned to c_{ij} according to the Saaty scale lie usually in the interval of 1–9 or their reciprocals. In our case, Table 17.1 presents the linguistic decision maker's preferences in the pairwise comparison process (Saaty, 1989).

It can be shown that the number of judgments (L) needed in the upper triangle of the matrix is

$$L = n(n-1)/2 \qquad (17.1)$$

where n is the size of the matrix C.

The vector of weights is the eigenvector corresponding to the maximum eigenvalue "λ_{max}" of the matrix C. The traditional eigenvector method of estimating weights in AHP yields a way of measuring the consistency of the referee's preferences arranged in the comparison matrix. The consistency index (CI) is given by $CI = (\lambda_{max} - n)/(n-1)$.

If the referee shows some minor inconsistency, then $\lambda_{max} > n$ and Saaty proposes the following measure of the consistency ratio: $CR = CI/RI$, where RI (random index) is the average value of CI (Table 17.2) for random matrices using the Saaty scale (Forman, 1990; Alonso and Lamata, 2006) and Saaty only accepts a matrix as a consistent one if $CR < 0.1$.

In AHP problems, where the values are fuzzy, not crisp, instead of using λ as an estimator of the weight, we will use the geometric normalized average, expressed by the following expression:

TABLE 17.1

Fuzzy Scale of Valuation in the Pairwise Comparison Process

Labels	Verbal Judgments of Preferences between Criterion i and Criterion j	Triangular Fuzzy Scale and Reciprocals
(II)	C_i and Cj are equally important	(1, 1, 1)/(1,1,1)
(M+I)	C_i is slightly more important than C_j	(2, 3, 4)/(1/4,1/3,1/2)
(+I)	C_i is strongly more important than C_j	(4, 5, 6)/(1/6,1/5,1/4)
(Mu+I)	C_i is very strongly more important than C_j	(6, 7, 8)/(1/8,1/7,1/6)
(Ex+I)	C_i is extremely more important than C_j	(8, 9, 9)/(1/9,1/9,1/8)

Source: Saaty, T.L., *Group Decision Making and the AHP*, Springer, New York, 1989.

TABLE 17.2

Random Index for Different Matrix Orders

n	1–2	3	4	5	6	7	8	9	10
RI	0.00	0.5247	0.8816	1.1086	1.2479	1.3417	1.4057	1.4499	1.4854
n	11	12	13	14	15				
RI	1.5140	1.5365	1.5551	1.5713	1.5838				

Source: Alonso, J.A. and Lamata, M.T., *International Journal of Uncertainty, Fuzziness and Knowledge-Based Systems*, 14, 445–459, 2006.

$$w_i = \frac{\left(\prod_{j=1}^{n}\left(a_{ij},b_{ij},c_{ij}\right)\right)^{1/n}}{\sum_{i=1}^{m}\left(\prod_{j=1}^{n}\left(a_{ij},b_{ij},c_{ij}\right)\right)^{1/n}} \tag{17.2}$$

where $\left(a_{ij},b_{ij},c_{ij}\right)$ is a fuzzy number.

Also, to obtain the weight vector, the normalizing operation must be used; this will be achieved through Equation 17.3:

$$\left(w_{c_{ia}},w_{c_{ib}},w_{c_{ic}}\right) = \left[\frac{c_{ia}}{\sum_{i=1}^{n}c_{ic}}, \frac{c_{ib}}{\sum_{i=1}^{n}c_{ib}}, \frac{c_{ic}}{\sum_{i=1}^{n}c_{ia}}\right] \tag{17.3}$$

17.5   Obtaining the Weight

The extraction of knowledge was performed by a group of experts who filled out a survey based on the application of the methodology described. The group of experts involved in the decision process consisted of those who specialize in this type of environmental management.

17.5.1   FAHP Survey

To determine the weights of the criteria, an FAHP survey was carried out. This survey included three questions:

Q1: Do you believe that the 10 criteria considered have the same weight?
If the answer was yes, then $w_i = w_j = 1/n$ for all *i,j*. Thus, it is not necessary to do anything else to obtain the weights of the criteria since these will all have the same value. Otherwise, if an expert considers that not all the criteria have equal importance, then it becomes appropriate to proceed to the next question of the survey.

The next step will be to find out which criterion is more important than another. This degree of importance will be analyzed to be able to assign a weight to each criterion. For example, when indicating that a particular criterion has a higher weight than the rest of the criteria, it is declared that this is the most important criterion. Forthwith, the weights of the criteria will be used to quantify their importance.

The four experts have considered that certain criteria should have a greater weight than others. Therefore, those weights need to be determined.

TABLE 17.3

Order of Importance of Criteria for Each Expert.

E_1	$C_9 = C_{10} > C_4 > C_5 = C_6 > C_8 > C_3 = C_2 > C_1 > C_7$
E_2	$C_2 = C_3 > C_8 > C_{10} > C_6 > C_7 > C_1 > C_5 > C_4 > C_9$
E_3	$C_2 = C_8 > C_4 = C_6 = C_{10} > C_1 = C_5 > C_3 = C_7 = C_9$
E_4	$C_6 = C_9 = C_{10} > C_2 > C_3 = C_7 > C_8 > C_4 = C_5 > C_1$

Q2: List the criteria in descending importance
According to Table 17.3, the four experts believe that the importance is different, although they differ in the order of importance of the criteria. Once the expert has indicated the order of importance, the next question would be considered.

Q3: Compare the approach to be considered first with respect to that considered secondly and successively, using the following tags {(II), (M+I), (+I), (Mu+I), (Ex+I)} that correspond to the scale of valuation in the pairwise comparison process (Table 17.1)

To determine the weights of the criteria, as has been discussed, a pairwise comparison has been made. Using Expert 4 as an example, in Figure 17.2, his or her appreciation by pairwise comparison is shown. The meaning is as follows: criterion C_6 is very strongly more important than both C_5 and C_1; with respect to C_8 and C_4, it is strongly more important, with respect to C_2, C_3, and C_7, it is slightly more important; and with respect to C_9 and C_{10}, it is equally important.

This, translated into the fuzzy numbers according to Table 17.1, gives Figure 17.3.

Taking into account both (Zadeh and Kacprzyt, 1999) and the operation (2), the weights of the criteria (provided in Figure 17.4) are obtained.

The information detailed above for E_4 would also be carried out by other experts. The normalized weights associated with the corresponding criterion C_j, $j=1,2,\ldots,10$, given by each expert can be seen in Table 17.4.

From the above table, it is observed that there is not a criterion that stands out above all the others, that is, certain differences can be appreciated among them. In order to use the weights of the criteria provided by the experts afterward, it is necessary to unify them. To deal with this, we will carry out a

$$C_6 \quad C_9 \quad C_{10} \quad C_2 \quad C_3 \quad C_7 \quad C_8 \quad C_4 \quad C_5 \quad C_1$$
$$C_6 \ [II \quad II \quad II \quad M+I \quad M+I \quad M+I \quad +I \quad +I \quad Mu+I \quad Mu+I]$$

FIGURE 17.2
Valuations given by E_4.

$$C_6 \quad C_9 \quad C_{10} \quad C_2 \quad C_3 \quad C_7 \quad C_8 \quad C_4 \quad C_5 \quad C_1$$
$$C_6 \ [(1,1,1) \ (1,1,1) \ (1,1,1) \ (2,3,4) \ (2,3,4) \ (2,3,4) \ (4,5,6) \ (4,5,6) \ (6,7,8) \ (6,7,8)]$$

FIGURE 17.3
Matrix of decision making for E_4.

$$
\begin{array}{c}
\\
C_6 \\
C_9 \\
C_{10} \\
C_2 \\
C_3 \\
C_7 \\
C_8 \\
C_4 \\
C_5 \\
C_1
\end{array}
\begin{bmatrix}
C_6 \\
(1, 1, 1) \\
(1, 1, 1) \\
(1, 1, 1) \\
(1/4, 1/3, 1/2) \\
(1/4, 1/3, 1/2) \\
(1/4, 1/3, 1/2) \\
(1/6, 1/5, 1/4) \\
(1/6, 1/5, 1/4) \\
(1/8, 1/7, 1/6) \\
(1/8, 1/7, 1/6)
\end{bmatrix}
=
\begin{bmatrix}
(0.188, 0.213, 0.231) \\
(0.188, 0.213, 0.231) \\
(0.188, 0.213, 0.231) \\
(0.047, 0.071, 0.115) \\
(0.047, 0.071, 0.115) \\
(0.047, 0.071, 0.115) \\
(0.031, 0.043, 0.058) \\
(0.031, 0.043, 0.058) \\
(0.023, 0.030, 0.038) \\
(0.023, 0.030, 0.038)
\end{bmatrix}
$$

FIGURE 17.4
Criteria weight for E_4.

TABLE 17.4

Weights of Criteria by the Four Experts

	Expert 1 (E_1)	Expert 2 (E_2)	Expert 3 (E_3)	Expert 4 (E_4)
C_1	[0.027, 0.037, 0.048]	[0.031, 0.040, 0.050]	[0.037, 0.052, 0.072]	[0.023, 0.030, 0.038]
C_2	[0.036, 0.051, 0.071]	[0.245, 0.280, 0.303]	[0.222, 0.261, 0.289]	[0.047, 0.071, 0.115]
C_3	[0.036, 0.051, 0.071]	[0.245, 0.280, 0.303]	[0.028, 0.037, 0.048]	[0.047, 0.071, 0.115]
C_4	[0.055, 0.086, 0.143]	[0.027, 0.031, 0.038]	[0.056, 0.087, 0.145]	[0.031, 0.043, 0.058]
C_5	[0.055, 0.086, 0.143]	[0.031, 0.040, 0.050]	[0.037, 0.052, 0.072]	[0.023, 0.030, 0.038]
C_6	[0.055, 0.086, 0.143]	[0.041, 0.056, 0.076]	[0.056, 0.087, 0.145]	[0.188, 0.213, 0.231]
C_7	[0.027, 0.037, 0.048]	[0.041, 0.056, 0.076]	[0.028, 0.037, 0.048]	[0.047, 0.071, 0.115]
C_8	[0.036, 0.051, 0.071]	[0.061, 0.093, 0.151]	[0.222, 0.261, 0.289]	[0.031, 0.043, 0.058]
C_9	[0.218, 0.257, 0.286]	[0.027, 0.031, 0.038]	[0.028, 0.037, 0.048]	[0.188, 0.213, 0.231]
C_{10}	[0.218, 0.257, 0.286]	[0.061, 0.093, 0.151]	[0.056, 0.087, 0.145]	[0.188, 0.213, 0.231]

homogeneous aggregation (considering that all experts are equally important in the decision problem) by the arithmetic average. The weights of each criterion (Table 17.5) and their graphical representation (Figure 17.5) were as follows.

Table 17.5 shows that according to the experts, the criteria that have the greatest importance are criteria C_{10} (level of protection), C_2 (distance to towns or villages), and C_9 (desertification risk). By contrast, those with a lower weight are criteria C_1 and C_7 (distance to game preserves, and areas contaminated by nitrates, respectively). Once the weights of criteria were obtained, our results were corroborated by the group of experts who validated the obtained results. It is worth mentioning that several studies have demonstrated the relevant role of such main criteria and their influence in climate change, with the aim of preserving the habitats of the Natura 2000 network (Barredo et al., 2016; Araújo et al., 2011; Vos et al., 2008; Martínez et al., 2006; Araújo et al., 2004). According to their point of view, it is not only necessary to have a suitable level of protection to avoid the extinction of the species

TABLE 17.5

Weights of Criteria through Experts'
Homogeneous Aggregation

C_1	[0.032, 0.045, 0.063]
C_2	[0.129, 0.154, 0.180]
C_3	[0.090, 0.111, 0.137]
C_4	[0.075, 0.093, 0.117]
C_5	[0.035, 0.050, 0.072]
C_6	[0.082, 0.107, 0.141]
C_7	[0.035, 0.049, 0.069]
C_8	[0.079, 0.100, 0.128]
C_9	[0.114, 0.133, 0.148]
C_{10}	[0.128, 0.159, 0.196]

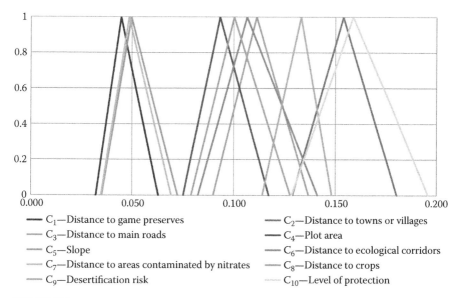

—— C_1—Distance to game preserves
—— C_3—Distance to main roads
—— C_5—Slope
—— C_7—Distance to areas contaminated by nitrates
—— C_9—Desertification risk

—— C_2—Distance to towns or villages
—— C_4—Plot area
—— C_6—Distance to ecological corridors
—— C_8—Distance to crops
—— C_{10}—Level of protection

FIGURE 17.5
Graphical representation of weights.

located inside but also to take into account the proximity to urban lands and
the aridity of these areas.

In order to verify the consistency of the AHP method, the consistency ratio
was calculated by each expert. This value was less than 0.1 and so it was possible to demonstrate the consistency of the method, and therefore, it was not
necessary to revise the judgments of the experts.

17.6 Conclusions

The current study has shown that a number of criteria must be taken into account when carrying out environmental actions in the Natura 2000 network areas. Moreover, such criteria do not equally influence the decision making, so it is very important to know beforehand the weights of these criteria when implementing such actions.

The highest rated criteria for carrying out environmental actions in the Natura 2000 network areas correspond to the level of protection (C_{10}), distance to towns or villages (C_2), and desertification risk (C_9). It is also interesting to highlight that there are a number of criteria whose importance is similar: these are distance to main roads (C_3), distance to ecological corridors (C_6), distance to crops (C_8), and plot area (C_4). Although excessive differences are not observed for the rest of the criteria, their values are far away of the rest.

With respect to the applied methodology, it is worth highlighting that carrying out the extraction of knowledge through an experts' group in the environmental field has allowed the combination of fuzzy logic techniques with a well-known decision-making tool like AHP methodology. Furthermore, it has been possible to not only select and define a criteria list that influences the prioritization problem, but also to obtain their coefficients of importance through the AHP methodology.

Therefore, this study shows the usefulness of fuzzy approaches of a well-known multicriteria method (FAHP in this case), which enables the solving of a trending decision problem such as obtaining of the weights of the criteria that must be taken into account when carrying out environmental actions.

Finally, it should be emphasized that this work consists of a decision problem taking into account significant technical criteria and the experience of an advisory group of experts, as well. To extend this work, a further study regarding additional relevant criteria, such as economic aspects or even institutional factors, may be carried out. Furthermore, once the weights of the criteria have been obtained, a second step allowing the prioritization of specific areas in any territory could be put into effect.

References

Alonso, J.A. and Lamata, M.T. (2006). Consistency in the analytic hierarchy process. A new approach. *International Journal of Uncertainty, Fuzziness and Knowledge-Based Systems*, 14, 445–459.

Araújo, M.B., Alagador, D., Cabeza, M., Nogués-Bravo, D., and Thuiller, W. (2011). Climate change threatens European conservation areas. *Ecology Letters*, 14, 484–492.

Araújo, M.B., Cabeza, M., Thuiller, W., Hannah, L., and Williams, P.H. (2004). Would climate change drive species out of reserves? An assessment of existing reserve-selection methods. *Global Change Biology*, 10, 1618–1626.

Aretano, R., Petrosillo, I., Zaccarelli, N., Semeraro, T., and Zurlini, G. (2013). People perception of landscape change effects on ecosystem services in small Mediterranean islands: A combination of subjective and objective assessments. *Landscape and Urban Planning*, 112, 63–73.

Barredo, J.I., Caudullo, G., and Dosio, A. (2016). Mediterranean habitat loss under future climate conditions: Assessing impacts on the Natura 2000 protected area network. *Applied Geography*, 75, 83–92.

Basim, A-N. and Alsyouf, I. (2003). Selecting the most efficient maintenance approach using fuzzy multiple criteria decision making. *International Journal of Production Economics*, 84, 85–100.

Bellman, R.E. and Zadeh, L.A. (1970). *Decision-Making in a Fuzzy Environment.* Washington, DC: National Aeronautics and Space Administration (NASA CR-1594).

Brans, J.P., Mareschal, B., and Vincke, Ph, (1984). *PROMETHEE: A New Family of Outranking Methods in Multicriteria Analysis.* In J.P. Brans, editor. North-Holland, Amsterdam: Operational Research, pp. 477–490.

Commission of the European Communities (1979). Council Directive 79/409/EC of 2 April 1979 on the conservation of the wild birds. Brussels, European Commission 79/409/EC.

Commission of the European Communities (2003). Council Directive 92/43/CEE of 21 May 1992 on the conservation of natural habitats and of wild fauna and flora. As amended by the Accession Act of Austria, Finland and Sweden (1995) and the Accession Act of the Czech Republic, the Republic of Estonia, the Republic of Cyprus, the Republic of Latvia, the Republic of Lithuania, the Republic of Hungary, the Republic of Malta, the Republic of Poland, the Republic of Slovenia and the Slovak Republic (2003). Official Journal of the European Union L 236 33 23.9.2003. Brussels. European Commission 1992/95/2003.

Commission of the European Communities (2009). Directive 2009/147/EC of the European Parliament and of the Council of 30 November 2009 on the conservation of wild birds, Brussels, European Commission 2009/147/EC.

Drius, M., Carranza, M.L., Stanisci, A., and Jones, L. (2016). The role of Italian coastal dunes as carbon sinks and diversity sources. A multi-service perspective. *Applied Geography*, 75, 127–136.

EU Council Directive (1992). Natura 2000 Network on the Conservation of Natural Habitats and of Wild Fauna. Council Directive 92/43/EEC (1) of 21 May 1992. The Council of the European Communities.

Foresta, M., Carranza, M.L., Garfì, V., Febbraro, M.D., Marchetti, M., and Loy, A. (2016). A systematic conservation planning approach to fire risk management in Natura 2000 sites. *Journal of Environmental Management*, 181, 574–581.

Forman, E.H. (1990). Random indices for incomplete pairwise comparison matrices. *European Journal of Operational Research*, 48, 153–155.

García-Cascales, M.S. and Lamata, M.T. (2011). Multi-criteria analysis for a maintenance management problem in an engine factory: Rational choice. *Journal of Intelligent Manufacturing*, 22, 779–788.

Geitzenauer, M., Hogl, K., and Weiss, G., (2016). The implementation of Natura 2000 in Austria—A European policy in a federal system. *Land Use Policy*, 52, 120–135.

Hwang, C.L. and Yoon, K. (1981) *Multiple Attribute Decision Methods and Applications.* Berlin Heidelberg: Springer.

IUCN. (2012). *IUCN Red List Categories and Criteria: Version 3.1.* Second edition. Gland, Switzerland and Cambridge, UK: IUCN. iv + 32pp.

Keeney, R. and Raiffa, H. (1976). *Decisions with Multiple Objectives: Preferences and Value Tradeoffs.* New York, NY: Wiley.

Klir, G.J. and Yuan, B. (1995). *Fuzzy Sets and Fuzzy Logic: Theory and Applications.* Upper Saddle River, NJ: Prentice Hall.

Kosmas, C., Kirkby, M., and Geeson, N. (1999). The MEDALUS project: Mediterranean desertification and land use. Brussels: European Commission. Available at http://www.kcl.ac.uk/projects/desertlinks/downloads/publicdownloads/ESA%20Manual.pdf

Ladisa, G., Todorovic, M., and Liuzzi, G.T. (2012). A GIS-based approach for desertification risk assessment in Apulia region, SE Italy. *Physics and Chemistry of the Earth*, 49, 103–113.

Lung, T., Lavalle, C., Hiederer, R., Dosio, A., and Bouwer, L.M. (2013). A multi-hazard regional level impact assessment for Europe combining indicators of climatic and non-climatic change. *Global Environmental Change*, 23, 522–536.

Malczewski, J. and Rinner, C. (2015). *Multicriteria Decision Analysis in Geographic Information Science. Series: Advances in Geographic Information Science.* New York, NY: Springer.

Martínez, I., Carreño, F., Escudero, A., and Rubio, A. (2006). Are threatened lichen species well-protected in Spain? Effectiveness of a protected areas network. *Biological Conservation*, 133, 500–511.

Mücher, C.A., Hennekens, S.M., Bunce, R.G.H., Schaminée, J.H.J., and Schaepman, M.E. (2009). Modelling the spatial distribution of Natura 2000 habitats across Europe. *Landscape and Urban Planning*, 92, 148–159.

Opricovic, S. and Tzeng, G.H. (2004). The compromise solution by MCDM methods: A comparative analysis of VIKOR and TOPSIS. *European Journal of Operational Research*, 156, 445–455.

Riccioli, F., Fratini, R., Boncinelli, F., El Asmar, T., El Asmar, J.-P., and Casini, L. (2016). Spatial analysis of selected biodiversity features in protected areas: A case study in Tuscany region. *Land Use Policy*, 57, 540–554.

Roy, B. (1968). Classement et choix en présence de points de vue multiples. La méthode ELECTRE. *La Revue d'Informatique et de Recherche Opérationelle (RIRO)*, 8, 57–75.

Russo, P., Carullo, L., Riguccio, L., and Tomaselli, G. (2011). Identification of landscapes for drafting Natura 2000 network management plans: A case study in Sicily. *Landscape and Urban Planning*, 101, 228–243.

Saaty, T.L. (1980). *The Analytic Hierarchy Process.* New York, NY: McGraw Hill.

Saaty, T.L. (1989). *Group Decision Making and the AHP.* New York, NY: Springer.

Santini, M., Caccamo, G., Laurenti, A., Noce, S., and Valentini, R. (2010). A multi-component GIS framework for desertification risk assessment by an integrated index. *Applied Geography*, 30, 394–415.

Semeraro, T., Mastroleo, G., Aretano, R., Facchinetti, G., Zurlini, G., and Petrosillo, I. (2016). GIS fuzzy expert system for the assessment of ecosystems vulnerability to fire in managing Mediterranean natural protected areas. *Journal of Environmental Management*, 168, 94–103.

Vos, C.C., Berry, P., Opdam, P., Baveco, H., Nijhof, B., O'Hanley, J., Bell, C., and Kuipers, H. (2008). Adapting landscapes to climate change: Examples of climate-proof ecosystem networks and priority adaptation zones. *Journal of Applied Ecology,* 45, 1722–1731.

Votsi, N.-E., Zomeni, M.S., and Pantis, J.D. (2016). Evaluating the effectiveness of Natura 2000 network for wolf conservation: A case-study in Greece. *Environmental Management,* 57, 257–270.

Zadeh, L.A. (1965). Fuzzy sets. *Information and Control,* 8, 338–353.

Zadeh, L.A. and Kacprzyt, J. (1999). *Computing with Words in Information/Intelligent Systems.* Part 1. Heidelberg and New York: Physica-Verlag (Springer-Verlag).

Żmihorski, M., Kotowska, D., Berg, Å., and Pärt., T. (2016). Evaluating conservation tools in Polish grasslands: The occurrence of birds in relation to agri-environment schemes and Natura 2000 areas. *Biological Conservation,* 194, 150–157.

Index

Printed and bound by CPI Group (UK) Ltd, Croydon, CR0 4YY

24/10/2024

01778306-0008